Soil Resources and its Mapping Through Geostatistics Using R and QGIS

Soil Resources and its Mapping Through Geostatistics Using R and QGIS

Editors

Priyabrata Santra

Mahesh Kumar

N.R. Panwar

C.B. Pandey

CRC Press is an imprint of the
Taylor & Francis Group, an **informa** business

NEW INDIA PUBLISHING AGENCY

New Delhi – 110 034

Figure 1.5, 3.1, 6.9, 6.10, 6.11, 6.12, 9.1, 9.2, 9.3, 9.4, 9.5, 9.6, 9.7, 9.8, 9.9, 9.10, 17.1, 17.2, 20.1, 22.1, 22.2 are originals of the respective editor/authors/contributors.

CRC Press
Taylor & Francis Group
6000 Broken Sound Parkway NW, Suite 300
Boca Raton, FL 33487-2742

First issued in paperback 2023

ISBN-13: 978-1-03-265404-1 (pbk)
ISBN-13: 978-0-3673-4052-0 (hbk)

Publisher's Note
The publisher has gone to great lengths to ensure the quality of this reprint but points out that some imperfections in the original copies may be apparent.

Print edition not for sale in South Asia (India, Sri Lanka, Nepal, Bangladesh, Pakistan or Bhutan)

Library of Congress Cataloging-in-Publication Data
A catalog record has been requested

Visit the Taylor & Francis Web site at
http://www.taylorandfrancis.com

and the CRC Press Web site at
http://www.crcpress.com

Preface

With the great explosion in computation and information technology has come vast amounts of data and tools in all fields of endeavour. Soil science is no exception, with the ongoing creation of regional, national, continental and worldwide databases. The challenge of understanding these large stores of data has led to the development of new tools in the field of statistics and spawned new areas such as data mining and machine learning. In addition to this, in soil science, the increasing power of tools such as geographical information systems (GIS), geographic positioning system (GPS), remote and proximal sensors and data sources such as those provided by digital elevation models (DEMs) are suggesting new ways forward. Fortuitously, this comes at a time when there is a global clamour for soil data and information for environmental monitoring and modelling. Consequently, worldwide, organisations are investigating the possibility of applying the new spanners and screwdrivers of information technology and science to the old engine of soil survey. The principal manifestation is soil resource assessment using geographic information systems (GIS), i.e., the production of digital soil property and class maps with the constraint of limited relatively expensive fieldwork and subsequent laboratory analysis. The production of digital soil maps ab initio, as opposed to digitised (existing) soil maps, is moving inexorably from the research phase to production of maps for regions and catchments and whole countries.

Rapid and reliable assessment of soil characteristics is an important step in agricultural and natural resource management. In general, soils are opaque to most sensing methods. For example, microwave radiations penetrate only a few centimetres of the topsoil; visible (VIS) and infrared radiations can barely penetrate through the soil surface. Consequently, most soil assessments are performed under laboratory conditions. Laboratory methods used for estimating soil chemical properties are based on wet chemistry with tedious and time-consuming sample preparation and analyses steps. Assessment of soil physical attributes generally takes a longer time than chemical attributes. Soil properties widely vary both in time and space. Consequently, rapid and in situ assessment of soil properties even in near-real time remains a formidable task despite decades of research and development in soil testing. Over the past few decades, remote sensing approaches provide some solution for rapid soil assessment. These approaches are fast, nondestructive and have large spatial coverage. There are four factors that influence the remote sensing (especially optical) signature of soil–mineral composition, organic matter, soil moisture and texture. Remote sensing data have been used for soil classification, soil resources mapping, soil moisture assessment and soil degradation (salinity) mapping among many others. Particularly, hyperspectral remote sensing (HRS) is emerging as a promising tool for its capability to measure the reflectance of earth surface features such as soil, water, vegetation, etc. at hundreds of contiguous and narrow wavelength bands. Availability of such a large pool of spectral information offers an opportunity to estimate multiple soil attributes from the same reflectance spectra with greater specificity than their multispectral counterpart.

Keeping in mind the requirement of digital soil maps for efficient management of natural resources, the book on "Soil Resources and its Mapping Through Geostatistics Using R and QGIS" is written. This book will provide an exposure to recent developments in the field of geostatistical modeling, spatial variability of soil resources, and preparation of digital soil maps using R and GIS and potential application of it in agricultural resource management. Specifically following major areas are covered in the book.

- Fundamentals of spatial processes and geostatistics
- Field soil sampling using GIS/GPS and spatial data handling
- Spatial variability analysis using semivariogram approach
- Introduction to 'GSTAT' package of R and QGIS
- Kriging technique and other advanced geostatistical approaches
- Procedures of preparing digital soil maps in R
- Assessment of uncertainty and error of digital soil maps
- Case studies on digital soil mapping

Editors

Contents

1

Fundamentals of Geostatistics

***Priyabrata Santra*[1] *and Gerard Heuvelink*[2]**

[1]*ICAR-Central Arid Zone Research Institute, Jodhpur, Rajasthan, India*
[2]*ISRIC-World Soil Information/ ESG Wageningen UR, The Netherland*

Introduction

Geostatistics is a branch of statistics that deals with the analysis and modelling of geo-referenced data. Its main aims are to quantify spatial variability and to create maps from point observations. Of course there is much more to geostatistics than just that and indeed some of the other uses of geostatistics are addressed in this chapter too, but the emphasis is on the assessment of spatial variability and geostatistical interpolation.

The first step of geostatistical interpolation is to quantify the spatial correlation structure of the variable of interest. This can be done by examining the observations and how these vary in space. Next spatial interpolation makes use of the quantified spatial correlation to derive optimal predictions at unobserved locations and create a map. The interpolation error is quantified as well, which helps to design optimal spatial sampling schemes that balance data collection costs and map accuracy. All this will be explained in this chapter, but in order to do so we first need to explain the statistical theory that underlies geostatistical interpolation followed by representation of the spatial correlation structure, the basics of geostatistical interpolation ('kriging'), kriging extensions and spatial stochastic simulation.

The Random Field Model

Geostatistical methods are based on a statistical model of reality. This model treats reality as if it were a 'realisation' of a stochastic spatial process. In order to describe this properly and explain what these abstract terms mean we first need to repeat some basic material from probability theory and statistics.

Random Variables

Random variables are variables whose values depend on the outcome of a probabilistic experiment. A typical example is the throw of a (fair) die. If we denote this outcome by D, then D is a *random variable* that has six possible outcomes: 1, 2, 3, 4, 5 and 6. Each outcome has equal probability (namely 1/6), which means that D has a *uniform* probability distribution. Now suppose that we perform the experiment and throw the die. Let the outcome be 5. Then $d = 5$ is the *realisation* of D. Note that we introduced notation in which random variables are written in upper case and realisations in lower case. Realisations are just numbers, they are not *stochastic* but *deterministic*.

Another example of a random variable is the number of tails ('successes') of ten tosses with a coin. The possible outcomes are all whole numbers between zero and ten. The probability distribution is no longer a uniform distribution because the probabilities are not equal for all outcomes. For instance, the probability of zero successes is much smaller than that of three successes, and this in turn is smaller than that of five successes. If we denote this random variable by K, then we have probability $P(K = k) = \binom{10}{k}\left(\frac{1}{2}\right)^{10}$ for all k between zero and ten. The probability distribution of K is an example of the *binomial* distribution. The probability distributions of D and K are given in Fig. 1.1.

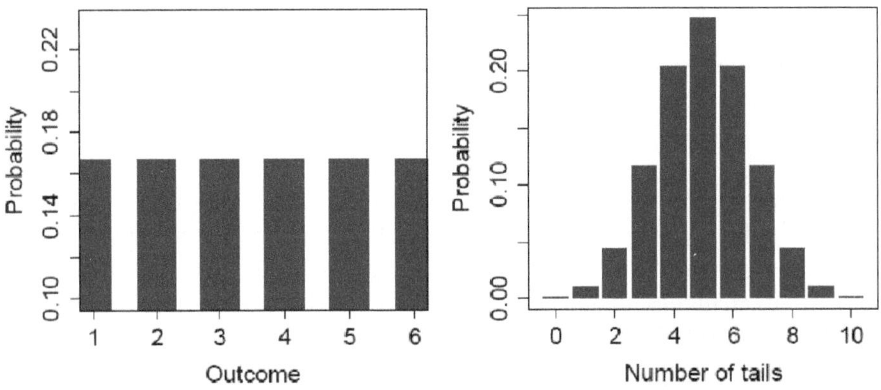

Fig. 1.1: Probability distribution of the throw with a die (left) and of the number of tails of ten tosses with a coin (right)

Key properties of a random variable are its *mean* and *variance*. The mean is also known as the *expected value*. It is calculated by taking a weighed sum of all possible outcomes of the random variable, using the probability of each outcome as a weight. Thus, we can calculate the mean of a throw with a die as:

$$\text{E[D]} = \sum_{d=1}^{6} d.P(D=d) = \sum_{d=1}^{6} d.\frac{1}{6} = \frac{1}{6}\sum_{d=1}^{6} d\frac{1}{6}(1+2+3+\ldots+6) = 3.5 \quad (1)$$

Here, E[....] stands for expected value. It is custom to denote the mean of a random variable by the Greek letter μ. The mean may be interpreted as the average outcome of a very large (infinite) number of repetitions of the same probability experiment. In other words, if we would throw the die 1 million times, then the average of these 1 million throws would be very close to 3.5.

The *variance* is the expected value of the square of the difference between the random variable and its mean. It is typically denoted by σ^2. For the throw with a die we get:

$$\sigma^2 \text{Var (D)} = \text{E [(D} - \text{E[D])}^2] = \sum_{d=1}^{6} (d-3.5)^2 .P(D=d) \frac{1}{6}\sum_{d=1}^{6} (d-3.5)^2 \quad (2)$$

The *standard deviation* is the square-root of the variance *(i.e.,* σ). It is a measure of the *spread* or *variability* of the random variable, while the mean is a measure of *centrality*. Other measures of centrality are the *median* and *mode*. The median is the value that separates the probability distribution of a random variable in two halves of equal probability mass (*e.g.*, if *X* is a random variable then (X < *median*) = *X* > *median*) = 0.5). The mode is that outcome of a random variable that has the largest probability.

The two examples discussed so far are examples of a *discrete* random variable. These are random variables that have a finite number of outcomes (or countably infinite). The opposite are *continuous* random variables, which have an (uncountable) infinite number of outcomes. For instance, let *U* be a randomly chosen real number between 0 and 1. Because there are an infinite number of outcomes, each outcome has equal probability and because all probabilities must sum to one, we have $P(U=u)=0$ for any value of u. This implies that we cannot characterise a continuous random variable by the probabilities associated with each possible outcome. Instead, we must use the concept of *probability density*. We can calculate the probability that *U* takes on a value within a given interval [*a*, *b*] by mathematical integration:

$$P(a \le U \le b) = \int_a^b f(u)\,du \quad (3a)$$

Here, *f* represents probability density. If you are not familiar with mathematical integration, then it may help to know that the integration sign is nothing else than a twisted letter 'S', which signifies that we *sum* from *a* to *b*, in very much the same way as we used the Σ symbol to represent summation over the outcomes of a discrete random variable. For random variable *U*, which has a uniform continuous distribution between 0 and 1, we have $f(U) = 1$ for all u in the interval

[0,1] and so P($a \le$ U $\le$ b)= $\frac{1}{b-a}$ (assuming $0 \le a \le$ b ≤ 1). (3b)

The normal distribution

The most common continuous probability distribution is the *normal distribution.* Let X be a normally distributed random variable, then its probability density is given by:

$$f(x) = \frac{1}{\sqrt{2\pi\sigma^2}} e^{\frac{(x-\mu)^2}{\sigma^2}} \tag{4}$$

where μ and σ^2 are parameters. If we calculate the mean and variance of X in a similar way as was done for D above (but then using integration instead of summation), we get that the mean of X equals μ while its variance equals σ^2 (this also explains why we used these symbols). The shape of the normal probability density is the well-known bell-shaped curve and an example is given in Fig. 1.2. Beware not to interpret the numbers on the y-axis as probabilities. In case of continuous random variables, probabilities can only be associated with the surface area below the curve. The total area below the curve must always be equal to one.

The importance of the normal distribution is due to the *Central Limit Theorem* from statistics, which states that the distribution of averages of realisations that are randomly and independently drawn from whatever distribution, will tend to the normal distribution. For instance, the height of a person is the cumulative effect of many factors, such as genetic material, food supply and illnesses during childhood, and so whenever we plot the frequency distribution of a person's height from a sufficiently large sample of adults drawn from a population, this

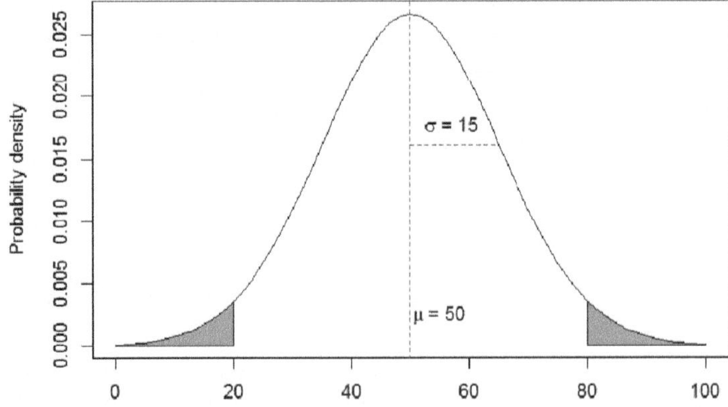

Fig. 1.2: Probability density of a random variable X that has a normal distribution with mean μ=50 and standard deviation σ = 15. The probability of values that deviate more than two standard deviations from the mean (grey area) is about 0.05.

distribution will be remarkably normal. This is illustrated also in Fig. 1.3, which shows that the number of tails with repeated tossing of a coin converges to the normal distribution as the number of tosses increases. Indeed, many variables encountered in the environmental sciences, such as air temperature, clay content of the soil or river water flux, fit the normal distribution curve fairly well. There are also many variables that follow better a *lognormal* distribution, such as the concentration of pollutants in the ground- or surface water. Such variables typically result from *multiplication* instead of *summation* of underlying random variables. The lognormal distribution is asymmetric and has a non-zero *skewness*. A random variable is lognormally distributed if its log-transform is normally distributed.

In geostatistics, we mainly work with normally distributed random variables. Not only because the normal distribution is often encountered in the real world, but also because the *statistical inference* associated with normally distributed random variables is much easier than with others. Whenever we come across a variable that deviates substantially from normality, we will look for a *mathematical transformation* (*e.g.*, logarithm, square-root) such that the transformed variable is (approximately) normal.

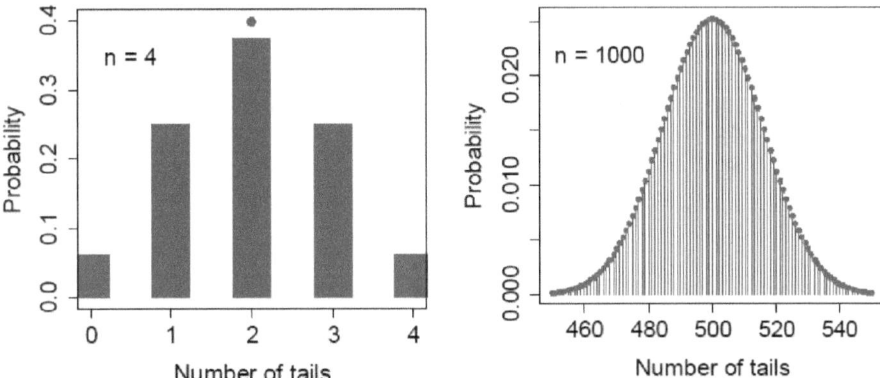

Fig. 1.3: The probability distribution of the number of tails when tossing a coin progressively approaches the normal distribution with increasing number of tosses (n). Bars represent the factual binomial distribution; dots represent the normal distribution.

Probability distribution versus frequency distribution

Above we explained that random variables are characterised by a probability distribution, such as the uniform and normal distribution. The attentive reader might ask, how can the observed data of an environmental variable have a probability distribution? It is not a random variable, it is a data set (i.e., a characteristic measured on a *population* of objects or on a *sample* from such population). Indeed there is a subtle but important difference between the two.

Whereas a random variable has a *probability distribution*, a data set has a *frequency distribution*. A frequency distribution shows how often a certain value (or range of values) occurs within the dataset, while a probability distribution shows the probability of a certain outcome of a probability experiment. Now if we define a random variable as the outcome of a *random draw* from the dataset, then it turns out that the probability distribution of that random variable equals the frequency distribution of the data.

Covariance and correlation

Before we address the spatial extension of random variables in the next section we must first briefly discuss the concept of *correlation*. The correlation between two random variables X and Y is usually denoted by the Greek symbol ρ and is defined as:

$$\rho(X,Y) = \frac{\mathrm{cov}(X,Y)}{\sqrt{\mathrm{var}(X).\mathrm{var}(Y)}} = \frac{E\left[(X-\mu_X).(Y-\mu_Y)\right]}{\sqrt{E\left[(X-\mu_X)^2.E(Y-\mu_Y)^2\right]}} \tag{5}$$

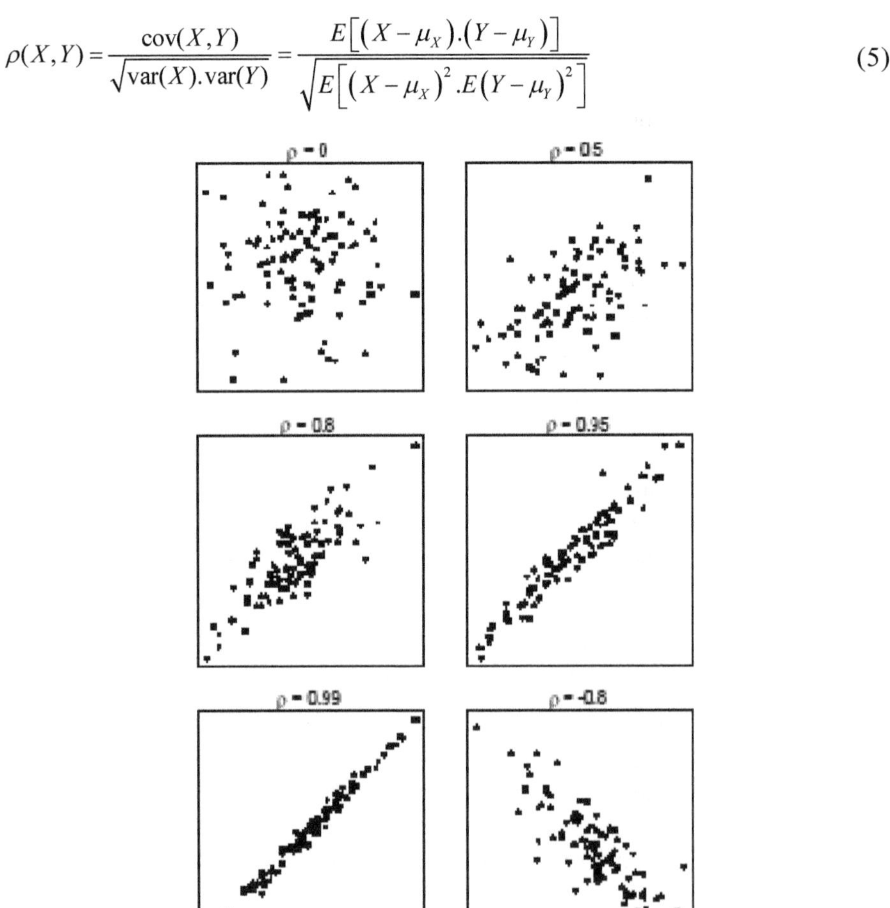

Fig. 1.4: Scatter plots of 100 paired draws from a bivariate normal distribution with different correlations.

Here, *Cov* (X,Y) is the *covariance* between X and Y. The correlation is a standardised or normalised covariance. It is dimensionless and always a number between -1 and 1. It is a measure of the joint variability of random variables: if the correlation is positive then highs (lows) of X often coincide with highs (lows) of Y. If the correlation is negative then it is the opposite. This is illustrated in Fig. 1.4 which shows *scatter plots* of 100 random draws from the joint probability distribution of X and Y for six different correlations. The absolute value of the correlation measures the strength of the (linear) relationship between X and Y. For instance, a zero correlation means that X and Y are not related to each other. In other words, knowing Y does not provide any information about X, and vice versa. In the environmental sciences, most variables are correlated to some degree because everything is related to everything else, but the strength of the correlation is often weak.

Random fields

In geostatistics we deal with variables that vary in space and for this reason we need to extend the concept of a random variable to a *random field*. A random field is defined as a collection of random variables indexed by a geographic coordinate. Let us explore this by considering the soil database of the arid western database as given in the appendix of this book. If we denote the soil properties at any location x in the geographic domain of interest A (*i.e.*, hot arid ecosystem of India) by (x), then in geostatistics we treat (x) as a random variable, while the collection of all random variables Z= {Z(x) x ε A} is a random field. Similar to random variables, a random field Z is fully characterised by its probability distribution, but such probability distribution can be very complex. It must specify the probability distribution of $Z(x)$ for all locations x in A, and it must also specify the correlations between $Z(x_1)$ and $Z(x_2)$ for all paired locations x_1 and x_2 in A. Now if we also assume that Z is normally distributed, then we have completely characterised Z by the (marginal) probability distributions and the correlations. In practice, we will not be able to specify different distributions for all locations x $\in A$ and different correlations for any pair of locations $x_1, x_2 \in A$, simply because we cannot glean that information from only a sample of observations at measurement locations, and so we have to make certain simplifying assumptions to be able to estimate the distributions and correlations from the available observations. You may already now agree that it makes sense to assume that the correlation between $Z(x_1)$ and $Z(x_2)$ only depends on the separation distance $|x_1 - x_2|$ between the two locations. In such case, the correlation becomes a function of just the separation distance and is called *spatial correlation*.

Fig. 1.5 shows twelve realisations of a random field whose probability distribution at every location is a normal distribution with a mean of 50 and a standard deviation of 15. The realisations within each row in Fig. 1.5 have the same spatial correlation, but the spatial correlation is different between rows. It is strongest in the top row and smallest (in fact completely absent) in the bottom row. As you can see the same random field yields different realisations, for the same reason that repeated throws with a die are unlikely to yield the same outcome every time. So none of the realisations in a row are the same, but you may notice that their 'spatial structure' is. This is because this is controlled by the spatial correlation.

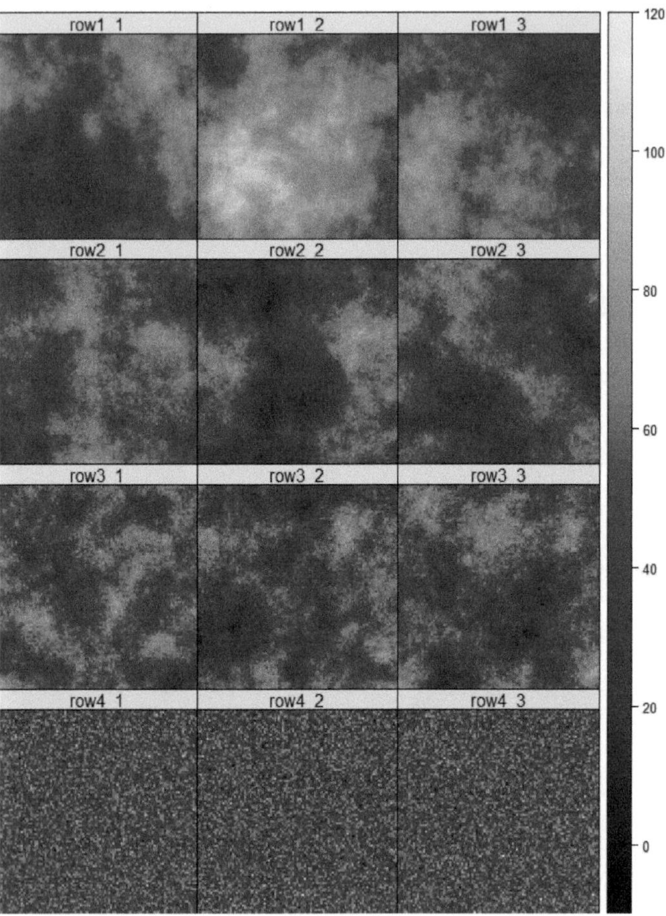

Fig. 1.5: Twelve realisations of a random field with a constant mean of 50 and standard deviation 15. The degree of spatial correlation is strongest in the top row and decreases from the top to the bottom row. Realisations within a row are all different but have the same spatial structure.

Statistical measure of spatial variation: The variogram

We concluded the previous section with a figure showing different realisations of random fields. The spatial structure of realisations within a row was quite similar, while between rows they were different. This was because the degree of spatial correlation that was used to generate these realisations was different between rows. The top row showed fairly large patches of similar value which reflects strong spatial correlation, the bottom row showed spatial 'noise' which means no spatial correlation at all, and the middle rows showed an in-between situation. What measure can we use to characterise spatial correlation? And how can we estimate this measure from a limited number of point observations? These are the two main questions that we address in this section.

The variogram

In geostatistics, the degree of spatial correlation is characterised by the *semivariance*. Let Z be a random field defined on a geographical domain D. The semivariance is defined as follows:

$$\gamma(h) = \frac{1}{2} E\left[\left(Z(x) - Z(x+h)\right)^2\right] \tag{6}$$

Here, E stands for mathematical expectation, as before. Thus, the semivariance $\gamma(h)$ is half the expected squared difference between the value of Z at two locations separated by the distance vector h. A graph of the semivariance against distance is called a *variogram* (sometimes also called: semivariogram). Clearly, $\gamma(h)$ cannot be negative because it is the average of a square. Taking a closer look also reveals that it typically increases with distance, because according to Tobler's first law of geography, "Everything is related to everything else, but near things are more related than distant things". In other words, since the difference between $Z(x)$ and $Z(x + h)$ will usually be small for small h, $\gamma(h)$ will also be small for small h.

The typical shape of the variogram is given in Fig. 1.6. Many real-world variables have a variogram with such shape. A few observations can be made. First, the semivariance increases with distance. However, note also that at some point the increase comes to a hold and the variogram stabilises. This happens at a distance which is called the *range*. It identifies the distance up to which there is still spatial correlation. In other words, for distances beyond the range there is no longer any spatial correlation and the semivariance has reached its maximum value. The maximum value is known as the *sill* of the variogram. In fact, it can be proven that the sill is equal to the *variance* of Z, which we introduced before in Section *Random Variables*. Finally, the third key parameter of the variogram is the *nugget*. This characterises the spatial variability at short distances, also known as the 'micro-scale' variation. Note from the definition of the semivariance that it must be zero for h=0, but apparently it can be greater than zero for distances close to zero. The term 'nugget' stems from mining: if you are lucky you may find a gold nugget at some location in a geologic deposit, while right next to the location there may be no gold at all. Thus, there is much spatial variation at short distances. Apart from micro-scale variation, random *measurement error* also contributes to the nugget variance. This is because repeated measurements of the same variable may yield different outcomes, due to measurement error, and so when calculating half the squared difference, an outcome bigger than zero will result.

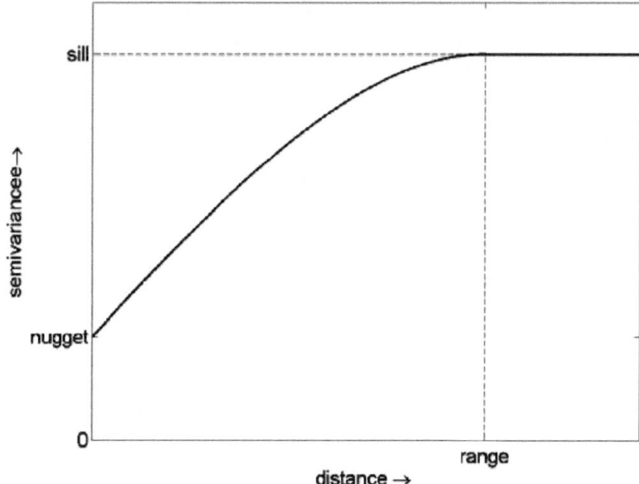

Fig. 1.6: Typical shape of the variogram with its three key parameters nugget, sill and range

The variograms of three out of four random fields that were used to generate the realisations shown in Fig. 1.5 are given in Fig. 1.7. It is important that you can link a spatial structure or 'pattern' as shown in Fig. 1.5 with a variogram. You should understand how the shape of the variogram influences the spatial pattern. A variogram with a small or zero nugget and large range yields large spatial patterns and no 'noise', while a variogram that has no spatial correlation ('pure nugget') creates patterns that are complete noise.

When we defined the semivariance with Eq. (6), we implicitly assumed that it only depends on the distance between the locations, and not on the locations themselves. This is known as the *stationarity* assumption. To be more precise, the stationarity assumption has two components, namely that the mean (expected value) of Z is constant and that the semivariance of Z only depends on the distance between locations. Perhaps the stationarity assumption is not realistic in all cases. For example, we know that air temperature is negatively correlated with elevation, so it would be unwise to assume a constant mean for air temperature in mountainous terrain. Another example: we know that spatial variability of soil organic matter depends on land use, so it would be unwise to assume a single variogram for organic matter in an area with cropland, grassland and forest. There are possibilities in geostatistics to relax the stationarity assumption, but for now we will assume that the assumption holds. Moreover, in Fig. 1.6 we also implicitly assumed that the semivariance only depends on the 'Euclidean' distance between locations, and not on the distance *vector* that also takes account of direction. This is the *isotropy* assumption. The opposite is *anisotropy*. For instance, the variogram of the soil clay content of a river floodplain may be different in a

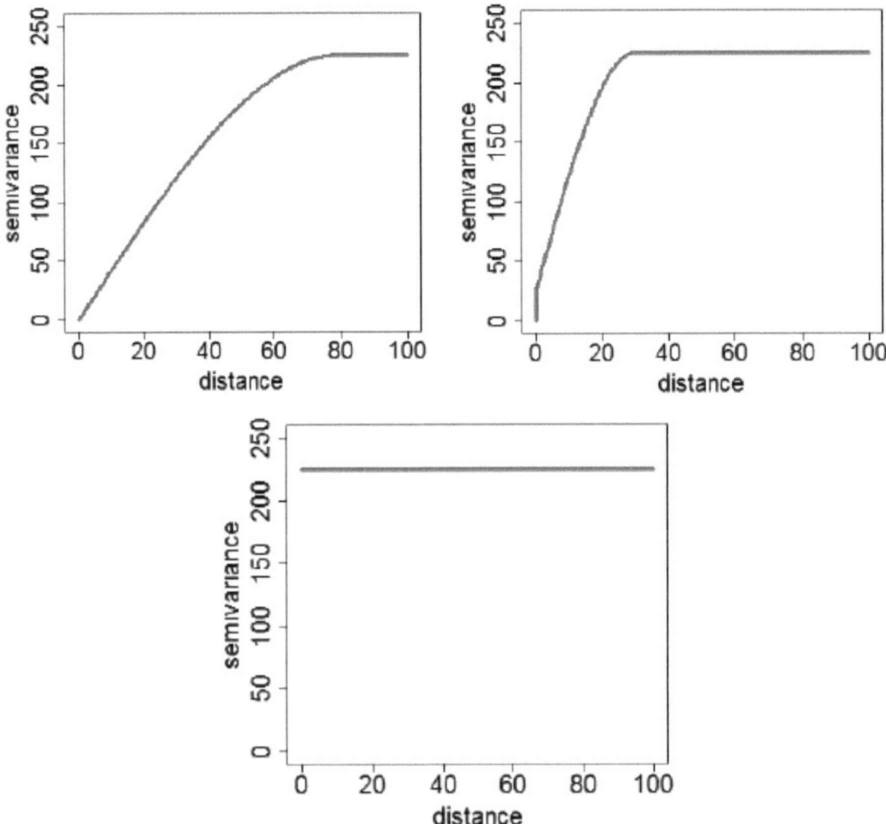

Fig. 1.7: The three variograms that correspond with realisations shown in Fig. 1.5 (left: top row; middle: third row; right: bottom row).

direction parallel to the river than in a direction perpendicular to it. In such case the isotropy assumption is not realistic and an anisotropic variogram should be used instead.

In practice we need to estimate the variogram from a set of observations. The procedure that is used for this is known as the *structural analysis*. We will not explain in detail how the structural analysis works but merely mention that it is based on evaluating the squared difference between observations for each pair of observation locations in the dataset. By averaging these squared differences over multiple distance intervals, a so-called sample variogram is obtained. Using curve fitting techniques a variogram model (mathematical function) is fitted through the sample variogram. The procedure works well but requires a sufficiently large set of observations. It is generally agreed that stable estimation of the variogram requires at least one hundred observations, which ideally are spread out through the entire study area, and include a few clusters of observations to be able to estimate the nugget reliably.

Relationship between the variogram, covariance function and correlogram

The variogram measures the degree of spatial variation as a function of the separation distance between two locations. We have seen that it typically is a function that increases with distance, due to Tobler's first law of geography. Unlike the variogram, the *covariance function* measures the covariance between the variable of interest at two locations as a function of the distance between the locations. The covariance was defined in Section *Covariance and correlation*, where it was explained that it measures the degree of co-variation of two random variables. In the case of a covariance function of a random field Z, it measures the degree of co-variation of $Z(x)$ and $A(x+h)$ and is defined as $C(h)=E[(Z(x)–E[Z(c)]).(Z(x+h)–E[Z(x+h)])]$. The covariance function tends to decrease with distance. This is confirmed by the mathematical relationship between the variogram γ and the covariance function C: $C(h) = \gamma(\infty) = (h)$ and $\gamma(h) = C(0) - C(h)$. In other words, the covariance function is a 'mirrored' version of the variogram, see Fig. 1.8.

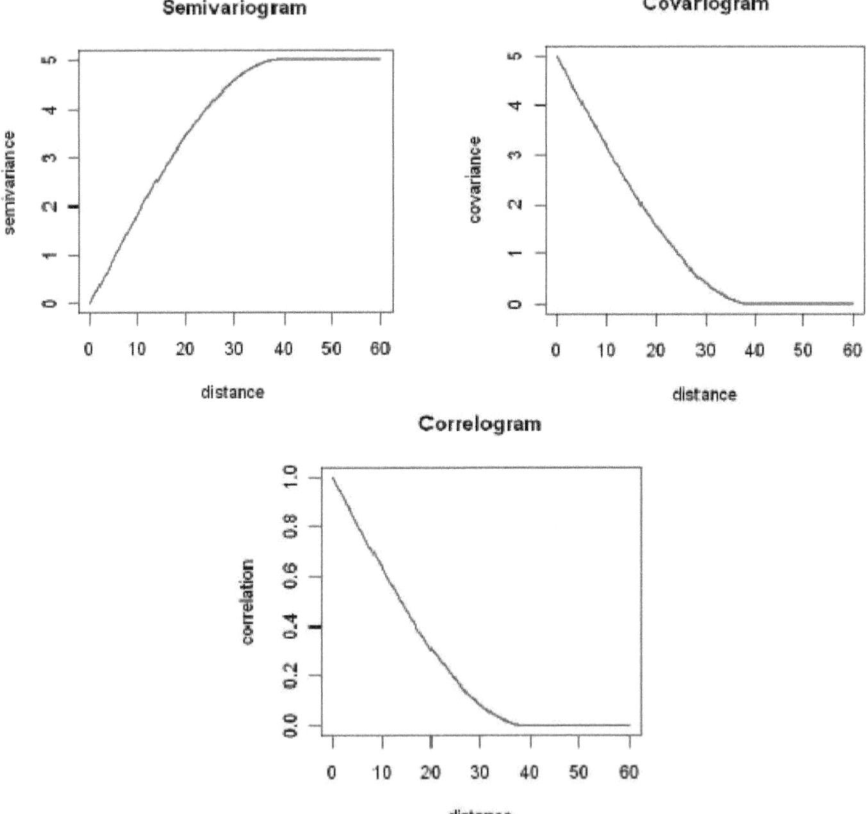

Fig. 1.8: Variogram, covariance function and correlogram of the same random field.

The *correlogram*, which plots the correlation as a function of distance, is simply a normalised version of the covariance function. It is obtained by dividing the covariance function by the overall variance (i.e., the sill of the variogram). Thus, the correlogram is dimensionless and cannot be greater than +1. In Fig. 1.8, the correlogram starts at 1, but note that it would be a value smaller than 1 if there were a nugget variance greater than 0.

Kriging

The section *'Statistical measure of spatial variation: the variogram'* explained how spatial variability can be characterised with a variogram. The variogram is a very useful tool because it provides insight into the degree and structure of spatial variation, which may help researchers understand the processes behind spatial variation, but it is also very useful for another reason. This is that it provides the information needed for geostatistical interpolation, also known as *kriging* (after the late South-African mining engineer Danie Krige). There are many versions of kriging. We will discuss some of these in this chapter, beginning with *ordinary kriging*.

Ordinary kriging equations

The idea of ordinary kriging is to predict $z(x_0)$ at a location x_0 where z was not measured as a linear combination of the observations:

$$\hat{z}(x_0) = \sum_{i=1}^{n} \lambda_i . z(x_i) \tag{7}$$

Here, the λ_i are so-called *kriging weights*. We would like to choose these weights such that the *interpolation error* $Z(x_0) - Z(x_0)$ is as small as possible. Now the problem is that we do not know the interpolation error, because in order to calculate it, we need to know $z(x_0)$, which we were aiming to predict in the first place. If we knew it, we would not have an interpolation problem. The solution that is used in geostatistics is that we characterise the interpolation error by a probability distribution. This means that we replace the realisation z by its geostatistical model Z and predict $Z(x_0)$ instead of $z(x_0)$. Given the geostatistical model that we have assumed for Z (including its variogram) we can derive the probability distribution of $Z(x_0) - z(x_0)$ for any choice of kriging weights, and next we choose the weights such that the variance of the prediction error is the smallest among all possible choices, while also ensuring that the expected value (*i.e.*, mean) of the prediction error is zero. In other words, we choose the kriging weights such that the expected squared prediction error is minimised under the condition that the prediction error is *unbiased*:

$$E\left[\left(\hat{Z}(x_0) - Z(x_0)\right)^2\right] = E\left[\left(\sum_{i=1}^{n} \lambda_i . Z(x_i) - Z(x_0)\right)^2\right] \tag{8}$$

$$E\left[\hat{Z}(x_0)-Z(x_0)\right]=E\left[\sum_{i=1}^{n}\lambda_i.Z(x_i)-Z(x_0)\right]=0 \tag{9}$$

The unbiasedness condition implies that the kriging weights must sum to one: $\sum_{i=1}^{n}\lambda_i=1$. This also makes intuitively sense: if the weights would add up to a number larger than one, then we would tend to overpredict the value of Z at x_0, while we would on average end up with too small predictions if the weights add up to a value smaller than one.

It was mentioned several times in this chapter that in geostatistics, we pose a statistical model Z of reality and assume that the single, true reality that we are interested in is a realisation Z of the statistical model Z. We characterise the model and its spatial variability by a variogram. Once this is done, we can derive the probability distribution of the kriging prediction error at any location x_0 in terms of a probability distribution. That distribution will look similar to the one shown in Fig. 1.2, only with different values for the mean and standard deviation. The 'real' prediction error at x_0 (which we do not know) may now be interpreted as a realisation from that distribution. Thus, the ideal shape of the distribution would be the one that has a zero mean and zero standard deviation, because in that case we would be certain that the interpolation error is zero. In practice, it will be impossible to choose the kriging weights such that this is achieved (unless the prediction location is an observation location and there are no measurement errors), so that we try to get as close as possible to it by ensuring that the distribution is centred around zero and has the smallest standard deviation possible. In this way, we achieve that on average, for many locations and for the many kriging exercises that we execute during our lifetime, the actual prediction errors made will be closer to zero compared with any other interpolation method. Of course all this is true under the assumption that the geostatistical model is a correct description of reality. But given this model, kriging is indeed an *optimal interpolator*.

Minimisation of Eq. (8) is achieved by working out Eq. (8), treating it as a function of the kriging weights λ_i, and minimising it by setting its mathematical derivative with respect to each of the λ_i to zero. This involves some calculus and linear algebra, which is not that difficult but will not be presented here. A slight complication is that the unbiasedness condition Eq. (9) must also be satisfied. For this, a mathematical 'trick' is applied in which an extra unknown, the so-called *Lagrange parameter*, is introduced. For us, it is sufficient to know that finally we end up with n+1 linear equations with n+1 unknowns:

$$\sum_{i=1}^{n}\lambda_i.\gamma\left(h_{ij}\right)+\varphi=\gamma\left(h_{i0}\right) \text{ for all } J=1...n \tag{10a}$$

$$\sum_{i=1}^{n} \lambda_i = 1 \tag{10b}$$

Here, h_{ij} is the geographic distance between x_i and x_j, h_{i0} the distance between x_i and x_0, and φ is the Lagrange parameter.

Eq. (10a and 10b) is known as the *kriging system*. It has a unique solution that is not difficult to calculate, because all equations are linear in the unknowns. However, calculation by hand becomes tedious and error-prone for values $n = 3$ and larger, and so in practice we use computers for that, also because we need to repeat the calculation for each and every prediction location x_0 (which are usually the nodes of a fine grid laid out over the area of interest). The resulting values for the λ_i obviously depend on the variogram and on the spatial configuration of the observation and prediction points, but the general picture is that observations nearby the prediction location get larger weights (because they are more strongly correlated with the variable at the prediction location), and that observations in clusters get smaller weights than more isolated observations (because of redundancy). In general, the smaller the nugget-to-sill ratio, the larger the diversity in kriging weights.

Once the kriging weights are calculated it is easy to calculate the ordinary kriging prediction, because it just requires substitution of the weights and the actual observations in Eq. (7). An attractive property of kriging is that the accuracy of the prediction is also quantified, by means of the *ordinary kriging variance*. This is nothing else than the expected squared prediction error Eq. (8), which can be worked out into:

$$\sigma_{OK}^2(x_0) = E\left[\left(\hat{Z}(x_0) - Z(x_0)\right)^2\right] = \sum_{i=1}^{n} \lambda_i.\gamma(h_{ij}) + \varphi \tag{11}$$

The square-root of Eq. (11) gives the *kriging standard deviation*, which is easier to interpret than the kriging variance because it has the same measurement units as the kriging prediction. One may also calculate the ratio of the kriging standard deviation and kriging prediction to get the *relative error*, or compute the lower and upper boundaries of an approximate 95% *prediction interval* by calculating $\hat{Z}(x_0) \pm 2 \cdot \sigma_{OK}(x_0)$ (see also Fig. 1.2).

Cross-validation

Although the kriging standard deviation map produces a measure of the accuracy of the kriging predictions, it is only valid given the kriging assumptions, which include the used variogram. After all, it is derived from the geostatistical model that was assumed. Instead, a more objective measure of map accuracy would be obtained if an independent data set were available. If this were the case, then we could compare the kriging predictions at the validation locations with the independent observations, and compute summary measures from these, such as the *Mean Error* (*ME*), *Root Mean Squared Error* (*RMSE*) and *Standardised Root Mean Squared Error* (*SRMSE*). These are defined as follows:

$$ME = \frac{1}{m}\sum_{i=1}^{m}\left[\hat{z}(x_i) - z(x_i)\right] \tag{12}$$

$$RMSE = \sqrt{\frac{1}{m}\sum_{i=1}^{m}\left[\hat{z}(x_i) - z(x_i)\right]^2} \tag{13}$$

$$SRMSE = \sqrt{\frac{1}{m}\sum_{i=1}^{m}\left[\frac{\hat{z}(x_i) - z(x_i)}{\sigma_{OK}(x_i)}\right]^2} \tag{14}$$

where m is the number of validation observations and the x_i are in this case the validation locations. The *ME* should be close to zero, the *RMSE* as small as possible, and the *SRMSE* should be close to one. If SRMSE is greater than one, then the 'observed' errors are bigger than the 'anticipated' errors. If SRMSE is smaller than one, then the observed errors are smaller than the anticipated errors. Both are undesirable outcomes because we want the kriging standard deviation to be a realistic measure of interpolation error.

Often we cannot afford splitting the dataset into one set for interpolation and one for validation, because we want to use all available data for interpolation to obtain the most accurate map possible. An often-used alternative then is to take out one or multiple observation(s) at a time for validation purposes and use all other data for prediction. This is called *cross-validation.* If one observation is removed at a time, it is called *leave-one-out cross-validation.*

Kriging Extensions

Regression kriging

Optimal spatial prediction using ordinary kriging only makes use of information contained in observations of the variable of interest. However, in many practical situations there is additional information that is useful too and can help improve prediction. For instance, we know that air temperature is correlated with elevation and since we have a DEM of most parts of the world, we might try to include elevation data to help predict air temperature. One way of doing that is through *regression kriging,* which is a combination of regression and kriging. It is also sometimes called *kriging with external drift* or *universal kriging,* although there are subtle differences that we need not discuss here. Before we explain regression kriging, let us first look at linear regression.

In (*multiple*) *linear regression*, we consider a dependent variable Y that is assumed to be linearly related to a number of independent (explanatory) variables X_i:

$$Y = \beta_0 + \sum_{k=1}^{p} \beta_k . X_k + \varepsilon \tag{15}$$

Here, the β_k are regression coefficients (β_0 is known as the *intercept*), p is the number of independent variables and ε is a zero-mean, normally distributed stochastic residual with constant variance σ^2. The regression coefficients and variance of the residual are estimated from paired observations of the dependent and independent variables. The standard deviation of ε is derived from the spread of the points around the fitted line. It conveys how well we can predict Y with the X_i. If the independent variables are informative about Y, then the variance of ε will be substantially smaller than that of Y. The variance reduction is neatly captured by the so-called R^2 (*R-square* or *goodness-of-fit*), which is a number between zero and one that expresses the *amount of variance explained* by the regression. In case of simple linear regression (i.e., $p = 1$), R^2 equals the square of the correlation between X and Y.

Prediction of Y given the X_i is done as follows:

$$Y = \beta_0 + \sum_{k=1}^{p} \beta_k . X_k + \varepsilon \tag{16}$$

In other words, we use the fitted regression line for prediction. This makes sense because the $\hat{\beta}_\kappa$ are our 'best' estimates of the true regression coefficients, and because ε has zero mean, and hence our 'best' estimate of it is simply zero. However, both the estimation errors of the regression coefficients and the stochastic residual of the regression model cause that the prediction differs from the true value of Y. In other words, there will be a prediction error, which has a mean of zero (*i.e.*, the prediction is unbiased), but whose standard deviation is bigger than zero. Multiple linear regression as explained above can also be used to predict a spatially distributed variable (such as air temperature) from other spatially distributed variables (such as elevation). However, the regression model assumes that the stochastic residual is *statistically independent*, while we learnt in previous sections of this chapter that many spatially distributed variables are spatially correlated. This implies that the assumption of statistical independence of the regression residual is often not realistic in case of spatially distributed variables, and this is where regression *kriging* comes in. Under the regression kriging model, the dependent and independent variables are made spatially explicit and the spatial correlation of the regression residual is modelled with a variogram. Since we had characterised the target variable by z and had used letter x for geographic location, we now write the spatial analogue of Eq. (15) as:

$$Z(x) = \beta_0 + \sum_{k=1}^{p} \beta_k . f_k(x) + \varepsilon(x) \tag{17}$$

Here, the f_k are spatially distributed explanatory variables, such as elevation, slope, vegetation index, land cover and geology. In fact, it could be any variable that is spatially exhaustively available and that is correlated with the dependent variable Z. Note that it must be spatially exhaustive, because otherwise we could not predict at locations where the explanatory variables are not available. Note also that the explanatory variables need not be numeric variables such as elevation and slope, but can also be categorical variables, such as land use and geology. Such variables can still be included in linear regression by treating them as *factors*, effectively replacing them with as many binary dummy variables as there are categories. In Eq. (17), the term $\beta_0 + \sum_{k=1}^{p} \beta_k . f_k(x)$ is known as the *trend.*

The only real difference between Eqns. (15) and (17) is that we now allow for spatial correlation of the stochastic residual ε. We typically assume stationarity, so that its spatial correlation is fully characterised by a variogram. Regression kriging then works as follows:

i). *Fit regression model.* Select spatially exhaustive explanatory variables (often termed '*covariates*'), overlay these with the locations of observations of the dependent variable, and fit a regression model on the resulting dataset.

ii). *Variography regression residual.* Subtract predictions of the regression model from the true observations at observation locations, and calculate a variogram of the resulting residuals.

iii). *Apply regression model.* Apply the regression model at all locations in the study area to generate a regression map of the dependent variable.

iv). *Krige residuals.* Krige the residuals with ordinary kriging to the same area.

v). *Combine results.* Add the map with kriged residuals to the map of the regression predictions.

This shows that regression kriging is truly a combination of regression and kriging. If we would do only regression, then we would not include a kriging interpolation of the regression residual. If we would do only kriging, we would only do a spatial interpolation of the dependent variable, without taking the information in the explanatory variables into account. In fact, ordinary kriging is a special case of regression kriging, namely the one where there are no explanatory variables (ρ=0) and where the trend is assumed constant. From this, you can also see that regression kriging relaxes the assumption of a constant mean or trend, as described in Section *The variogram*. It poses a more realistic model of reality, and it can potentially provide a more accurate prediction map because it uses more information: not only the observations of the dependent variable, but also explanatory covariate information. Thus, if there is explanatory power in the covariates, then the regression kriging variance should be smaller than the ordinary kriging variance.

Indicator kriging

The previous section mentioned that categorical variables can be used in regression kriging. In that case the categorical variables were used as explanatory variables, but what to do if the dependent variable is measured on a categorical scale? For instance, what if we have observations of vegetation type or land cover at locations in an area, and want to interpolate these observations to create a map of the dependent variable? In such case we can make use of *indicator kriging*. This is done as follows. First, for each category the observations are transformed into observations of a binary 'indicator' variable, which is one if the categorical variable at a location equals the category and zero otherwise. As a result we get a set of point observations that are either zero or one. Next, a variogram is estimated from the indicator values and used in kriging. This is done in the same way as described in Sections *Kriging and Kriging extensions*. The resulting prediction map typically has values between zero and one, which is as expected because Eq. (7) shows that the prediction is a weighed average of zeroes and ones. The prediction is interpreted as the *probability* that the category will occur at the prediction location. This is intuitively sensible, because values close to one will result if there are many ones in the local neighbourhood, meaning that the category often occurs in the local neighbourhood, while values close to zero turn up in areas where the specific category is rare or absent. Indicator kriging is then repeated for all other categories, again transforming observations to a binary indicator variable, estimating the variogram and applying kriging.

In practice, it may occur that predictions are smaller than zero or greater than one. In such case a clipping to zero or one is done. Also, it usually happens that the sum of the predictions for all categories do not add up to one, while obviously the probabilities should. This is repaired by scaling the predictions such that they do sum to one. In a way, these corrections show that while indicator kriging may work satisfactory in practice, its theoretical foundation is weak. Many geostatisticians are therefore not in favour of it. Unfortunately, there are not many alternative geostatistical methods and those that do exist are quite involved. Apparently, spatial interpolation of numerical variables is much easier than spatial interpolation of categorical variables.

Sampling design optimisation

We have seen before that one of the attractive properties of kriging is that it quantifies the interpolation error, by means of the kriging variance. This implies that we can compare the interpolation accuracy of different spatial sampling designs, by comparing the associated kriging variance maps. When comparing sampling, it makes sense to select the sampling design that has the smallest average kriging variance, because this yields the smallest overall interpolation uncertainty. Any design is characterised by two components. The first is the total number of observations. Collecting more observations yields a more accurate interpolated map, but at the expense of greater costs. The second is the spatial configuration of the sampling locations. Here, one configuration or design might produce a more accurate map than another, while sampling costs remain the same. This section briefly explores these issues and presents some common sampling design approaches used in geostatistics.

While it makes sense to use the average kriging variance as a criterion to compare sampling designs and select the optimal design, there is a problem. This is that the kriging variance can only be computed after the observations have been taken and a variogram calculated, while the sampling design must be chosen prior to the actual sampling. Indeed, Eq. (11) shows that the kriging variance can only be computed if the kriging weights, the variogram and the Lagrange parameter are known. The kriging weights and Lagrange parameter are computed by solving Eq. (10a,b), which again requires the variogram. This implies that geostatistical sampling design optimisation can only be done given the variogram. Note, however, that the observations themselves are not required to compute the kriging variance. Thus, all that we need to be able to optimise the sampling design is the variogram. In practice, two approaches are used to get hold of it prior to sampling for interpolation. The first is to 'borrow' the variogram from a previous, similar study, where the same variable was measured in a comparable study area. The second approach is to conduct sampling in two stages: the first only aiming at estimation of the variogram, the (optimised) second to interpolate the variable of interest with kriging. The disadvantage of the first method is that extrapolation of variograms is risky, the disadvantage of the second is that one needs to do field work and data collection twice. Nonetheless, it may definitely pay off, particularly in large projects and/or when observations (*e.g.*, laboratory analyses) are expensive. It is beyond the scope of this chapter to explain how the actual sampling design optimisation is done. We only note that it tends to go for a fairly uniform distribution of observations points across the study area, with slightly higher sampling density in the boundary region of the study area.

In practice, it may frequently happen that we cannot 'borrow' a variogram from another study and cannot afford to do field work twice. In that case we must design a spatial sampling design that will be used both for variogram estimation and kriging. A uniform distribution of points across the study area is preferred for kriging, as we noted above. One practical way of doing that is to sample at the nodes of a regular grid, where the grid mesh is chosen such that the total number of observations stays within the field work budget. However, regular grid sampling is far from ideal for variogram estimation. This is because such design provides no information about the spatial variability at distances smaller than the grid mesh, while kriging is sensitive in particular to the behaviour of the variogram at short distances (i.e., the nugget variance). Therefore, it is recommended to supplement the regular grid with multiple small 'clusters' of points, such as shown in Fig. 1.9. As a rule of thumb, one might use two-thirds of the total number of observations for regular grid sampling, and the remaining one-third to assess short-distance spatial variation.

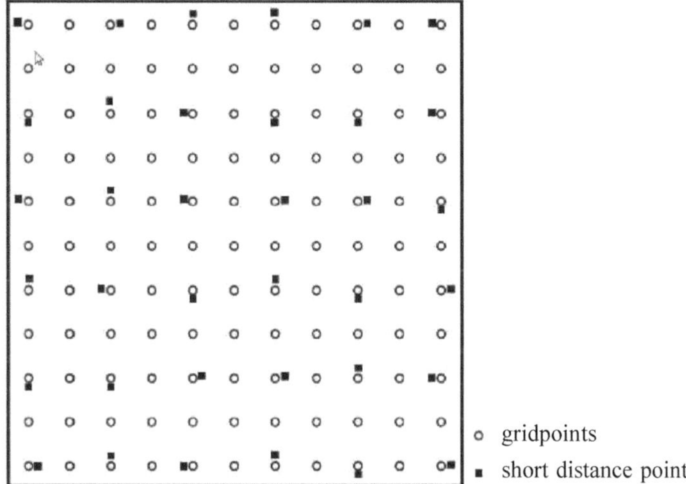

Fig. 1.9: Regular grid sampling supplemented with short-distance points to achieve a spatial sampling design that works well for variogram estimation and kriging.

Spatial stochastic simulation

Spatial stochastic simulation versus kriging

Kriging makes predictions at points, such that the expected squared prediction error is minimised. This is attractive because it means that the predicted value is on average closest to the true (unknown) value. *Spatial stochastic simulation* has an entirely different objective. Here, the goal is to generate possible realities from the probability distribution of the uncertain variable. This is done using a *pseudo-random number generator*, while accounting for the shape of the probability distribution, the spatial correlation structure and the observations. The result of a stochastic simulation exercise is not unique, because there are an infinite number of possible realities, from which just one or several are taken. To illustrate this, let us recall the example of the throw of a die discussed in Section *Random variables*. Optimal prediction would produce the value of 3.5, because on average this is the number closest to any of the outcomes 1 to 6. However, stochastic simulation would simply take one of the numbers 1 to 6, where each of the six values would have an equal chance of being selected.

Spatial stochastic simulation is useful for two main purposes. First, visualisation and comparison of multiple simulated realities nicely conveys the uncertainty about the mapped variable. Where uncertainty is large the differences between simulated values will be large, where there is little uncertainty the differences will be small. Also, the spatial structure in the simulated maps agrees with that of reality, whereas kriging produces a smoothed version of reality. Second, spatial stochastic simulation is also required in spatial uncertainty analysis studies. The aim of such studies is to analyse how errors and uncertainties in inputs to environmental models (such as interpolated maps) propagate through the model. This can be analysed using a Monte Carlo simulation approach, which requires that realisations of the uncertain inputs are generated using stochastic simulation. For spatially distributed inputs, spatial stochastic simulation must be used.

There are various ways for spatial stochastic simulation. Perhaps the most attractive method is *Sequential Gaussian Simulation*, which works as follows:

i). Define a grid of simulation locations across the study area (as also done in kriging).

ii). Visit a randomly selected simulation location and verify that it does not coincide with an observation or already simulated location (if it does, then select another simulation location).

iii). Krige to the simulation location, this yields a kriging prediction and kriging standard deviation.

iv). Use a pseudo-random number generator to sample from the normal distribution with mean equal to the kriging prediction and standard deviation equal to the kriging standard deviation.

v). Add the simulated value to the data set, in other words treat the simulated value as if it were another observation.

vi). Go back to step 2 and repeat the procedure until there are no more simulation locations left.

Note that adding previously simulated values to the dataset as required in step 4 causes the kriging system (i.e., the number of equations, see Eq. (10a & b) soon becomes very large so that it will be needed to set a maximum on the number of observations to be used in the kriging (step 3). Usually, one limits the observations to those located in a circular window surrounding the simulation location. If this is done, it is called *local kriging*. The opposite, in which all available observations are used, is called *global kriging*. Local kriging is attractive because it is fast (in the case of spatial stochastic simulation on a fine grid it is imperative for this reason, because global kriging would take 'forever'), and also because the kriging result is more robust against deviations from the stationarity assumption. The disadvantages are that the kriging variance will increase slightly, and that anomalies in the kriging prediction map may result if the local window is chosen too small.

2

Digital Soil Mapping

Amit Kumar[1]*, Probir Das*[1]*, Joydeep Mukherjee*[1]*, Anupam Das*[1]
Nilimesh Mridha[1]*, Priyabrata Santra*[2] *and Debashis Chakraborty*[1]

[1]*ICAR-Indian Agricultural Research Institute, New Delhi, India*
[2]*ICAR-Central Arid Zone Research Institute, Jodhpur, Rajasthan, India*

Introduction

A complex spatial soil properties pattern is directly controlled by soil internal factors and anthropogenic impacts. Hartemink et al. (2001) elaborate a strong decline in mineralogy, morphology and genesis of soil research than strong increase in pedometrics applications during 1967 to 2001. Hartemink and McBratney (2008) identified this change as "soil science renaissance". Basically, the soil science deeply rooted in agriculture, geology, and chemistry but the paradigm has been shifted from classification and inventorization to quantification of soil patterns in spatial as well as temporal patterns and their impact on the ecosystem health and hydrological cycle. The environmental keen approach explains that soil is in the center of the ecosystem and highly interacting with the biotic and abiotic factors of the ecosystem and a complex pattern and process has been co-evolved with the time.

In these days, the world economy is in the high urge of the quantitative Land Resource Assessment (LRA), which is mostly provided by from soil survey agencies over the globe. LRA provides necessary information on the spatial distribution pattern of soil and land attributes, which is used to predict the environmental quality by using physical and chemical models. In the present time, soil characterization (physical and chemical parameters) is very expensive and the information gathered by soil survey is not upto the required quality to use in the soil models. This is the ultimate limitation of most of the available soil and land resources information. Therefore, linking the presently available soil information to accurately measure soil functional properties and their simulation prediction to provide the easy, accessible, inexpensive techniques is the biggest challenge.

Characterization of soil spatial variation can be done by mapping the soil classes (Lagacherie 2005). Soil classification can be retrieved from local, national, global soil classification such as numerical classification (Odeh et al. 1992, Bragato 2004), the Australian Soil Classification System (Isbell 1996), or World Reference Base (IUSS Working Group WRB 2006). The preference of using these soil maps by scientist is due to their familiarity with the soil class concept and the properties linked to these classes. These soil classes can also be utilized to estimate the soil physico-chemical properties without doing laboratory exercise or in the unavailability of laboratory. In digital soil mapping, at each soil observation location, there is a set of co-located environmental parameters/variables, but the linking of these classes of soil with environmental variables is huge task.

In these days, there are great advances in the computation and information technology. The detail regional, national, continental and worldwide soil data is available (McBratney et al. 2003). Data mining is the emerging field by utilizing these data sets (Hastie et al. 2001). Geographic information system is used to produces soil class map and digital properties of soil without doing the expensive and exhaustive soil survey and then laboratory analysis.

Introduction of digital technologies like remote sensing, high speed of computer processing, spatial data management, and quantitative methodology to describe soil patterns, processes and scientific visualization methods provides new opportunities to predict soil properties and processes.The history of digital soil mapping and modeling (DSMM) is marked by adoption of new tools and techniques to analyze, integrate, and visualize soil and environmental datasets (Grunwald 2006).

What is digital soil mapping (DSM)?

Digital soil mapping (DSM) is "computer-assisted production of digital representations of soil type or soil properties, which involve the creation and population of spatially-explicit information by the use of field and laboratory methods, coupled with spatial and non-spatial soil inference systems". The digital soil map production is not the digitization of existing soil map (Skidmore et al. 1991, Favrot and Lagacherie 1993, Moore et al. 1993). In the last fifteen years, many published literature related to Digital Soil Mapping (DSM) define the creation and population of spatial soil information using field and laboratory observations which are coupled with spatial and non-spatial soil inference system (McBratney et al. 2003, Lagacherie and McBratney 2007). These important techniques have been accepted and used by the soil science community at a large scale (Lagacherie et al. 2007).

The outputs of DSM are soil properties/classes derived by a spatial inference system. Spatial inference system is based on a number of predictive approaches involving environmental covariates, prior soil information (point & map form), survey and laboratory observational methods linked together with spatial and non-spatial soil inference systems (Lagacherie et al. 2007). This allows the prediction of soil properties/classes by utilizing soil information and environmental properties (Dobos et al. 2006). The important advantages of DSM are costs, consistency and documentation, and it can also be easily updated when new data become excisable along with the capability of uncertainties estimation for predicted outcomes thus allowing the tracking of error propagation through the whole process (Carré et al. 2007). DSM can be used for quantitative modelling of difficult-to-measure soil attributes, necessary for assessing threats to soil, *i.e.* erosion, decline of organic matter, compaction, salinization, landslides, sealing, floods, decline of biodiversity and the soil functions, i.e. biomass production, environmental interactions, physical support, production of raw material, cultural heritage, carbon pool, source of biodiversity (Carré et al. 2007). DSM can also be used for quantitative evaluation of soil-related scenarios for providing policy guidance using from digital soil assessment plus socioeconomic data outputs along with the general environment information (Carré et al. 2007).

According to McBratney et al. (2000) there are three resolutions of interest, viz. >2 km, 20 m –2 km and <20 m corresponding to national to global, catchment to landscape and local extents. The details are given in Table 2.1. The D3 survey, which deals with sub-catchments, catchments and regions, is the most important. In digital soil maps (Bishop et al. 2001), scale is replaced by resolution and spacing. D3 surveys have a scale of 1:20,000 down to 1:200,000; have cell size from 20-200 m, a spacing of 20–200 m with a nominal spatial resolution of 40–400 m (Table 2.1).

How to do digital soil mapping

Major Approaches of Soil Spatial Prediction

State Factor or clorpt Model

The state factor model is expressed by the following equation:

$$S = f\left(cl, o, r, p, t, \ldots\right) \tag{1}$$

where soil (S) is considered to be a function of climate (cl), organisms (o), relief (r), and parent material (p) acting through time (t) (Jenny 1941, 1980). The ellipsis (.) in the model is reserved for additional unique factors that may be locally significant, such as atmospheric deposition. The 'clorpt' equation illustrates that by correlating soil attributes with observable differences in one or more of the state factors, a function (f) or model can be developed that explains the relationship between the two, which can be used to predict soil attributes at new locations when the state factors are known. An important distinction of the Clorpt model is that "The factors are not formers, or creators, or forces; they are variables

Table 2.1: Suggested resolutions and extents of digital soil maps

Name	Approximate USDA survey Order[a]	Pixel size and spacing[b]	Cartographic Scale[b]	Resolution	Nominal spatial Resolution[b]	Extent[d] (km)	Cartographic Scale[b]
D1	0	< (5 × 5) m	> 1:5000	< (25 × 25) m	< (10 × 10) m	< (50 × 50)	> 1:5000
D2	1, 2	(5 × 5) m	1:5000 -	(25 × 25) m	(10 × 10) m	(500 × 500) to	1:5000 -
		(20 × 20) m	1:20,000	(100 × 100) m	(40 × 40) m	(200 × 200)	1:20,000
D3	3, 4	(20 × 20) m	1:20,000 -	(100 × 100) m	(40 × 40) m	(2 × 2) to	1:20,000 -
		(200 × 200) m	1:200,000	(1 × 1) km	(400 × 400) m	(2000 × 2000)	1:200,000
D4	5	(200 × 200) m	1:200,000 -	(1 × 1) km	(400 × 400) m	(20 × 20) to	1:200,000 -
		(2 × 2) km	1:2,000,000	(10 × 10) km	(4 × 4) km	(20,000 × 20,000)	1:2,000,000
D5	5	> (2 × 2) km	< 1:2,000,0000	> (10 × 10) km	> (4 × 4) km	> (200 × 200)	< 1:2,000,0000

(Source: McBratney et al., 2003)

[a]Soil Survey Staff (1993).

[b]Digital soil maps are partly defined by their block size and spacing (Bishop et al. 2001), which, here, we equate with pixelsize. The cartographic scale, calculated as 1 m/(side length of 1000 pixels), assumes that the smallest area discernible 1×1 mm. Conversely, the pixel size (*p*) of a 1:100,000 conventional map can be calculated as $p=\chi*\lambda=$ 100,000×0.001 = 100 m if we consider the smallest are solvable on a map (λ), with representative fraction ÷ to be 1×1 mm. Following notions in microscopy, and the Nyquist frequency conceptfrom signal processing, it may be argued that the minimum resolution is the size of 2×2 pixels. We define this here as the nominal spatialresolution. Small pixel sizes correspond to fine resolutions and large pixel sizes correspond to coarse resolutions.

[c]According to Boulaine (1980), the smallest area discernible on a map is 0.5×0.5 cm or one quarter of a cm^2, hence, the term'loi du quart'. The USDA Soil Survey Field Handbook (Soil Survey Staff 1993) quotes 0.6×0.6cm. Both of these really refer toconventional map delineations, and resolution estimates based on these minimum areas should be regarded as very conservative.

[d]Calculated as minimum resolution times 100 (pixels) up to maximum resolution times 10,000 pixels.

[e]This order was suggested by Dr. Pierre C. Robert, University of Minnesota, for applications in precision agriculture.

(state factors) that define the state of a soil system" (Jenny 1961). This means that the factors do not constitute pedogenic processes, but are factors of the environmental system which condition processes.

Numerous researchers have taken the quantitative path and have tried to formalise this equation largely through studies of cases where one factor varies and the rest are held constant. Therefore, quantitative climofunctions, topofunctions, etc., have been developed. Much of this work was done before sophisticated numerically intensive statistical methods became available. Here are some brief examples.

Jones (1973) found relationships between carbon, nitrogen and clay and annual rainfall and altitude in West African savanna using linear and multiple linear regression. Simonett (1960) found a power–function relationship between mineral composition of soil developed on basalt in Queensland and annual rainfall.There seems less development of organofunctions, many believing that the principal organofunction or biofunction, that of vegetation, is dependent on soil rather than the converse. Noy-Meir (1974) found relationships between vegetation and soil type in South Eastern Australia. The other principal organofunction, the anthropofunction, has only been recently quantified. Much of the work on soil degradation and soil quality are evidence of the effect of humanity on soil. The classic work of Nye and Greenland (1960) is an early quantitative example.

The relationship between soil and topographic factors has been evident since Milne's (1935) paper. Quantitative topofunctions are manifold. For example, Furley (1968) and Anderson and Furley (1975) found a piece-wise linear relationship between organic carbon, nitrogen and pH of surface horizons, and slope angle for profiles developed on calcareous parent materials around Oxford in England. Quantitative lithofunctions have not been developed often, perhaps due to a difficulty in recognising and quantifying the dependent and independent variables. Barshad (1958) quantified mean clay content as a function of rock type.

Some consider this the only truly independent variable. Chronofunctions are often theoretical or hypothetical rather than observed. Hay (1960), however, found an exponential relationship between clay formation and time for soil developed in volcanic ash on the island of St. Vincent, as would be expected from first-order kinetics.

1.2. Scorpan Model – The Digital Soil Mapping Formula

Recently, McBratney et al. (2003) have suggested a revised formalization of the state factor model. This revised formula is expressed by the following equation:

$$S = f(s, c, o, r, p, a, n) \tag{2}$$

where S, a set of soil attributes (Sa) or classes (Sc), is considered a function of other known soil attributes or classes (s), climate (c), organisms (o), relief (r), parent materials (p), age or time (a), and spatial location or position (n). The scorpan equation also explicitly incorporates space (x,y coordinates) and time(~t). Thus, the scorpan equation can be expanded as follows:

$$S[x,y,\sim t]=f\left(s[x,y,\sim t],c[x,y,\sim t],o[x,y,\sim t],r[x,y,\sim t],p[x,y,\sim t],a[x,y],[x,y]\right) \quad (3)$$

This expansion indicates that scorpan is a geographic model, where the soil and factors are spatial layers that can be represented in a geographic information system.

The Scorpan model deviates from clorpt in that it is intended for quantitative spatial prediction, rather than explanation (McBratney et al. 2003). This distinction justifies the inclusion of soil and space as factors, because soil attributes can be predicted from other soil attributes and spatial information. For example, many soil attributes are correlated and can thus be reasonably predicted from each other. Also, Tobler's first law of geography (Tobler 1970) tells us that near things are spatially correlated, and can thus be predicted by their distance from their neighbors. To account for the soil factor, prior soil information can come from either published soil maps or the expert knowledge of soil surveyors. The space factors can come from indices of relative position, or by incorporating spatial auto correlation (e.g. co-kriging or regression kriging). While prior soil information is undoubtedly not an independent factor, the independence of spatial information is dubious. In most situations, spatial information likely accounts for relationships not captured by other factors (McBratney et al. 2003). Otherwise, from a metaphysical perspective, spatial information accounts for the random diffusion of particles trying to achieve a uniform state within a system (Hengl 2009). The addition of space is not a new idea, as Jenny himself spent a great deal of time validating spatial soil relationships (Hudson 1992). However, the revised formulation of scorpan recognizes the importance of prior soil information and spatial relationships to help explain soil spatial variation.

STEP-AWHB Model – A space-time modeling framework

Given the importance of anthropogenic forcings in determining observed soil properties, Grunwald et al. (2011) have proposed a new conceptual model for understanding soil properties for a pixel (p_x) of size x (width = length = x) at a specific location on earth, at a given depth (z), and at the current time (t_c):

$$SA(z,p_x,t_c)=f\left\{\sum_{j}^{n}\left[S_j((z,p_x,t_c),T_j(p_x,t_c),E_j(p_x,t_c),P_j(p_x,t_c)\right]\right\};$$

$$\int_{i=0}^{m}\left\{\sum_{j}^{n}\left[A_j((p_x,t_i),W_j(p_x,t_i),B_j(p_x,t_i),H_j(p_x,t_{ci})\right]\right\} \quad (4)$$

where the soil property of interest (SA) is a function of a number (j = 0, 1, 2, ..., n) of relatively static environmental factors (only at t_c): ancillary soil properties (S), topographic properties (T), ecological properties (E), and parent material properties (P), as well as a number (j = 0, 1, 2, ., n) of dynamic environmental conditions (with values representing dynamics through time t_i with i= 0, 1, 2,., m): atmospheric properties (A), water properties (W), biotic properties (B), and human-induced forcings (H). The model is spatially explicit because it constrains all properties included in equation to a specific pixel location. The model is temporally explicit as indicated by the inclusion of (time) in the above equation. This recognizes the spatial variation and temporal evolution of STEP-AWBH properties that covary and coevolve with the target soil property SA. Similar to the 'scorpan' model, the STEP-AWBH model reflects the new emphasis on existing soil information and spatial location as key attributes capable of providing predictive power in soil models.

The STEP-AWBH model separates hydrologic properties (W) from topographic (T) and climatic (A) factors, and formally includes anthropogenic properties (H). In previous factorial soil models, topography (relief) indirectly expressed the effects of hydrology on soil genesis. Yet, this is a simplification of reality that the STEP-AWBH model tries to overcome, where T represents topographic properties (e.g. elevation, slope gradient, slope curvature, and compound topographic index) that have been shown to be correlated with soil properties (Grunwald 2009). However, water flow and transport processes in soil depend on the interplay of A (atmospheric properties, such as precipitation arriving at the soil surface); soil surface conditions (e.g. salt crusts, residues, and density and composition of vegetation cover); internal soil characteristics that determine infiltration, percolation, and lateral flow processes (such as soil texture and soil organic matter); parent material (geologic formations) that control flow into the surface and deep aquifer; and topographic properties that enhance or subdue water flow at or close to the soil surface. The AWBH variables can be entered into the equation using spatially and temporally explicit sets (e.g. temperatures at a discrete time, *e.g.* Jan. 1, 2011) or condensed/aggregated data over a period of time (*e.g.* mean annual temperatures 2000–2010). The H factor represents different anthropogenic forcings that can act across shorter or longer periods of time on $SA(z,p_x,t_c)$ to shift SA into a different state, such as greenhouse gas emissions, contamination (*e.g.* an oil spill), disturbances, overgrazing and others.

Currently available digital soil data at global scales

Mostly Earth's land area is covered by existing soil maps at of different scales, from low-resolution (1:5,000,000 FAO-UNESCO, Soil Map of the World), to moderate resolution (1:24,000, NRCS soil survey maps), to high resolution (1:5000 Belgium pedologic map). Although, some of them with their associated databases have been digitized and are available in digital format yet digital soil mapping is more than digitizing existing soil maps. Finke (2007) told how to assess the accuracy of digital soil maps such as producer accuracy and user accuracy. When evaluating presently available digital soil maps, or when considering the potential for new digital soil map products, data quality can be perceived as a function of positional quality, attribute quality, completeness, semantic quality, currency, logical consistency, and lineage (Finke 2007).

The FAO-UNESCO Soil Map of the World (Nachtergaele 1999), originally published as paper maps between 1971 and1981, has been digitized, generalized, modified, and updated later to produce several global digital soil databases. While there is no current alternative to these digital versions of the FAO-UNESCO Soil Map of the World at the global scale, some fixed shortcomings to this map are recognized and identified (Sanchez et al. 2009) such as the map does not adequately represent the present soil condition and not the current state of knowledge of soils and soil classification. All of the digital versions of the Soil Map of the World are at coarse scales (1:25,000,000 or 1:5,000,000) and represent information from soil classes. The very first versions - the World Soil Resources Map was produced in 1990 and then World Reference Base (WRB) World Soil Resources Map were produced in 2003 and digitized as generalized paper map versions, which are presented at a scale of 1:25,000,000. The primary difference between these two databases is that the legend for the WRB World Soil Resources Map was updated to conform to the WRB classification system. The more recent Digital Soil Map of the World was produced in 2007 which, provides a digital rendering of the FAO-UNESCO map at the original resolution of 1:5,000,000 as its predecessors, represents the dominant soil types within each soil map unit polygon. The soil types/classes were only indirectly related to soil-environmental change induced by the Anthropocene. The other permutation is the global data set of derived soil properties from 2005 and 2006. These were created by the International Soil Reference and Information Center (ISRIC), which are combined spatial from the digital soil map of the world with measured soil property data from the World Inventory of soil emission potentials global soil profile set (Table 1). Both databases are of coarse spatial resolution (at the equator, the 0.5° resolution raster of the global data set of derived soil properties is approximately equivalent to a horizontal resolution of 55 km, whereas the 5 arc-min resolution of the derived soil properties database is approximately 9 km). Both datasets provide about 20 soil properties for 2-5 depth intervals, with available water capacity, base saturation, bulk density, cation-exchange capacity, coarse fragment content, drainage class, electrical conductivity, organic C, particle size distribution, pH, and total N. The digital harmonized world soil database (FAO/IIASA/ISRIC/ISS-CAS/JRC 2009) is the recently derived spatial data set from the soil map of the world. It is a raster data set with a 30 arc-sec resolution (~1 km), which provides data on soil classes and 13 selected soil properties. While the harmonized world soil database is also derived from the FAO-UNESCO map, which is having regional and national soil information from around the globe to update both the spatial and tabular data in the database to produce a more consistent representation of world soil resources (FAO/IIASA/ ISRIC/ISS-CAS/JRC 2009).

There are many digital soil map products accessible at regional to continental scales but does not having non-digital products such as atlas publications i.e. Soil Atlas of the Northern Circumpolar Region (eusoils.jrc.ec.europa.eu/library/maps/ Circumpolar/; verified 9 May 2011), they also does not include non-map products such as soil profile databases i.e. the NRCS (National Soil Characterization Database). These digital map products were produced by digitization of older paper maps without sampling of current soil resources, and most of them represent a compilation of multiple paper maps created at different times, by different individuals, at different scales, and for different purposes. These compilations have varied currency and lineage and often contain logical inconsistencies (most evident where two or more separate maps have been joined, e.g., at political boundaries), which completely diminish the data quality (Finke 2007). This is the urgent need to improve attribute quality and the semantic quality of the resulting digital soil map simultaneously. Canada and the United States digital soil survey databases are the exception, in which, all of the regional and continental databases are at coarse scales (1:1,000,000 or coarser). The most detailed data the soil landscapes of Canada (version 3.1.1) and the SSURGO database have incomplete spatial coverage.

The Soil and Terrain Digital Database (SOTER) project has been a source for multiple national and regional digital soil maps and databases. The aim of the SOTER project was to develop global soil database coverage at 1:1,000,000 scales (Batjes 1990). Cooperation among United Nations Environmental Program (UNEP), FAO, and ISRIC, soil class maps that represent standardized soil and terrain attributes have been developed for South America, southern and Central Africa, and eastern and central Europe at scales of 1:2,000,000 to 1:5,000,000. The map units of the SOTER databases delineate land areas with distinct patterns of soils and associated land forms and parent materials. These SOTER products are expected to have higher positional accuracy, better attribute accuracy, higher semantic accuracy, and better logical consistency than the Digital Soil Map of the World. However, they are still at relatively coarse scales.

Finally, the available digital global soil data do not meet the profound needs of the Anthropocene. If such coarse-scale soil data are included in global ecological biodiversity models, global climate change simulation models and ecosystem service assessments, major uncertainties arise with unknown outcomes. Global soil monitoring networks describing soil change have not been implemented at this point in time. Assessing the impact of anthropogenic impact on soil health, quality, services, degradation, and change are in need of higher spatial and temporal-resolution soil data, which are presently not available at continental and global scales.

Future trends of DSM

Recent research papers have dealt with Digital Soil Mapping (DSM), defined the creation and population of spatial soil information by the use of field and laboratory analysis/observational methods coupled with spatial and non-spatial soil inference systems (McBratney et al. 2003, Lagacherie and McBratney 2007), which are accepted and Widely used by the soil science community (Lagacherie et al. 2007). The DSM can be used for Digital Soil Assessment (DSA) and Digital Soil Risk Assessment (DSRA), both of which are downstream application of DSM. The DSA is the quantitative modelling of less easily measured soil attributes, essential for assessing threats to soil (erosion, decline of organic matter (OM), compaction, salinization, landslides, sealing, floods and biodiversity decline) and soil functions (biomass production, environmental interactions, physical support, production of raw material, cultural heritage, carbon pool, biodiversity sources) (European Commission 2006b), using DSM outputs. The DSRA is the quantitative evaluation of soil-related scenarios for providing policy guidance using the outputs from DSA along with socio-economic data and more general environment information.

The scope of DSA

DSA comprises two main processes, which are:

(1) A soil attribute space inference system.

(2) An evaluation of the soil functions and the threats to soils.

A description of each these two processes follow together with the suggested advice appropriate to policy makers.

The soil attributes space inference system

Basically, the primary data is loaded into a soil spatial inference system to build up soil property maps which are used as input to the soil attribute space inference system to extend the range of available properties/characters. The soil attribute space inference system is composed of pedo-transfer function that predicts diffcult to measure soil properties essentially required by the user (Lagacherie and McBratney 2007). One important precaution for using the pedo-transfer functions is that they should not be used without their uncertainty evaluated (McBratney et al. 2002).

Different approaches can be used for uncertainty quantification of the particular model. One is that the error can be allotted to the pedo-transfer functions and evaluate the impact on the output (Vereecken et al. 1992, Christiaens and Reyen 2001). Secondly, error evaluated by summation of the function error and the input data uncertainties (Minasny et al. 1999). Second approach is mostly preferred due to its advantage of considering all the possible uncertainties associated with the global process and can be achieved by Monte-Carlo simulation (Heuvelink 1998). Further, the validity domains of each pedo-transfer function should be evaluated to ensure that they are appropriate for the intended use.

There are three major approaches for extracting inferred attributes (hard class, fuzzy class and continuous class) from different types of input data (fuzzy class or continuous class) with pedo-transfer functions (McBratney et al. 2002). Each pedo-transfer function can then be calculated from several suitable models. The inference system is a very powerful tool within DSA as it enables the calculation of a much wider range of soil properties/classes that can be provided by traditional soil surveys in which laboratory analyzed data are often restricted to a few soil properties/characteristics. Many models assessing soil functions and threats as well as ecosystem services require a much larger number of input variables, particularly dynamic models.

Digital assessment of soil functions and threats to soil

Soil functions represent the various ecological and socio-economic roles of soils (COM231, European Commission 2006a) and threats to soil are the soil degradation processes arising principally from anthropogenic pressure (COM232, the European Commission 2006b), the assessment processes essentially considered into account the different approach methodologies. Whether for actual or potential, for functions or threats an individual assessment approach may be required. Applied to soil functions, potential is taken to mean the capacity of soil to perform a function, whereas assessment of actual function is the occurrence of the soil function. For instance, potential soil biomass production is modeled through different soil properties, climate data and land management information, whereas actual biomass production can be estimated through yield measurement.

The assessment of the actual threat is the occurrence of the threat. For example, potential erosion is generally modelled by combining soil erodibility (i.e. resultant of texture, structure, soil moisture, roughness and organic matter content), parent material, climate and relief. Actual soil erosion can be re-examined/assessed using relationships between in situ measured suspended soil particles concentration and atmospherically corrected spectral reflectance obtained from satellite remote-sensing data (Carpenter and Carpenter 1983, Harington et al. 1992, Nellis et al., 1998, Vrieling 2006, Vrieling et al. 2007).

The assessment of the potential involves the inference of soil properties, whereas actual assessment requires some degree of measurement. Since, DSA is a quantitative model with spatially inferred soil information and other environmental and socio-economic data as input parameters, it refers only to potential assessment. Hence, the discussion that follows is restricted to the assessment of potential functions and threats.

The digital assessment of soil functions

The seven most important soil functions identified in COM232 (the European Commission 2006b) are: (1) biomass production along with in agriculture and forestry, (2) storage, filtration & transformation of nutrients, substances and water, (3) biodiversity pool at genetic, specific and habitat levels, (4) physical and cultural environmental development for human and anthropogenic activity, (5) source of raw materials, (6) serve as carbon pool, and (7) archive of geological and archeological heritage. However, each function encompasses several sub-functions for example soil biomass productivity may be used separately for production of arable, grasses, energy and forest production. Then, all sub-functions must have to be estimated independently and separately.

For biomass production of soil and its interactions with environment, some functions are on high priority than others. For example after the Second World War, governments paid particular attention to land suitability and to crop production since food production for the need of population was on high merit and challenges and so biomass production of soil can be quantitatively evaluated whereas cultural heritage production is tedious to assess.

Crop models can be considered in a same trend to pedo-transfer functions. Therefore, estimation of the overall accuracy of such crop models can be done. However, most of crop models, soil variables/parameters play a secondary role to weather and climate information. This results in a lack of appreciation of inaccuracy of soil data. Accuracy can be evaluated in some cases, for example, Lagacherie et al. (2000) evaluated empirical information accuracy by using fuzzy logic for modeling. However, in case of legacy data use, then the accuracy is not easy to assess.

Digital assessment of threats to soil

The major threats to European soil are (i) erosion, (ii) decreased inorganic matter, (iii) local and diffuse contamination, (iv) sealing, (v) compaction, (vi) decline in biodiversity, (vii) salinisation and (viii) landslides (COM179, European Commission 2002, Eckelmann et al. 2006). Beside these, different part of the world also having additional threats such as desertification and acidification, and these may assume greater importance for the respective region. This is because of the most of the risk related to soil are highly diverse, so, there is no single model can be function as cosmopolitan in nature. For example, soil erosion is a physical phenomenon resulting in the removal/trasporatation of soil particles by water/wind, to elsewhere. Thus there are different erosion types such as water, wind, snow, bank erosions etc. Each type of erosion must then be quantified in order to assess all aspects of erosion.

Digital assessment of soil processes affecting the ecosystem functions

DSA also allows integrating soil functions with other variables/ covariates like air, biosphere and water in order to assess ecosystem functions. In this way, soil can be a threat for the ecosystem, whereas in the previous section it was itself under threat. For example, soil compaction can be a more or less intensive source of emission of greenhouse gases (Ruser et al. 2006). Due to compaction, restricted rooting can decrease crop yields by making soil moisture and nutrients unavailable (Sriramam 1968, Lipiec and Hatano 2003, Choi et al. 2005).

In these days, mapping of soil functions and its related threats is largely empirical. When quantitative relations are constructed to correlate soil parameters to the intensity of the functions or its related threats, a complete assessment of the total accuracy is rarely identified. However, as relationships between soil parameters, soil environmental variables/covariates and soil functions or its related threats to soil can usually be quantified, they can be assimilated into pedo-transfer functions. The process is the same as the attribute inference system described above. The accuracy of the full process can also be calculated as for the attribute inference system. Soil processes have also to be consider as they may themselves present some threat to the overall ecosystems.

The scope of DSRA

Digital Soil Risk Assessment (DSRA) integrates political, environmental, social, economical characters and the DSA outputs. The purpose is then to build, model and test soil related environmental issues, along with it measurement of the accuracy of the predicted risk is also done by which the modeler can provide a degree of uncertainty to decision-makers. The below section briefly state that how to build scenarios? How to evaluate the accuracy of the risk assessment? How decision-makers can use this accuracy for making strategic plans, such as developing policies for soil protection?

How to make and model a scenario?

The theme, the level of detail (the resolution of analysis) and the time-scale required are the three key points for developing a scenario. The basic step for developing a scenario is to: (i) identify what are the central attributes to consider, dependent upon the driving factors being tested, (ii) identify all the driving factors and quantitatively model the relations between the central attributes and the driving factors from reviews, existing models or new models and (iii) testing the accuracy of such relations according to: the study area, the precision of the input parameters, and the timescale.

Tzilivakis et al. (2005) proposed a method for assessing threats to soil functions for assessment of confidence and communication of uncertainty. It consists of building and updating a ‘knowledge base’ system to store descriptions of how changes in soil properties might have consequences relating to the soil stability to perform its various functions. But the uncertainty evoked here is qualitative and depends on the expert who built the scenario.

Some models consider the change in one parameter can affect the function of whole ecosystem such as Clue model, which deals with the effects of land use on the water quality and living resources of estuaries from socio-economic and biophysical driving factors (Verburg et al. 2001, Veldkamp et al. 2002, et al. 2002). Some study areas have been tested in order to estimate the accuracy of such a model but for the moment, the model does not present a validation of the application domain and its robustness has not been tested. For example, SWAT model is a river basin scale model developed to quantify the impact of land management practices in large, complex watersheds (Allen et al. 1997). It can be used then to assess soil erodibility in certain conditions. Mostly, the longer the scenario, the more complex the dynamics and the modelling processes.

Soil assessment is a dynamic process. The important issues are: (i) to identify starting points in the overall process; (ii) how to model all the feedback mechanisms that exist between the soil parameters and the other features of ecosystems to take account of the constantly changing interactions.

Human factors are the most influential in producing soil changes (European Commission 2006a), is the basic hypothesis. Therefore, land use is the first parameter to change and then soil and environmental variables change as a consequence. The process needs to be studied in depth and then quantified to allow for uncertainty estimation of the whole process.

Monte-Carlo simulation estimate the uncertainty of the full process but for this, uncertainties of each process is much needed (Heuvelink 1998, Finke 2007, Heuvelink 2007). The resulting uncertainty needs to be taken into account by decision-makers.

Soil information needs for decision-makers

(i) A digital soil assessment map with large uncertainties should not be used without discuss. However, it may be good enough start point. This methodology allows the quantification of accuracy improvement costs and it is then up to the decision-makers to make the value judgment to what extent further investment is made.

(ii) For a digital soil map made with fuzzy logic, which is the more important parameter to take into account: the membership or the uncertainty? Uncertainty is established according to memberships, it is better to give the membership values priority and preference over the uncertainty. So, large membership values are more important than low uncertainties.

References

Allen, P.M., Arnold, J.G. and Jakubowski, E. 1997. Design and testing of a simple submerged jet device for field determination of soil erodibility. Environmental and Engineering Geoscience 3: 579–584.

Anderson, K.E. and Furley, P.A. 1975. An assessment of the relationship between surface properties of chalk soils and slope form using principal component analysis. Journal of Soil Science 26: 130– 143.

Barshad, I. 1958. Factors affecting clay formation. 6th National Conference on Clays and Clay Mineralogy. pp. 110– 132.

Batjes, N.H. 1990. Macro-scale land evaluation using the 1:1M world soils and terrain digital database: Identification of a possible approach and research needs. SOTER Report 5. International Soil Science Society, Wageningen, the Netherlands.

Bishop, T.F.A., McBratney, A.B. and Whelan, B.M. 2001. Measuring the quality of digital soil maps using information criteria. Geoderma 103: 95–111.

Bragato, G., 2004. Fuzzy continuous classification and spatial interpolation in conventional soil survey for soil mapping of the lower Piave plain. Geoderma 118: 1–16.

Carpenter, D.S. and Carpenter, S.M. 1983. Monitoring inland water quality using Landsat data. Remote Sensing of Environment 13: 345–352.

Carré, F., McBratney, A.B., Mayr, T. and Montanarella, L. 2007. Digital soil assessments: Beyond DSM. Geoderma 142: 69-79.

Choi, W.J., Chang, S.X., Curran, M.P., Ro, H.M., Kamaluddin, M. and Zwiazek, J.J. 2005. Foliar delta C-13 and delta N-15 response of lodgepole pine and Douglasfir seedlings to soil compaction and forest floor removal. Forest Science 51: 546–555.

Christiaens, K. and Reyen, J. 2001. Analysis of uncertainties associated with different methods to determine soil hydraulic properties and their propagation in the distributed hydrological MIKE SHE model. Journal of Hydrology 246: 63–81.

Dobos, E., Carre, F., Hengl, T., Reuter, H. and Tóth, G. (Eds.) 2006. Digital Soil Mapping as a Support to Production of Functional Maps. Office for Official Publications of the European Communities, Luxembourg. EUR22123 EN, 68 p.

Eckelmann, W., Baritz, R., Bialousz, S., Bielek, P., Carre, F., Houskova, B., Jones, R.J.A., Kibblewhite, M.G., Kozak, J., Le Bas, C., Toth, G., Varallyay, G., YliHalla, M. and Zupan, M. 2006. Common criteria for risk area identification according to soil threats. European Soil Bureau Research Report n°20, EUR22185 EN, 94 pp. Office for Official Publications of the European Communities, Luxembourg.

European Commission 2006a. Thematic strategy for soil protection. Communication from the Commission to the Council, the European Parliament, the Economic and Social Committee of the Regions. European Commission COM, Brussels. 2006. 231 final.

European Commission 2006b. Proposal for a Directive of the European Parliament and of the Council Establishing a Framework for the Protection of Soil and Amending Directive 2004/35/EC. European Commission COM, Brussels. 2006. 232 final.

FAO/IIASA/ISRIC/ISS-CAS/JRC. 2009. Harmonized world soil database (version 1.1). IIASA, Laxenburg, Austria.

Favrot, J.C. and Lagacherie, P. 1993. La cartographieautomatisee des sols: une aide a la gestionecologique des paysagesruraux. ComptesRendus de L'Academied'Agriculture de France 79: 61– 76.

Finke, P. 2007. Chapter 39. Quality assessment of digital soil maps: producers and users perspectives. In: Lagacherie, P., McBratney, A.B., Voltz, M. (Eds.), Digital Soil Mapping: An Initial Perspective. Developments in Soil Science 31.Elsevier, Amsterdam, p. 250.

Furley, P.A. 1968. Soil formation and slope development: 2. The relationship between soil formation and gradient angle in the Oxford area. Zeitschriftfu¨ r Geomorphologie 12: 25- 42.

Grunwald, S. (Ed.) 2006. Environmental Soil-Landscape Modeling — Geographic Information Technologies and Pedometrics. CRC Press, New York.

Grunwald, S. 2009. Multi-criteria characterization of recent digital soil mapping and modelling approaches. Geoderma 152: 195–207.

Grunwald, S., Thompson, J.A. and Boettinger, J.L. 2011. Digital soil mapping and modeling at continental scales – finding solutions for global issues. Soil Science Society of America Journal (SSSA 75th Anniversary Special Paper) 75: 1201–1213.

Harington Jr., J.A., Schiebe, F.R. and Nix, J.F. 1992. Remote sensing of LakeChicot, Arkansas: monitoring suspended sediments, turbidity, and Secchidepth with Landsat MSS data. Remote Sensing of Environment 39: 15–27.

Hartemink, A.E. and McBratney, A.B. 2008. A soil science renaissance. Geoderma 148: 123–129.

Hartemink, A.E., McBratney, A.B. and Cattle, J.A. 2001. Developments and trends in soil science 100 volumes of Geoderma (1967–2001). Geoderma 100: 217–268.

Hastie, T., Tibshirani, R. and Friedman, J. 2001. The elements of statistical learning: data mining, inference and prediction. Springer Series in Statistics. Springer-Verlag, New York.

Hay, R.L. 1960. Rate of clay formation and mineral alteration in a 4000-years-old volcanic ash soil on St. Vincent, B.W.I. American Journal of Science 258: 354– 368.

Hengl, T. 2009. A Practical Guide to Geostatistical Mapping, second ed. University of Amsterdam. www.lulc.com. p. 291.

Heuvelink, G.B.M. 1998. Error Propagation in Environmental Modelling with GIS. Taylor & Francis, London.

Heuvelink, G.B.M. 2007. Chapter 2.3. Accuracy assessment. In: Dobos, E., Carre,F., Hengl, T., Reuter,H., Tóth, G. (Eds.), In Digital Soil Mapping as a Support to Production of Functional Maps. Office for Official Publications of the EuropeanCommunities, Luxembourg. Luxembourg, EUR 22123 EN, 68.

Hudson, B.D. 1992. The soil survey as paradigm-based science. Soil Science Society of America Journal 56: 836–841.

Isbell, R.F. 1996. The Australian Soil Classification. CSIRO Publishing, Collingwood, Victoria.

IUSS Working Group WRB 2006. 2nd edition. World Soil Resources Reports, vol. 103. FAO, Rome.

Jenny, H. 1941. Factors of Soil Formation: A System of Quantitative Pedology. McGraw-Hill, New York.

Jenny, H. 1961. Derivation of state factor equations of soils and ecosystems. Soil Sci. Soc. Am. Proc. 25, 385–388. Jenny, H., 1980. The Soil Resources. Spring-Verlag, ew York.

Jones, M.J. 1973. The organic matter content of the savanna soils of West Africa. Journal of Soil Science 24: 42– 53.

Lagacherie, P. 2005. An algorithm for fuzzy pattern matching to allocate soil individuals to pre-existing soil classes. Geoderma 128: 274–288.

Lagacherie, P., Cazemier, D.K., Martin-Clouaire, R. and Wassenaar, T. 2000. A spatial approach using imprecise soil data for modelling crop yields over vast areas. Agriculture, Ecosystems & Environment 81: 5–16.

Lagacherie, P. and McBratney, A.B. 2007. Chapter 1. Spatial soil information systems and spatial soil inference systems: perspectives for digital soil mapping. In: Lagacherie, P., McBratney, A.B., Voltz, M. (Eds.), Digital Soil mapping: An Initial Perspective. Developments in Soil Science 31. Elsevier, Amsterdam, p. 250 pp.

Lagacherie, P., McBratney, A.B. and Voltz, M. (Eds.) 2007. Digital Soil Mapping: An Introductory Perspective. Developments in Soil Science, vol. 31. Elsevier, Amsterdam.

Lipiec, J. and Hatano, R. 2003. Quantification of compaction effects on soil physical properties and crop growth. Geoderma 116: 107–136.

McBratney, A., Mendonça Santos, M.L. and Minasny, B. 2003. On digital soil mapping. Geoderma 117: 3–52.

McBratney, A.B., Odeh, I.O.A., Bishop, T.F.A., Dunbar, M.S. and Shatar, T.M. 2000. An overview of pedometric techniques for use in soil survey. Geoderma 97: 293–327.

Milne, G. 1935. Some suggested units of classification and mapping particularly for East African soils. Soil Research 4: 183–198.

Moore, I.D., Gessler, P.E., Nielsen, G.A. and Peterson, G.A. 1993. Soil attribute prediction using terrain analysis. Soil Science Society of America Journal 57: 443–452.

Noy-Meir, I. 1974. Multivariate analysis of the semiarid vegetation in south-eastern Australia: II. Vegetation catena and environmental gradients. Australian Journal of Botany 22: 115–140.

Nye, P.H. and Greenland, D.J. 1960. The Soil under Shifting Cultivation. Commonwealth Bureau of Soils, Harpenden, UK.

Odeh, I.O.A., McBratney, A.B. and Chittleborough, D. 1992. Fuzzy-c-means and kriging for mapping soil as a continuous system. Soil Science Society of America Journal 56: 1848–1854.

Rusler, R., Flessa, H., Russow, R., Schmidt, G., Buegger, F. and Munch, J.C. 2006. Emission of N_2O, N_2 and CO_2 from soil fertilized with nitrate: effect of compaction, soil moisture and rewetting. Soil Biology and Biochemistry 38: 263–274.

Simonett, D.S. 1960. Soil genesis in basalt in North Queensland. Transactions of the 7th International Congress of Soil Science, Madison, Wisconsin. pp. 238–243.

Skidmore, A.K., Ryan, P.J., Dawes, W., Short, D. and O'Loughlin, E. 1991. Use of an expert system to map forest soils from a geographical information system. International Journal of Geographical Information Science 5: 431–445.

Soil Survey Staff 1993. Soil Survey Manual. Handbook No. 18. USDA, Washington, DC.

Sriramam, G. 1968. Effect of soil compaction on plant growth and nutrientuptake. Proceedings of the National Academy of Sciences India Section A Physical Science 3, 71.

Tzilivakis, J., Lewis, K.A. and Williamson, A.R. 2005. A prototype framework for assessing risks to soil functions. Environmental Impact Assessment Review 25: 181–195.

Veldkamp, A., Verburg, P.H., Kok, K., de Koning, G.H.J. and Soepboer, W. 2002. Spatial explicit land use change scenarios for policy purposes: some applications of the CLUE framework. In: Walsh, S.J., Crews-Meyer, K.A.(Eds.), Linking People, Place, and Policy. A GIScience Approach. Kluwer Academic Publishers, Boston/Dordrecht/London. pp. 317–341.

Verburg, P.H., de Koning, H.G.J., Kok, K., Veldkamp, A. and Priess, J. 2001. The CLUE modelling framework: an integrated model for the analysis of land use change. In: Singh, R.B., Fox, Jefferson, Himiyama, Yukio (Eds.), Land Use and Cover Change. Science Publishers, Enfield, NH.

Vereecken, H., Diles, J., van Orshoven, J., Reyen, J. and Bouma, J. 1992. Functional evaluation of pedotransfer functions for the estimation of soil hydraulic properties. Soil Science Society of America Journal 56:1371–1378.

Vrieling, A. 2006. Satellite remote sensing for water erosion assessment: a review. Catena 65: 2-18.

Vrieling, A., Rodrigues, S.C., Bartholomeus, H. and Sterk, G. 2007. Automatic identification of erosion gullies with ASTER imagery in the Brazilian Cerrados. International Journal of Remote Sensing 28: 2723–2738.

3

Soils of Arid Region of India, Their Taxonomic Classification and Fertility Characteristics

Mahesh Kumar, N.R. Panwar and Priyabrata Santra

ICAR-Central Arid Zone Research Institute, Jodhpur, Rajasthan, India

The arid region in India is spread over in 38.7 million hectare land. Out of the total 31.7 m hectare lies in hot arid region and remaining 7 m hectare comes under cold arid region. The hot arid region occupies major part of northwestern India (28.7 m ha) and a small pocket (3.13 mha) in southern India. The north western arid region occurs between 22°30' and 32°05' N latitude and 68°05' to 75°45' E, covering western part of Rajasthan, north western Gujarat and south western parts of Haryana and Punjab. About 62% area of arid region falls in western Rajasthan and 20% in Gujarat. Haryana and Punjab together constitute 7% area of arid region. The state wise distribution of arid region is shown in Table 3.1. Further our discussion is based on the majority area of arid region, belonging to Rajasthan, Gujarat, Punjab and Haryana.

Table 3.1: State wise distribution of area of hot arid region in India

State	Area (ha)	% Area of arid zone
Rajasthan	196150	61.8
Gujarat	62180	19.6
Punjab	14150	5.0
Haryana	12840	4
Maharashtra	1290	0.4
Karnataka	8570	2.7
Andhra Pradesh	21550	6.9
Total	316730	100

The soils

The distribution of soils in the arid region of India is shown in Table 3.2. The soils of this tract have been mapped in Entisols, Aridisols and Alfisols soil orders. The Entisols cover maximum 17134.26 (54.21%) thousand hectares, followed by Aridisols and Alfisols. The area of the latter two is 14254.32 and 213.10 thousand hectares, comprising 45.1 and 0.67% of hot arid India. Haplocalcids, Haplocambids, Haplosalids, Petrocalcids and Haplogypsids constitute Aridisols at the great group level, while Torripsamments, Torriorthents and Torrifluents are the part and parcel of Entisols. The Natrargids, Pleargids and Haplargids are the great groups mapped in Alfisols. The spatial distribution of these great groups is shown in Fig. 3.1. Haplocambids dominantly occur in the arid part of Haryana, Gujarat and Punjab, covering around 77, 42 and 39.6% of delineated area in their respective state. Torripsamments dominantly mapped in 47.1% area of western Rajasthan, while the share of these soils to the marked area is around 43, 14.3 and 9.7% in Punjab, Haryana and Gujarat, respectively. Haplocalcids are the second dominant soils of Gujarat after Haplocambids with 11.1% of allocated area under arid land. Petrocalcids and Haplocalcids together constitute 8.4% area of western Rajasthan. The contribution of Haplosalids to the delineated arid land is higher in Gujarat than Rajasthan. Torriorthents, the shallow skeletal soils associated with hills are mapped in 7.8 and 4.3% area of arid part of Rajasthan and Gujarat, respectively. Torrifluvents fine textured deep soils along the course of river constitute around 2% area of arid zone both in Rajasthan and Gujarat. Haplogypsids and Haploargids represent relict soils in Rajasthan and Gujarat, while Natrargids and Pleargids are the paleosols mapable only in the latter.

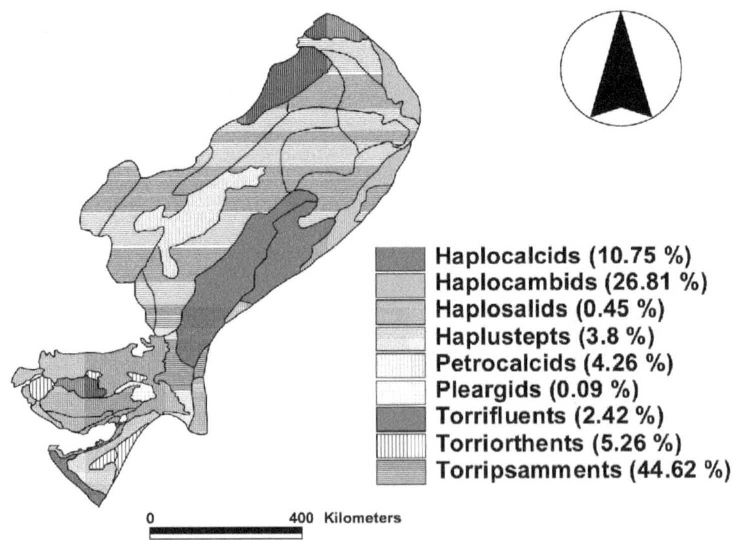

Fig. 3.1: Dominant great groups of arid region of India

Table 3.2: Dominant Soil of Arid Region of India (000, ha)

Order / Great Group	Area	Rajasthan	Gujarat	Haryana	Punjab
Alfisol	213.10 (0.67)				
Natrargids		-	43.32 (0.14)		
Pleagragids		-	28.45 (0.09)		
Haplargids		136.27 (0.43)	5.00 (0.01)		
Aridisol	14254.32 (45.10)				
Haplosalids		96.23 (0.30)	49.04 (0.15)		
Petrocalcids		1349.20 (4.27)	-		
Haplogypsids		287.28 (0.90)	-		
Haplocalcids		996.28 (3.15)	651.50 (2.06)	1155.44 (3.65)	594.82 (1.8)
Haplocambids		6706.7 (21.22)	2366.81 (7.48)		
Entisol	17134.26 (54.21)				
Torripsamments		13198.9 (41.76)	568.56, (1.8)	214.50 (0.67)	720.00 (2.3)
Torrifluvents		639.63 (2.02)	127.98 (0.40)		
Torriorthents		1207.77 (3.82)	456.88 (1.44)		

Thematic mapping in western Rajasthan

Deep to very deep soils constitute 73.5%, while moderately shallow and moderately deep together cover 21.5% area. Shallow and very shallow soils together mapped in 4.7% area of western Rajasthan. Sandy and coarse loamy soils constitute 52.4 and 32.7% area, respectively. Sandy-skeletal and loamy-skeletal together constitute 3.3%, while loamy and fine soils jointly cover 0.5% area of delineated arid part of Rajasthan. Fine loamy soils alone cover another 8% area of arid Rajasthan (Shyampura et al. 2002).

Fertility status of arid zone soils

The major and micro nutrient contents of typical arid soils of Rajasthan are presented in Table 3.3. The arid zone soils are low in organic carbon and deficient in nitrogen. Organic carbon in the soils below 300 mm rainfall zone ranged between 0.05 to 0.2% in sandy soils, 0.2 to 0.3% in coarse loamy soils and 0.3 to 0.4% in fine loamy soils. The soils of stabilized dunes contained higher organic carbon than their unstabilized counterparts (Aggrawal and Lahiri 1977 Aggrawal et al. 1978). With increase in clay content there is increase in soil organic carbon content. The nitrogen content ranges from 0.021 to 0.056%, major part of which is in organic form and mineralized to ammonium and then to nitrate. Under alkaline condition, part of ammonium N is converted to ammonia and escapes to the atmosphere and part of the ammonical N converted to nitrate N is utilized by crop plants and remaining is leached to lower horizons.

Available P in medium and fine textured alluvial soils is well provided but sandy soils are generally medium to low (<10 kg ha^{-1}). Applied P gets immobilized into insoluble forms as Ca phosphate in alkaline soils. Use efficiency of P seldom exceeds 20%. Significant amount of P is lost through runoff and erosion resulting in eutrophication of water bodies.

There is large variation in available K content of different arid soils and most soils are provided with sufficient available K. Potassium is more mobile than P and susceptible to loss by leaching, runoff and erosion but carries no environmental hazards (Dutta and Joshi, 1993).

In Jhunjhunu district, organic carbon content in soils under irrigated crops was slightly higher (0.22%) followed by those rainfed and grazing land (0.13 and 0.15%) and sand dunes soils (0.07%). P deficiency in soils (2.7-18.4 kg ha^{-1}) was spread all over the district. Mean values of soil available K were 212, 161, 156 kg ha^{-1} respectively under grazing lands, agriculture (Rainfed and irrigated) and sand dunes (Mahesh Kumar et al. 2011). In Churu district among nine soil series studied the Dune, Molasar, Modasar, Devas and Chirai are coarse textured associated with sand dunes, sandy plain with scattered sand dunes and sand hummocks where as Sarupdesar, Masitawali and Naurangpura are medium to fine textured. All the studied soils were low in organic carbon (0.05-0.23%). Available P was low in 67.5%, medium in 27.5% samples. Available K was low in 18% and medium in 64.5% samples (Mahesh Kumar et al. 2011).

Table 3.3: Fertility characteristics of arid soils of Rajasthan

Soils	Org. (%)	Avail. P (kg ha^{-1})	Avail. K (kg ha^{-1})	Available (mg kg^{-1})			
				Fe	Mn	Zn	Cu
Typic Torripsamments (Dune and interdune)	0.05- 0.16	10-15	116-392	3.5-27.5	1.1-16.1	0.27-1.91	0.12- 1.52
Coarse loamy Typic Haplocambids/calcids (Sandy hummocky plain)	0.15- 0.25	9-15	78-482	3.4-8.6	4.0-13.4	0.41-1.32	0.36 -0.89
Coarse loamy Typic Haplocambids/calcids (Medium textured alluvial plain)	0.15- 0.25	15-20	300-800	1.6- 16.0	5.4-25.0	0.27-3.28	0.44- 0.93
Fine loamy Typic Haplocambids/calcids (Fine textured alluvial plain)	0.25- 0.80	11-39	105-890	3.0- 10.4	2.2- 16.0	0.36-2.36	0.36- 1.80
Typic Petrocalcids	0.10- 0.15	9-15	70-300	3.4-8.6	4.0-13.4	0.41-1.32	0.36- 0.89
Typic Haplosalids	0.08- 0.73	9-25	67- 440	2.6- 16.0	6.0-29.0	0.56-3.37	0.55- 2.08

Available Fe content in all studied soils was found sufficient with about 40% samples contained 2-5 mg kg^{-1} and 54% samples has 5-10 mg kg^{-1}. Available Mn content was deficient in 23% samples which were from dune and interdune soils (Joshi and Dhir 1983a, b). DTPA soluble Zn and Cu ranged from 0.27- 2.36 mg kg^{-1} and 0.28-1.25 mg kg^{-1}, respectively (Joshi and Dhir 1982, Sharma et al. 1985). Samples showing low ranges were from dune and interdune soils in extremely arid part of Rajasthan. Deficiency of Zn was encountered in 13.5 and 9.6% samples in Barmer and Jaisalmer districts respectively. All studied soils are well provided with available Cu. It can be broadly inferred that arid soils are well provided with Cu and Mn micronutrients to meet the needs of rainfed cropping. Soils of Jhunjhunu district of Rajasthan were adequate in available Fe (4.2- 30.4 mg kg^{-1}), Mn (5.5- 40.5 mg kg^{-1}), and Cu (0.22-7.4 mg kg^{-1}). Zn content varied from 0.28 – 5.90 mg kg^{-1} with about 30% samples deficient in Zn (Mahesh Kumar and Sharma 2011). In Churu district all soils were well provided with available Fe, Mn and Cu but Zn was deficient in 38% and marginal in 57% samples (Mahesh Kumar et al. 2010, 2011).

Gujarat: In Jamnagar, Kachchh, Banaskantha and Mahesana districts 8-23% samples were deficient and 45-65% samples marginal in available Fe and only <5% samples were deficient in available Mn. Zinc deficiency has been reported in 44, 39 and 13% samples respectively in Banaskantha, Jamnagar and Kachchh districts. In Banaskantha district 22% samples were deficient in Cu (Dangarawala et al. 1983).

Punjab and Haryana: Sharma et al. (1992) for arid soils of Punjab reported that soils were low in organic carbon and nitrogen, available phosphorus in medium range and sufficient with respect to available potassium. All soils were deficient in Fe, Zn and Mn. Fe deficiency has also been reported in Bhatinda (38%), Faridkot (40%) and Sangrur (11%) districts of Punjab. In Sangrur and Gurdaspur district respectively 22 and 25% samples were deficient in Cu (Takkar et al. 1976). In Haryana, soils of Mahendragarh district were sufficient in available Fe but soils of Sirsa (25%) and Hisar (51.6%) districts were found deficient. About 80% samples in Hisar, Mahengragarh and Bhiwani districts of Haryana and 25% similar soils in Punjab were also found deficient in available Zn (Anonymous 1983).

References

Aggrawal, R.K., Gupta, J.P., Saxena, S.K. and Muthana, K.D. 1978. Studies on soil physico-chemical and ecological changes under twelve years old five desert tree species of western Rajasthan. Indian Forester 102, 863-872.

Aggrawal. R.K. and Lahiri, A.N. 1977. Influence of vegetation on the status of the organic matter and nitrogen of the desert soils. Science and Culture 43: 533-534.

Anonymous 1983. All India co-ordinated scheme on micronutrient in soils and plants. Annual report 1982-83, ICAR, New Delhi.

Dangarawala, R.T., Trivedi, B.S., Patel, M.S. and Mehta, P.M. 1983. Micronutrient research in Gujarat, 137 pp. GAU, Anand, Gujarat.

Dutta, B.K. and Joshi, D.C. 1993, Studies on potassium fixation and release in arid soils of Rajasthan.Transaction. ISDT 18: 191-199.

Joshi, D.C. and Dhir, R.P. 1983a. Distribution of micronutrient forms along dune landscape. Annals of Arid Zone 22: 135-141.

Joshi, D.C. and Dhir, R.P. 1983b. Available forms of manganese and iron in some arid soils and their relationship with soil properties. Annals of Arid Zone 22: 7-14.

Joshi, D.C. and Dhir, R.P. 1982. Distribution of different forms of Cu and Zn in some soils of arid Rajasthan.Journal of the Indian Society of Soil Science 30: 547-549.

Mahesh Kumar and Sharma, B.K. 2010. Micronutrient status of soils irrigated with high residual sodium carbonate water in Jhunjhunu district of Rajasthan. Annals of Arid Zone 49:137-140.

Mahesh Kumar and Sharma, B.K. 2011. Soil fertilility approisal under dominant land systems in north eastern parts of arid Rajasthan. Annals of Arid Zone 50: 11-15.

Mahesh Kumar, Singh, S.K., Raina P. and. Sharma, B.K. 2011. Status of available major and micronutrients in arid soils of Churu district of western Rajasthan. Journal of the Indian Society of Soil Science 59: 188-192.

Sharma, B.K., Dhir, R.P. and Joshi, D.C. 1985. Available micronutrient status of some soils of arid zone. Journal of the Indian Society of Soil Science 33: 50-55.

Sharma, B.D., Sidhu, P.S. and Nayyar, V.K. 1992. Distribution of micronutrients in arid soils of Punjab and their relation with soil properties. Arid Soil Research and Rehabilitation 6: 233-242.

Shyampura, R.L., Singh, S.K., Singh, R.S., Jain, B.L. and Gajbhiye, K.S. 2002. Soil series of Rajasthan, NBSS Publication No. 95 NBSS& LUP, Nagpur, 364p.

Takkar, P.N., Bansal, R.L., Mann, M.S. and Randhawa, N.S. 1976. Micronutrient status of soils and wheat crop of Punjab. Fertilizer News 21: 47-51.

4

Basics of Soil Sampling Through Field Survey and Processing of Georeferenced Data for Geostatistical Analysis

N.R. Panwar and Priyabrata Santra

ICAR–Central Arid Zone Research Institute, Jodhpur, Rajasthan, India

Soil sampling is an integral and essential component of soil fertility evaluation and nutrient management research. The effectiveness of soil sampling is a prerequisite for soil testing to achieve its goals for efficient nutrient management and plant nutrition for improved yield and quality. The underlying basis for soil sampling is that a soil sample taken represents the "population" which may be a plot, field or a watershed. It further implies that nutrient status of the representative soil sample(s) determined in a laboratory would reflect nutrient status of a plot, field or watershed and is of interest for correcting nutrient disorders in the field or watershed.

The most important factor that influences the effectiveness of soil sampling is soil heterogeneity. However, in a relatively homogenous group of fields or plots, a small number of samples may be sufficient to represent the population to a more heterogenous group of fields that would require more number of samples to represent the soil population. Whenever, soil samples to be taken in field following points needs to be kept in mind.

Seasonal effects on soil test values

- Considerable seasonal influence on soil test values and every effort to maintain consistency within season when taking soil tests should be made.
- Two analytes most affected by seasonal influences are potassium and pH.
- Soils having medium to high clay contents, potassium soil test values have a tendency to be higher during the winter months.

- Soil pH values can also vary appreciably over the year depending on nitrogen and sulfur inputs, amounts of rainfall or irrigation and soil buffering capacity (amount and types of clay and free carbonates).
- Year to year variation of soil test values can be appreciable as well, depending on the amount and timing of rainfall, and the duration of freezing and thawing over the winter months and wetting and drying in summer months.

Cropping history and its effects on soil test values

- Soil sampling events should be consistent as much as possible as significant differences in total nutrient uptake between crops or crop specific nutrient inputs exist that can impact on soil test values.
- Irrigation requirements vary between crops, leading to possible soil test variations following the irrigation season in the areas of nitrate – nitrogen, sulphate – sulphur, soil pH, sodium, carbonates and electrical conductivity as a function of soluble salts.
- Effect of a given crop on seasonal nutrient uptake and crop specific nutrient / irrigation requirements can help/explain a great deal of year to year soil test variation.

Effect of tillage

- Tillage systems have been demonstrated to cause significant layered, stratification of organic matter, pH and soil nutrients (especially where subsurface banding of fertilizer is not utilized).
- Under reduced tillage, ridge tillage and zero tillage condition soil samples should include some samples that are split into 0"-3" and 3"-7" depth increments, to properly assess to what extent stratification is occurring in order to modify fertilizer/soil amendment rates, timing and/or placement.
- While sampling for ridge till, it is recommended that the sample is taken halfway down the ridge at a 45° angle to the ridge.

Random composite sampling *vs.* point sampling

- Composite soil sampling consists of physical probes being taken at randomly chosen sites throughout an entire sampling area and combined into a single sample.
- Soil test results from the sample are used to represent the entire sampling area.
- A disadvantage to composite sampling is that it poorly characterizes field variability, creating coarse maps with distinct, sharp divisions between sampled areas.
- In point sampling, a sample location (point) is established and the physical sample is obtained within a specified radius from this point.

- Soil test results are linked to each sample point and interpolation methods are used to obtain values for the remaining unsampled areas of the field.
- Technically, point sampling can be considered a variation of composite sampling, but differs because it represents a single point, not an entire area.

Advantage and disadvantage to point and composite sampling

Types	Advantage	Disadvantage
Point sampling	Good for detecting various patterns of field variability	Accurate estimation requires close spacing
	Soil surface maps can be prepared by using soil test values	More expensive
Composite sampling	Somewhat less expensive	Part of field may be under or over fertilized
	Year to year results can be tracked easily	Coarse map with sharp divisions
	Highly reproducible	Good for uniform application

Grid soil sampling

Development of site-specific nutrient management via global positioning systems (GPS) and variable rate fertilization (VRF) demands that soil sampling be intensively organized into a systematic grid pattern. Grid soil samples should be taken at a specific point, either within the grid cell or at intersection points between grid cells, consisting of 8-10 cores per sample taken within a 10-foot radius.

Diagram showing the plan of a square grid and location from where soil cores would be collected

- To more correctly represent soil test variability within a field (especially for implementation of soil test mapping), the grid sample points should be organized into a systematic grid-diamond pattern

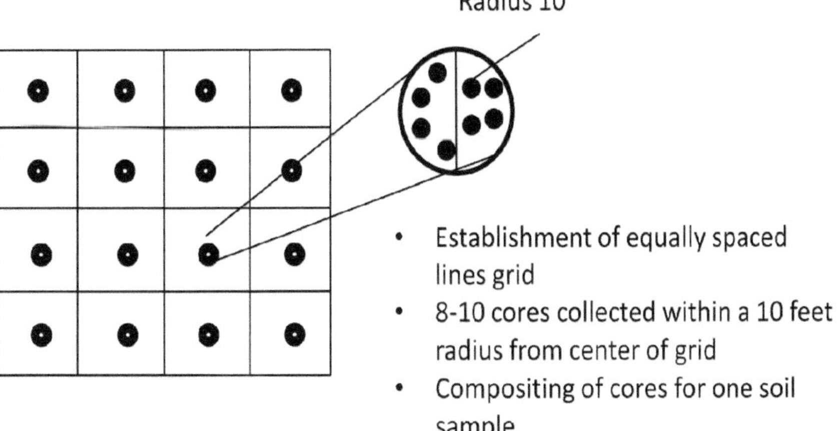

Radius 10'
• Establishment of equally spaced lines grid
• 8-10 cores collected within a 10 feet radius from center of grid
• Compositing of cores for one soil sample

- The grid-diamond pattern is accomplished by shifting the sample points to the left or right of the grid cell center in alternating rows perpendicular to the measurement pattern (established by GPS).

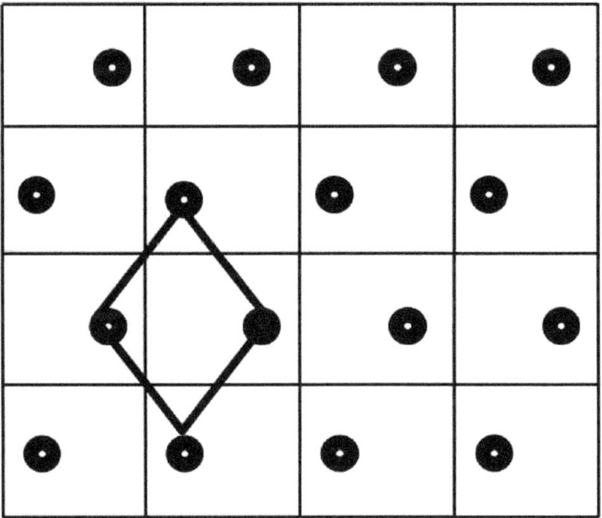

Grid-diamond sampling plan

Modification of a square grid where altering rows of sample points are shifted one half the distances from the cell center and edge.

The systematic unaligned sampling pattern is best utilized via GPS, following this procedure:

- Divide the field into cells by means of a coarse grid. Square cells are the norm but not mandatory.
- Superimpose a finer grid (reference grid) in each coarse cell. For example, if there are 5 rows and 5 columns in the coarse grid, you might choose to divide each coarse cell into 25 smaller cells.
- Choose a corner of the coarse grid, say top left, and randomly select a reference cell—in this sample, one of the 25 reference cells.
- Move horizontally to the next coarse cell in the top row and keep the X coordinate the same but randomly select a new Y coordinate.
- Repeat the process for all the coarse cells in the top row.
- Return to the upper left corner and repeat the process down the first column of cells, this time keeping the Y coordinate the same, but changing the X coordinate in each successively lower coarse cell.

- The remaining positions are determined by the X coordinate of the point in the left-hand square of its row and the Y coordinate of the point in the uppermost square of its column.

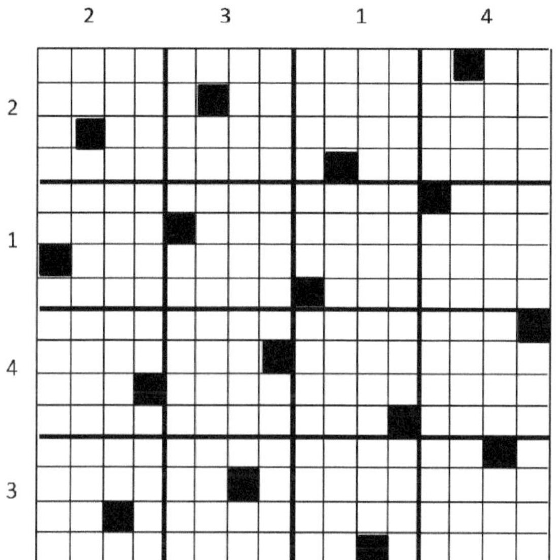

Systematic unaligned grid

Diagram showing layout of a systematic unaligned grid. The x, y coordinates were determined from a random number table.

Management zone sampling methods

- An alternative to grid soil sampling is management zone sampling (also called directed or smart sampling).
- Actual management zones are established using a variety of resources and/or datasets. Include soil surveys, past yield data, remote sensing imagery, landscape/topography, elevation, electrical conductivity, and/or past knowledge of field characteristics.
- Unlike grid sampling, the shape, size, and number of management zones will vary depending on field variability and the information derived from datasets.
- Method is less systematic and precise than grid sampling.
- Reliable results can be obtained if sample points and/or walk patterns are consistent between sampling events (utilizing row counts, distance measuring devices, or GPS).
- Specific points within the field are chosen based on soil type and yield data (if available), and 10-15 cores are taken within a 20-foot radius around each point.

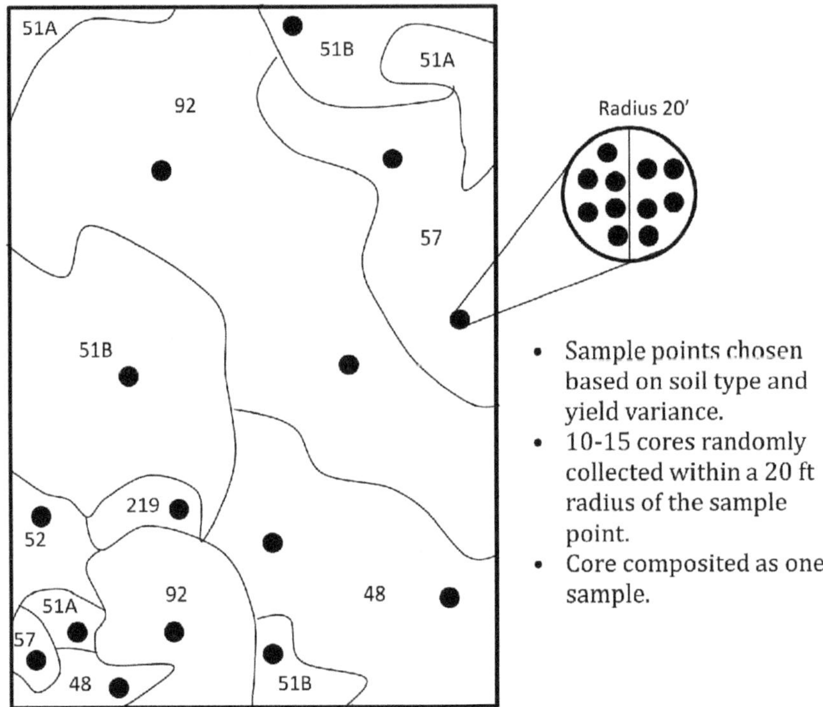

Diagram showing the layout of a specific sample point based soil type and yield variation

- Area represented by each sample should be no more than 20 acres depending on soil type, slope, drainage, old field boundaries and variation in cropping pattern.
- Variation on the grid-point sampling technique can be useful in developing more consistent, non-grid sample results.

Code	Soil Type	Acres	Bajra	Mung
51A	Silt loam, 0-2% slope	8.8	167	51
51B	Silt loam, 2-4% slope	12.1	165	50
52	Silty clay loam	2.6	136	44

Grid/Management zone hybrid soil sampling method

- A third option for soil sampling is the grid/management zone hybrid method. Management zones are created using various data sources as described previously.
- Management zones are used as a basemap and a grid is overlaid onto this basemap.
- Final sampling areas are defined according to basemap

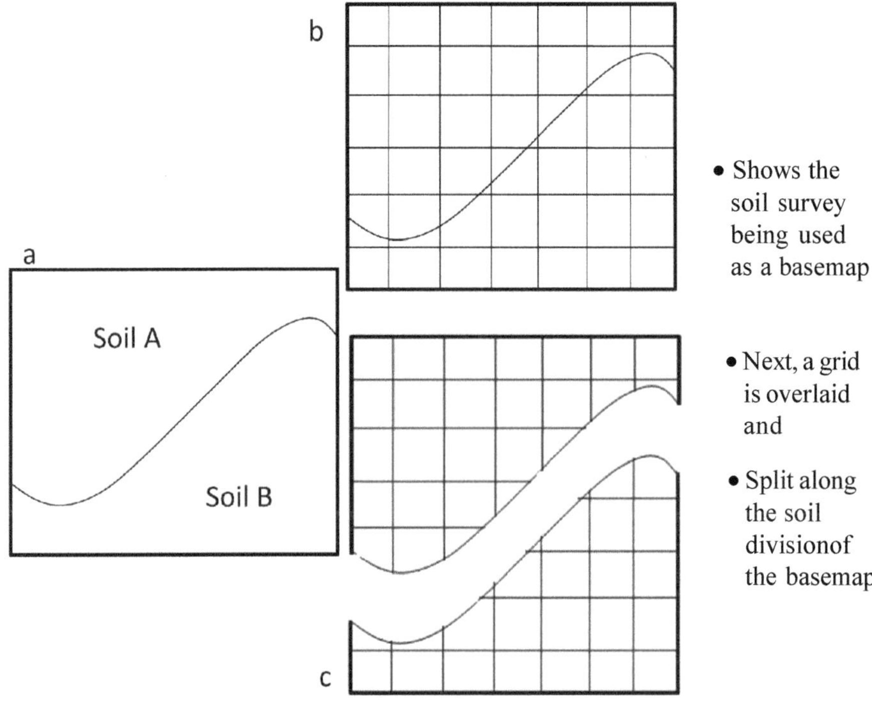

Hybrid grid or management layout

Conclusion

- Soil tests should be taken in such a manner to maximize their use as a soil fertility index based on comparison between sampling events.
- Soil sampling is mainly depend on your final goal/purpose.
- Consistency, in the areas of season, location (aided by GPS techniques), crop rotation, soil type and sampling depth must be maintained for proper soil test interpretation.
- Inconsistencies in any of these areas of soil sampling collection will lessen the interpretation value of soil test changes that occurred since the last soil sample was taken.
- Along with consistency, soil samples should reflect past soil and fertilizer/ amendment management of a given field, taking into account tillage, crop rotation, fertilizer/ amendment placement and also soil characteristics (texture, slope and drainage).
- Following these guidelines will allow soil tests to be used more effectively for nutrient management and also for effective natural resource management.

5

Introduction To R

Priyabrata Santra[1] and Uttam Kumar Mandal[2]

[1]ICAR-Central Arid Zone Research Institute, Jodhpur, Rajasthan, India
[2]ICAR-Central Soil Salinity Research Institute, Regional Station Canning Town, West Bengal, India

Introduction

R is one of the most popular platforms for data analysis and visualization currently available. It is free, open-source software, with versions for Windows, Mac OS X, and Linux operating systems. R is a language and environment for statistical computing and graphics, similar to the S language originally developed at Bell Labs. It's an open source solution to data analysis that's supported by a large and active worldwide research community. But there are many popular statistical and graphing packages available *e.g.* Microsoft Excel, SAS, IBM SPSS, Stata, and Minitab etc., however R has many interesting features so why it is mostly followed by researchers and academicians. Few of the major features of R are given below:

i). Most of the statistical software platforms are commercial with a cost of thousands of dollars. R is free! If you're a teacher or a student, the benefits are obvious.

ii). R is a comprehensive statistical platform, offering all manner of data analytic techniques. Just about any type of data analysis can be done in R.

iii). R has state-of-the-art graphics capabilities. If you want to visualize complex data, R has the most comprehensive and powerful feature set available.

iv). R is a powerful platform for interactive data analysis and exploration. For example, the results of any analytic step can easily be saved, manipulated, and used as input for additional analyses. Getting data into a usable form from multiple sources can be a challenging proposition. R can easily import data from a wide variety of sources, including text files, database management systems, statistical packages, and specialized data repositories. It can write data out to these systems as well.

v). R provides an unparalleled platform for programming new statistical methods in an easy and straightforward manner. It's easily extensible and provides a natural language for quickly programming recently published methods.

vi). R contains advanced statistical routines not yet available in other packages. In fact, new methods become available for download on a weekly basis. If you're a SAS user, imagine getting a new SAS PROC every few days.

vii). If you don't want to learn a new language, a variety of graphic user interfaces (GUIs) are available, offering the power of R through menus and dialogs.

viii). R runs on a wide array of platforms, including Windows, Unix, and Mac OS X. It's likely to run on any computer you might have (I've even come across guides for installing R on an iPhone, which is impressive but probably not a good idea).

Obtaining and installing R

R is freely available from the Comprehensive R Archive Network (CRAN) at http://cran.r-project.org. Precompiled binaries are available for Linux, Mac OS X, and Windows. RStudio allows the user to run R in a more user-friendly environment. It is open-source (*i.e.* free) and available at http://www.rstudio.com/. After downloading both these computer programs from internet, you need to install R program first and then RStudio. Advanced users straightway can work in R console after installing only R, however RStudio gives better visualization of the computing processes including inputs (data type), outputs (plots) and necessary packages etc. and thus are more user friendly.

Getting started with R

In Windows operating system, R can be launched from the Start Menu. On a Mac system, R is opened by double-click the R icon in the Applications folder. For Linux operating system, type R at the command prompt of a terminal window. Any of these will start the R interface or R console (Fig. 5.1).

R Studio can be opened in the Windows operating system by double click the R Studio icon in start menu. The R Studio window contains four sub-windows as shown in Fig. 5.2. As seen from the Fig. 5.2, R Studio contains additional three sub-windows besides R console. In the R script window, programming codes are written and run by pressing Control-Enter key. Single line of programming code can be run by placing the cursor anywhere on the selected line and then pressing the Control-Enter key. Multiple lines of programming code can be run by selecting them and then pressing the Control-Enter key. Whenever the codes are run in script window, the results are displayed in the R console window. Any error or warning messages while running the script is displayed in R console window. All inputs and outputs for the current session are displayed in the top right sub-window. It also displays the data type of inputs and outputs. Apart from these inputs and outputs, working history of programming codes, which were executed in recent past are displayed in the top-right sub window and can be accessed by clicking the history tab. The bottom right sub window contains the information on files, plots, packages, help and viewer. The files of the directory in which current session is running are displayed. Graphical outputs e.g. bar plot, histograms, scatter plot etc. are displayed in the plot tab when a script is run in script window. All installed packages in R are displayed in package tab, however, those packages which are being used in the current session can only be seen as check marked. Any help on R commands and packages can be obtained by searching in help tab.

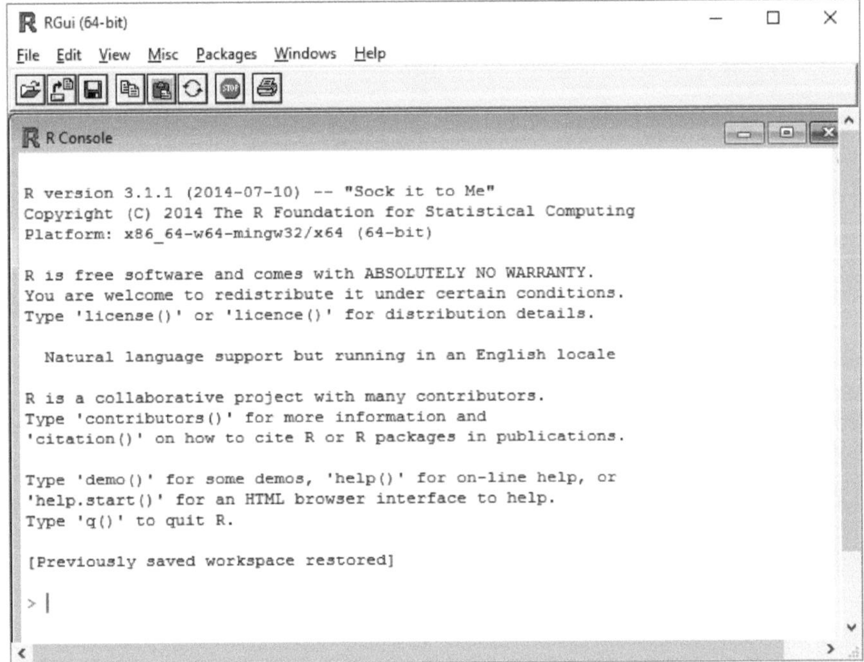

Fig. 5.1: R console interface on Windows

Working with R

R is a case-sensitive, interpreted language. Commands can be entered one at a time at the command prompt (>) or run a set of commands from a source file. Users typically access it through a command line interpreter. If a user types 2+2 at the R command prompt and presses enter, the computer replies with 4, as shown below:

> 2+2

[1] 4

There are wide variety of data types, including vectors, matrices, data frames (similar to datasets), and lists (collections of objects). Most functionality is provided

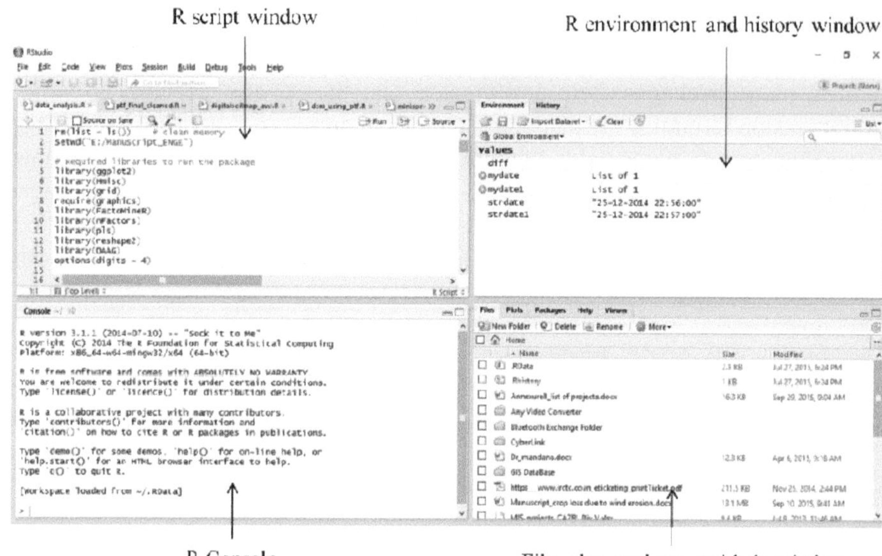

Fig. 5.2: R Studio interface on Windows

through built-in and user-created functions, and all data objects are kept in memory during an interactive session. Basic functions are available by default. Other functions are contained in packages that can be attached to a current session as needed.

Statements consist of functions and assignments. R uses the symbol <- for assignments, rather than the typical = sign. For example, the statement *x <- rnorm(5)* creates a vector object named x containing five random deviates from a standard normal distribution. *Comments* are preceded by the # symbol. Any text appearing after the # is ignored by the R interpreter.

Like other similar languages such as APL and MATLAB, R supports matrix arithmetic. R's data structures include vectors, matrices, arrays, data frames (similar to tables in a relational database) and lists. R's extensible object system includes objects for (among others): regression models, time series and geospatial coordinates. The scalar data type was never a data structure of R. Instead, a scalar is represented as a vector with length one.

Getting help in R

R provides extensive help facilities, and learning to navigate them will help significantly in programming efforts. The built-in help system provides details, references, and examples of any function contained in a currently installed package. Help is obtained using the functions listed in Table 5.1.

Table 5.1: R help functions (*adapted redrawn from Kabacoff, R.I. 2011.*)

Function	Action
help.start()	General help
help("*foo*") or?*foo*	Help on function *foo* (the quotation marks are optional).
help.search("*foo*") or??*foo*	Search the help system for instances of the string *foo*.
example("*foo*")	Examples of function *foo* (the quotation marks are optional).
RSiteSearch("*foo*")	Search for the string *foo* in online help manuals and archived mailing lists.
apropos("*foo*", mode="function")	List all available functions with *foo* in their name.
data()	List all available example datasets contained in currently loaded packages.
vignette()	List all available vignettes for currently installed packages.
vignette("*foo*")	Display specific vignettes for topic *foo*.

The function *help.start*() opens a browser window with access to introductory and advanced manuals, FAQs, and reference materials. The *RSiteSearch*() function searches for a given topic in online help manuals and archives of the R-Help discussion list and returns the results in a browser window. The vignettes returned by the *vignette*() function are practical introductory articles provided in PDF format. Not all packages will have vignettes. R provides extensive help facilities, and learning to navigate them will definitely aid in programming efforts.

R workspace

The workspace is the current R working environment and includes any user-defined objects (vectors, matrices, functions, data frames, or lists). At the end of an R session, user can save an image of the current workspace that's automatically reloaded the next time R starts. Commands are entered interactively at the R user prompt. Use the up and down arrow keys to scroll through your command history. Doing so allows to select a previous command, edit it if desired, and resubmit it using the Enter key.

The current working directory is the directory R will read files from and save results to by default. User can find out what the current working directory is by using the *getwd*() function. He can set the current working dirctory by using the *setwd*()function. If it is needed to input a file that isn't in the current working directory, use the full path name in the call. Always enclose the names of files and directories from the operating system in quote marks. Some standard commands for managing workspace are listed in Table 5.2.

Table 5.2: Functions for managing the R workspace (*adapted redrawn from Kabacoff, R.I. 2011.)*

Function	Action
getwd()	List the current working directory.
setwd("*mydirectory*")	Change the current working directory to *mydirectory*
ls()	List the objects in the current workspace.
rm(*objectlist*)	Remove (delete) one or more objects.
help(options)	Learn about available options.
options()	View or set current options.
history(#)	Display your last # commands (default = 25).
savehistory("*myfile*")	Save the commands history to *myfile*(default =.Rhistory).
loadhistory("*myfile*")	Reload a command's history (default = .Rhistory).
save.image("*myfile*")	Save the workspace to myfile (default = .RData).
save(*objectlist, myfile*)	Save specific objects to a file.
load("*myfile*")	load ("*myfile*") Load a workspace into the current session (default =.RData).
q()	Quit R. You'll be prompted to save the workspace

Some command actions of the functions are listed below as an example by which one can understand how the above listed functions work:

```
setwd ("C:/myprojects/project1")
options()
options(digits=3)
x <- runif(20)
summary(x)
hist(x)
savehistory()
save.image()
q()
```

First, the current working directory is set to C:/myprojects/project1, the current option settings are displayed, and numbers are formatted to print with three digits after the decimal place. Next, a vector with 20 uniform random variates is created, and summary statistics and a histogram based on this data are generated. Finally, the command history is saved to the file. Rhistory, the workspace (including vector x) is saved to the file. RData, and the session is ended.

It is to be noted here that forward slashes are used in the pathname of the *setwd*() command. R treats the backslash (\) as an escape character. Even when using R on a Windows platform, use forward slashes in pathnames needs to be used. It is also to be noted that the *setwd*() function will not create a directory that doesn't exist. If necessary, user can use the *dir.create*() function to reate a directory, and then use *setwd*() to change to its location.

It's a good idea to keep different projects in separate directories. Typically an R session is started by issuing the *setwd()* command with the appropriate path to a project, followed by the *load()* command without options. This helps to start up where the user left off in last session and keeps the data and settings separate between projects. In windows operating system, it can also be loaded by just navigating to the project directory and double-click on the saved image file. Doing so will start R, load the saved workspace, and set the current working directory to this location.

Input and output system in R

By default, launching R starts an interactive session with input from the keyboard and output to the screen. Otherwise, commands may also be processed from a previously created script file (a file containing R statements) and direct output to a variety of destinations.

The *source("filename")* function submits a script to the current session. If the file name doesn't include a path, the file is assumed to be in the current working directory. For example, *source("myscript.R")* runs a set of R statements contained in file myscript R. By convention, script file names end with an R extension, but this is not required.

The *sink("filename")* function redirects output to the file *filename*. By default, if the file already exists, its contents are overwritten. The option append=TRUE needs to be added in the command line to append text to the file rather than overwriting it. Including the option split=TRUE will send output to both the screen and the output file. Issuing the command *sink()* without options will return output to the screen alone.

Although *sink()* redirects text output, it has no effect on graphic output. To redirect graphic output, use one of the functions listed in Table 5.3. Use *dev.off()* to return output to the terminal.

Table 5.3: Functions for saving graphic output (*adapted redrawn from Kabacoff, R.I. 2011.)*

Function	Output
pdf("filename.pdf")	PDF file
win.metafile("filename.wmf")	Windows metafile
png("filename.png")	PNG file
jpeg("filename.jpg")	JPEG file
bmp("filename.bmp")	BMP file
postscript("filename.ps")	PostScript file

How input and output works in R environment is demonstrated below through some example. For example, there are three script files containing R code (script1.R, script2.R, and script3.R). Issuing the statement *source("script1.R")* will submit the R code from script1.R to the current session and the results will appear on the screen.

If then following statements are issued

sink("myoutput", append=TRUE, split=TRUE)

pdf("mygraphs.pdf")

source("script2.R")

the R code from file script2.R will be submitted, and the results will again appear on the screen. In addition, the text output will be appended to the file my output, and the graphic output will be saved to the file mygraphs.pdf. Finally, if you issue the statements

sink()

dev.off()

source("script3.R")

the R code from script3.R will be submitted, and the results will appear on the screen. This time, no text or graphic output is saved to files.

R Packages

R comes with extensive capabilities but its most exciting features are available as optional modules as packages that can be downloaded and installed. There are over 2,500 user-contributed modules called *packages* that user can download from http://cran.r-project.org/web/packages. All these packages are available for specific statistical and mathematical analysis. These packages provide a tremendous range of new capabilities, from the analysis of geostatistical data to protein mass spectra processing to the analysis of psychological tests!

R comes with a standard set of packages (including base, datasets, utils, grDevices, graphics, stats, and methods). They provide a wide range of functions and datasets that are available by default. Other packages are available for download and installation. Once installed, they have to be loaded into the session in order to be used. The command *search*() tells you which packages are loaded and ready to use.

There are a number of R functions that let you manipulate packages. To install a package for the first time, use the install.packages() command. For example, install.packages() without options brings up a list of CRAN mirror sites. Once you select a site, you'll be presented with a list of all available packages. Selecting one will download and install it. If you know what package you want to install, you can do so directly by providing it as an argument to the function. For example, the 'pls' package contains functions for partial least squares and principal component regression. It can be downloaded and installed with the command *install.packages*("pls").

To use a package in R, it needs to be installed just once. However, there is a requirement to update the package as it is often updated by authors. To update the installed package, *update.packages*() needs to be used. Details on the installed packages can be seen by the *installed.packages*() command. It lists the packages you have, along with their version numbers, dependencies, and other information.

Installing a package downloads it from a CRAN mirror site and places it in your library. It is to be noted here that the computer should be connected with internet to directly download and install a package from a CRAN mirror site. Otherwise, the package can be downloaded separately to a removable hard drive as 'package.rar' or 'package.zip' file from internet source and then needs to be installed in R environment. For installation of a package in R environment, install button from the package tab available in bottom right window of RStudio can be used (Fig. 5.3). As it is shown in the figure, source of package may be selected from repository or from package archive file (*.rar, *.zip, or *.tar) saved in a removable hard drive. Entering the correct name of the package in the package line will automatically search the packages and dependencies from the repository and will install it in R HOME Directory (e.g. C:/Users/Computer/Documents/R/Win-library/3.1.

To use an installed package in an R session, it should be loaded first using the *library*() command. For example, to use the packaged 'pls' issue the command *library*(*pls*). Of course, the packages should be available or installed first before it is loaded. The most common approach is to load the required libraries first in R script before running a command.

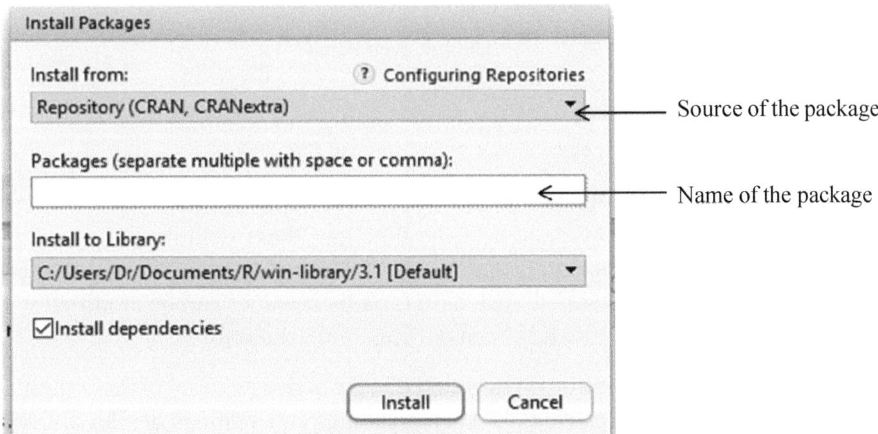

Fig. 5.3: Installation of packages in R

When a package is loaded, a new set of functions and datasets becomes available in the R workspace. Small illustrative datasets are provided along with sample code, allowing you to tryout the new functionalities. The help system contains a description of each function (along with examples), and information on each dataset included. Entering *help(package="package_name")* provides a brief description of the package and an index of the functions and datasets included. Using *help()* with any of these function or dataset names will provide further details. The same information can be downloaded as a PDF manual from CRAN.

How R works

The working principle of R is so simple that even a user who does not know the fundamentals of programming language can start to do work in R because of mainly two reasons. First, R is an interpreted language and not a compiled one, which indicates that all commands typed on the keyboard are directly executed without requiring to build a complete program like in most computer languages (*e.g.* C, Fortran, Pascal *etc*). Second, R's syntax is very simple and intuitive. For instance, a linear regression can be done with the command lm(y ~ x) which means "fitting a linear model with y as response and x as predictor". In R, in order to be executed, a function always needs to be written with parentheses, even if there is nothing within them (*e.g.*, *ls*()). If one just types the name of a function without parentheses, R will display the content of the function. When R is running, variables, data, functions, results, *etc*., are stored in the active memory of the computer in the form of objects which have a name. The user can do actions on these objects with operators e.g. arithmetic, logical, comparison etc and functions. An R function may be sketched as follows (Fig. 5.4):

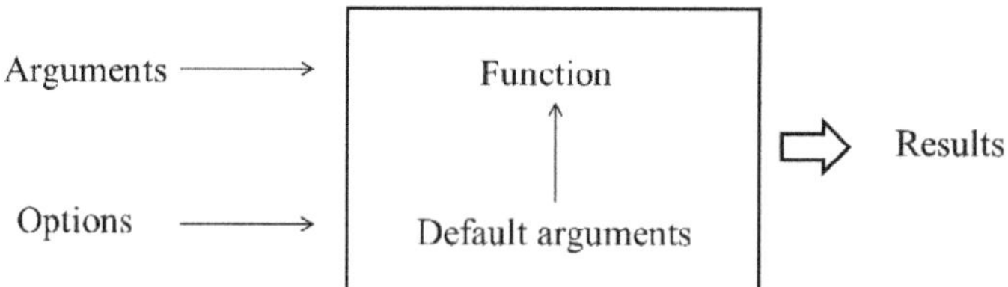

Fig. 5.4: Sketch diagram of working principle of R
(adapted redrawn from Kabacoff, R.I. 2011.)

The arguments can be objects ("data", formulae, expressions, . . .), some of which could be defined by default in the function; these default values may be modified by the user by specifying options. An R function may require no argument: either all arguments are defined by default (and their values can be modified with the options), or no argument has been defined in the function.

All the actions of R are done on objects stored in the active memory of the computer: no temporary files are used (Fig. 5.5). The readings and writings of files are used for input and output of data and results. The user executes the functions via some commands. The results are displayed directly on the screen, stored in an object, or written on the disk particularly for graphics. Since the results are themselves objects, they can be considered as data and analyzed as such. Data files can be read from the local disk or from a remote server through internet.

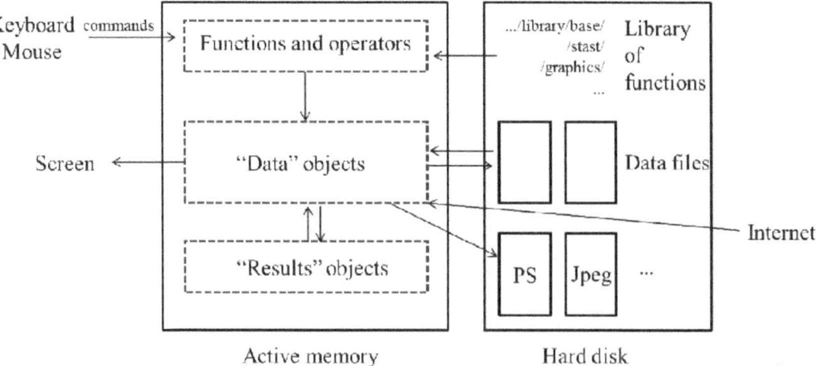

Fig. 5.5: A schematic view of how R works *(adapted redrawn from Kabacoff, R.I. 2011.)*

The functions available to the user are stored in a library localized on the disk in a directory called R HOME/library (R HOME is the directory where R is installed). This directory contains packages of functions, which are themselves structured in directories. The package named base is in a way the core of R and contains the basic functions of the language, particularly, for reading and manipulating data. Each package has a directory called R with a file named like the package (for instance, for the package base, this is the file R HOME/library/base/R/base). This file contains all the functions of the package.

One of the simplest commands is to type the name of an object to display its content. For instance, if an object n contents the value 10:

```
>n
[1] 10
```

The digit 1 within brackets indicates that the display starts at the first element of n. This command is an implicit use of the function print and the above example is similar to *print*(n) (in some situations, the function print must be used explicitly, such as within a function or a loop). The name of an object must start with a letter (A–Z and a–z) and can include letters, digits (0–9), dots (.), and underscores (_). R discriminates between uppercase letters and lowercase ones in the names of the objects, so that x and X can name two distinct objects.

Creating, listing and deleting the objects in memory

An object can be created with the "assign" operator which is written as an arrow with a minus sign and a bracket; this symbol can be oriented left-to-right or the reverse:

```
> n <- 10
>n
[1] 10
>6 -> n
>n
[1] 6
> x <- 2
> X <- 8
>x
[1] 2
> X
[1] 8
```

If the object already exists, its previous value is erased. The modification affects only the objects in the active memory, not the data on the disk. The value assigned this way may be the result of an operation and/or a function:

```
> n <- 10 + 2
>n
[1] 12
> n <- 3 + rnorm(1)
>n
[1] 2.208807
```

The function *rnorm*(1) generates a normal random variate with mean zero and variance unity. It is to be noted here that one can simply type an expression without assigning its value to an object and in such case the result will be displayed on the screen but is not stored in memory. For example, the arithmetic operation of (10+2) × 5 will result 60 will be displayed as follows but not stored in memory:

```
> (10 + 2) * 5
[1] 60
```

The function *ls* lists simply the objects in memory: only the names of the objects are displayed.

```
>name<- "Carmen"; n1 <- 10; n2 <- 100; m <- 0.5
```

>*ls*() [1] "m" "n1" "n2" "name"

It is noted here that semi-colon is used here to separate distinct commands on the same line. If we want to list only the objects which contain a given character in their name, the option pattern, which can be abbreviated with *pat*, can be used:

>*ls*(pat = "m")

[1] "m" "name"

To restrict the list of objects whose names start with this character:

>*ls*(pat = "^m")

[1] "m"

The function ls.str displays some details on the objects in memory:

>*ls.str*()

m :num 0.5

n1 :num 10

n2 :num 100

name :chr "Carmen"

To delete objects in memory, the function *rm*() is used. For example, *rm*(x) deletes the object x, rm(x,y) deletes both the objects x and y. The command *rm*(list=ls()) deletes all the objects in memory. Generally at the start of any R script, this command is used to clean the memory and then start loading the new objects and data. The same options mentioned for the function *ls*() can be used to delete selectively some objects, for example *rm*(list=*ls*(pat="^m")) command will delete those objects whose name starts with m.

Summary

The attractive features of R is shown here which are its basic strengths for which students, researchers, statisticians, and data analysts like to use it. How the R can be installed in a computer system is discussed. Basically both R and Rstudio need to be installed for easy user-interface options. Different options for installing, updating and loading additional packages of R are discussed besides the default packages. How the output file is stored specifically for graphics is discussed. The working system of R commands is also discussed here. Finally, how the objects are created, modified and deleted in R workspace is mentioned. Overall, it has been shown that R is very user-friendly and most importantly it is open source program and hence freely available.

References

Kabacoff, R.I. 2011. R in Action: Data analysis and graphics with R. Manning Publications Co. 20 Baldwin Road PO Box 261 Shelter Island, NY 11964.

Paradis, E. 2005. R for beginners. Institut des Sciences de l'Evolution Universite Montpellier II.F-34095 Montpellier cedex 05 France.

6

Spatial Data Handling and Plotting in R

Priyabrata Santra

ICAR - Central Arid Zone Research Institute, Jodhpur, Rajasthan, India

Introduction

Spatial data is different from a non-spatial data in the sense that it always has a spatial reference system which is nothing but a pair of coordinate values and a system of reference for these coordinates. For example, the locations of benchmark soil series of India can be defined as pairs of longitude and latitude in decimal degree values with respect to the prime meridian at Greenwich and zero latitude at the equator. The World Geodetic System (WGS84) is a frequently used system of reference on the Earth. The different types of spatial data are: (i) point, which is a single point location, such as a GPS reading or a geocoded address (ii) line, which is defined as a set of ordered points, connected by straight line segments (iii) polygon, an area, marked by one or more enclosing lines, possibly containing holes and (iv) grid, a collection of points or rectangular cells, organised in a regular lattice. Among the above said spatial data, the first three are vector data models and represent entities as exactly as possible, while the fourth data model is a raster data model, representing continuous surfaces by using a regular tessellation. In this chapter, we discuss how a spatial data can be handled in R. There are several packages available in R in which spatial data analysis can be handled e.g. sp, rgdal, raster etc. All the contributed packages for spatial data analysis in R address two broad areas: moving spatial data into and out of R, and analysing spatial data in R. In the following section, the sp package for spatial data handling is briefly discussed.

The sp package of R for handling spatial data

A group of R developers have developed the R package sp to extend R with classes and methods for spatial data (Pebesma and Bivand 2005). Before, we further go into details of sp package, we should know the definition of classes and methods in R. Classes generally specify a structure and define how spatial data are organised and stored in R environment. The spatial data classes implemented in R are points, grids, lines, rings and polygons. For example, Spatial Points is a spatial data without any attribute data whereas its extension SpatialPointsDataFrame is a spatial data where attribute information which is stored in a data.frame is available for spatial points entity. The available data classes in R are presented in Table 6.1.

Table 6.1: Available classes for spatial data in R

Data types	Class	Attributes
Points	SpatialPoints	No
Points	SpatialPointsDataframe	data.frame
Multipoints	SpatialMultipoints	No
Multipoints	SpatialMultipointsDataframe	data.frame
Pixels	SpatialPixels	No
Pixels	SpatialPixelsDataframe	data.frame
Full grid	SpatialGrid	No
Full grid	SpatialGridDataframe	data.frame
Line	Line	No
Lines	Lines	No
Lines	SpatialLines	No
Lines	SpatialLines	data.frame
Polygons	Polygon	No
Polygons	Polygons	No
Polygons	SpatialPolygons	No
Polygons	SpatialPolygonsDataframe	data.frame

In the table, we see SpatialPixels Dataframe and SpatialGridDataframe both represents a raster grid, however there is a fundamental difference between them. SpatialPixels can be viewed as SpatialPoints objects, but the coordinates in a SpatialPixel have to be regularly spaced. The coordinates of SpatialPixel are stored as grid indices. Therefore, SpatialPixelsDataFrame objects not only store attribute data where it is present, but need to store the coordinates and grid indices of those grid cells. SpatialGridDataFrame objects do not need to store coordinates, because they fill the entire defined grid, but they need to store NA values where attribute values are missing. The difference of the SpatialPixelDataframe and SpatialGridDataframe can be more understood from tutorial available in http://r-spatial.org/r/2016/03/08/plotting-spatial-grids.html and from the following example data of R.

```
>data(meuse.grid)                                  # only the non-missing valued cells
>coordinates(meuse.grid) = c("x", "y")             # promote to SpatialPointsDataFrame
>gridded(meuse.grid) <- TRUE                       # promote to SpatialPixelsDataFrame
>x = as(meuse.grid, "SpatialGridDataFrame")        # creates the full grid
```

In this example, meuse.grid is a DataFrame consisting of 3103 observations of 7 variables including x and y coordinates of regularly spaced locations. When the coordinates of this DataFrame is defined as x and y, it becomes a SpatialPointsDataframe. The gridded method of sp package converts this SpatialPointDataframe to a SpatialPixelDataframe. The structure of the SpatialPixelDataFrame can be viewed as follows by typing the code str(meuse.grid). Here, it may be noted that SpatialPixelDataFrame contains seven slots. Specially the @data slot contains 3103 observations for 5 variables for which data is available and thus the slot @grid.index contains 3103 grid indices.

```
> str(meuse.grid)
Formal class 'SpatialPixelsDataFrame' [package "sp"] with 7 slots
..@ data :'data.frame':   3103 obs. of 5 variables:
.. ..$ part.a: num [1:3103] 1 1 1 1 1 1 1 1 1 1 ...
.. ..$ part.b: num [1:3103] 0 0 0 0 0 0 0 0 0 0 ...
.. ..$ dist : num [1:3103] 0 0 0.0122 0.0435 0 ...
.. ..$ soil : Factor w/ 3 levels "1","2","3": 1 1 1 1 1 1 1 1 1 ...
.. ..$ ffreq : Factor w/ 3 levels "1","2","3": 1 1 1 1 1 1 1 1 1 ...
..@ coords.nrs : num(0)
..@ grid :Formal class 'GridTopology' [package "sp"] with 3 slots
.. .. ..@ cellcentre.offset: Named num [1:2] 178460 329620
.. .. .. ..- attr(*, "names")= chr [1:2] "x" "y"
.. .. ..@ cellsize : Named num [1:2] 40 40
.. .. .. ..- attr(*, "names")= chr [1:2] "x" "y"
.. .. ..@ cells.dim : Named int [1:2] 78 104
. .. .. ..- attr(*, "names")= chr [1:2] "x" "y"
..@ grid.index : int [1:3103] 69 146 147 148 223 224 225 226 301 ...
..@ coords : num [1:3103, 1:2] 181180 181140 181180 181220 ...
.. ..- attr(*, "dimnames")=List of 2
.. .. ..$ : chr [1:3103] "1" "2" "3" "4" ...
.. .. ..$ : chr [1:2] "x" "y"
..@ bbox : num [1:2, 1:2] 178440 329600 181560 333760
.. ..- attr(*, "dimnames")=List of 2
.. .. ..$ : chr [1:2] "x" "y"
.. .. ..$ : chr [1:2] "min" "max"
..@ proj4string:Formal class 'CRS' [package "sp"] with 1 slot
.. .. ..@ projargs: chr NA
```

However, when this SpatialPixelDataFrame is converted to a SpatialGridDataframe the structure has been changed as follows. It can be seen from below that the SpatialGridDataFrame has now 4 slots. The slots for @coords.nrs, @grid.index and @coords are missing in the SpatialGridDataframe. It is also noted that the @data slot now contains data for 8112 observations within the bounding box including the missing data as NA. From the dimension of grid as 78 × 104 it may be calculated as the total number of cells in the grid as 8112 out of which data is available for 3103 observations and rest of the cells are filled with NA.

> *str*(*x*)

```
Formal class 'SpatialGridDataFrame' [package "sp"] with 4 slots
..@ data :'data.frame': 8112 obs. of 5 variables:
.. ..$ part.a: num [1:8112] NA NA NA NA NA NA NA NA NA NA ...
.. ..$ part.b: num [1:8112] NA NA NA NA NA NA NA NA NA NA ...
.. ..$ dist : num [1:8112] NA NA NA NA NA NA NA NA NA NA ...
.. ..$ soil : Factor w/ 3 levels "1","2","3": NA NA NA NA NA NA ...
.. ..$ ffreq : Factor w/ 3 levels "1","2","3": NA NA NA NA NA NA ...
..@ grid :Formal class 'GridTopology' [package "sp"] with 3 slots
.. .. ..@ cellcentre.offset: Named num [1:2] 178460 329620
.. .. .. ..- attr(*, "names")= chr [1:2] "x" "y"
.. .. ..@ cellsize : Named num [1:2] 40 40
.. .. .. ..- attr(*, "names")= chr [1:2] "x" "y"
.. .. ..@ cells.dim : Named int [1:2] 78 104
.. .. .. ..- attr(*, "names")= chr [1:2] "x" "y"
..@ bbox : num [1:2, 1:2] 178440 329600 181560 333760
.. ..- attr(*, "dimnames")=List of 2
.. .. ..$ : chr [1:2] "x" "y"
.. .. ..$ : chr [1:2] "min" "max"
..@ proj4string:Formal class 'CRS' [package "sp"] with 1 slot
.. .. ..@ projargs: chr NA
```

Methods are instances of functions specialized for a particular data class. For example, the summary method for all spatial data classes may tell the range spanned by the spatial coordinates, and show which coordinate reference system is used (such as degrees longitude/latitude, or the UTM zone). A list of available methods in sp package of R is given in Table 6.2.

Table 6.2: List of standard methods available in sp package of R

Methods	**What it does**
[	select spatial items (points, lines, polygons, or rows/cols from a grid) and/or attributes variables
[[	extracts a column from the data attribute table
[[<-	assign or replace a column in the data attribute table.
$, $<-	retrieve, set or add attribute table columns
dimensions(x)	returns number of spatial dimensions
spsample(x)	sample points from a set of polygons, on a set of lines or from a gridded area
bbox(x)	get the bounding box
proj4string	get or set the projection (coordinate reference system)
spTransform(x,CRS())	For example, y = spTransform(x, CRS("+proj=longlat +datum=WGS84")) convert or transform from one coordinate reference system (geographic projection) to another (requires package rgdal to be installed)
Coordinates(x)	set or retrieve coordinates
coerce	convert from one class to another
overlay	combine two different spatial objects
gridded(x)	tells whether x derives from SpatialPixels, or when used in assignment, coerces a SpatialPoints object into a SpatialPixels object.
over(x, y)	retrieve index or attributes of y corresponding (intersecting) with spatial locations of x.
spplot(x)	plots attributes, possibly in combination with other types of data (points, lines, grids, polygons), and possibly in as a conditioning plot for multiple attributes
geometry(x)	strips the data.frame, and returns the geometry-only object

Other methods available in sp package of R are: plot, summary, print, dim and names (operate on the data.frame part), as.data.frame, as.matrix and image (for gridded data), lines (for line data), points (for point data), subset (points and grids), stack (point and grid data.frames). Apart from the basic classes and methods of sp package as discussed above, further details on spatial data analysis using sp packagae of R is available in https://cran.r-project.org/web/packages/sp/vignettes/intro_sp.pdf. Details on visualization of spatial data using sp package is available in https://cran.r-project.org/doc/contrib/intro-spatial-rl.pdf and http://r-spatial.org/r/2016/03/08/plotting-spatial-grids.html.

Other packages in R for spatial data analysis

One of the most important steps in handling spatial data with R is the ability to read in spatial data, such as shapefiles, a common geographical file format. Among several ways to read a shapefile in R, the most commonly used and versatile one is readOGR, which is available in rgdal package. The rgdal is R's interface to the "Geospatial Data Abstraction Library (GDAL)" which is used by other open source

GIS packages such as QGIS and enables R to handle a broader range of spatial data formats. The readOGR function automatically extracts the information from the spatial data. The following line of code is used to load a shapefile and assign it to a new object called "x".

>x <- readOGR(dsn = "data", layer = "layer1")

In principle, readOGR is a function which accepts two arguments: dsn which stands for "data source name" and specifies the location where the file is stored, and layer specifies the file name. It may be noted here that each new argument is separated by a comma and there is no need to specify the file extension (e.g. .shp) when providing the file name. Both arguments in this case are character strings indicated by quote marks. R functions have a default order for arguments, so "dsn =" and "layer =" do not actually have to be typed for the command to run, for example, readOGR("data", "layer1") would work just as well. The other way to read a shapefile and load in R environment is by the function readShapePoly. In the following example, "arid_western_india.shp" is loaded in R environment as studyarea by using readSpahePoly function.

>studyarea <- readShapePoly("arid_western_india.shp")

Some of the packages of R, which are commonly used for handling, visualizing and analyzing spatial data are mentioned below:

- **Classes for spatial data:** The 'sp' package is a base package for importing and using spatial data including storing of data and applying different functions for visualising it. Many other packages have become dependent on the 'sp' classes, including 'rgdal' and 'maptools'. The 'rgeos' package provides an interface to topology functions for 'sp' objects using 'GEOS'. The 'raster' package is a major extension of spatial data classes to virtualise access to large rasters, permitting large objects to be analysed, and extending the analytical tools available for both raster and vector data. Used with 'rasterVis', it can also provide enhanced visualisation and interaction. The 'spacetime' package extends the shared classes defined in 'sp' for spatio-temporal data. The 'Grid2Polygons' converts a spatial object from class SpatialGridDataFrame to SpatialPolygonsDataFrame.
- **Handling spatial data:** The 'raster' package introduces many GIS methods that now permit much to be done with spatial data without having to use GIS in addition to R. The 'spsurvey' package provides a range of sampling functions. The 'GeoXp' package permits interactive graphical exploratory spatial data analysis. The 'spcosa' provides spatial coverage sampling and random sampling from compact geographical strata. The 'magclass' offers a data class for increased interoperability working with spatial-temporal data together with corresponding functions and methods.

- **Reading and writing spatial data** - The 'rgdal' package provides bindings to GDAL-supported raster formats and OGR-supported vector formats. It contains functions to write raster files in supported formats. The package also provides PROJ.4 projection support for vector objects. Affine and similarity transformations on sp objects may be made using functions in the 'vec2dtransf' package. The'regeos' package provides functions for reading and writing well-known text (WKT) geometry, and the 'wkb' package provides functions for reading and writing well-known binary (WKB) geometry. There are a number of other packages for accessing vector data on CRAN. For example, 'maps' (with 'mapdata' and 'mapproj') provides access to the same kinds of geographical databases as S - RArcInfo allows ArcInfo v.7 binary files and *.e00 files to be read, and 'maptools' and 'shapefiles' read and write ArcGIS/ArcView shapefiles. The 'maptools' package also provides helper functions for writing map polygon files to be read by WinBUGS, Mondrian, and the tmap command in Stata. 'OpenStreetMap' gives access to open street map raster images, and 'osmar' provides infrastructure to access OpenStreetMap data from different sources, to work with the data in common R manner, and to convert data into available infrastructure provided by existing R packages. The 'rpostgis' package provides additional functions to the 'RPostgreSQL' package to interface R with a 'PostGIS'-enabled database, as well as convenient wrappers to common 'PostgreSQL' queries. The 'postGIStools' package provides functions to convert geometry and 'hstore' data types from 'PostgreSQL' into standard R objects, as well as to simplify the import of R data frames (including spatial data frames) into 'PostgreSQL'
- **Integration with GIS software:** The version 6.* and of the leading open source GIS, GRASS, is provided in CRAN package 'spgrass6', using 'rgdal' for exchanging data. For GRASS 7.*, use 'rgrass7.RPyGeo' is a wrapper for Python access to the ArcGIS GeoProcessor, and 'RSAGA' is a similar shell-based wrapper for SAGA commands. The 'RQGIS' package establishes an interface between R and QGIS, i.e. it allows the user to access QGIS functionalities from the R console. It achieves this by using the QGIS Python API via the command line.
- **Visualisation:** For visualization, the colour palettes provided in the 'RColorBrewer' package are very useful, and may be modified or extended using the colorRampPalette function provided with R. The 'classInt' package provides functions for choosing class intervals for thematic cartography. The 'tmap' package provides a modern basis for thematic mapping optionally using Grammar of Graphics syntax. Because it has a custom grid graphics platform, it obviates the need to fortify geometries to use with ggplot2. The 'mapview' package provides methods to view spatial objects interactively, usually on a web mapping base. The 'quickmapr' package provides a simple method to visualize 'sp' and 'raster' objects,

allows for basic zooming, panning, identifying, and labeling of spatial objects, and does not require that the data be in geographic coordinates. The 'cartography' package allows various cartographic representations such as proportional symbols, choropleth, typology, flows or discontinuities. The 'mapmisc' package is a minimal, light-weight set of tools for producing nice looking maps in R, with support for map projections. If the user wishes to place a map backdrop behind other displays, the 'RgoogleMaps' package for accessing Google Maps(TM) may be useful. The package 'ggmap' may be used for spatial visualisation with Google Maps and OpenStreetMap; 'ggsn' provides North arrows and scales for such maps. The 'plotGoogleMaps' package provides methods for the visualisation of spatial and spatio-temporal objects in Google Maps in a web browser. 'plotKML' is a package providing methods for the visualisation of spatial and spatio-temporal objects in Google Earth. A further option is 'leafletR', which provides basic web-mapping functionality to combine vector data files and online map tiles from different sources.

- **Geostatistics**: The 'gstat' package provides a wide range of functions for univariate and multivariate geostatistics, also for larger datasets, while 'geoR' and 'geoRglm' contain functions for model-based geostatistics. Variogram diagnostics may be carried out with 'vardiag'. Automated interpolation using 'gstat' is available in 'automap'. This family of packages is supplemented by 'intamap' with procedures for automated interpolation and 'psgp', which implements projected sparse Gaussian process kriging. A similar wide range of functions is to be found in the 'fields' package. The 'spatial' package is shipped with base R, and contains several core functions. The 'geospt' package contains some geostatistical and radial basis functions, including prediction and cross validation. Besides, it includes functions for the design of optimal spatial sampling networks based on geostatistical modelling. The 'geostatsp' package offers geostatistical modelling facilities using Raster and SpatialPoints objects are provided. Non-Gaussian models are fit using INLA, and Gaussian geostatistical models use Maximum Likelihood Estimation. The 'RandomFields' package provides functions for the simulation and analysis of random fields, and variogram model descriptions can be passed between 'geoR', 'gstat' and this package.

Understanding spatial data classes and objects in R

Spatial Objects in R

The foundation class of all spatial data in R is the Spatial class, with just two slots. The first is a bounding box (bbox), a matrix of numerical coordinates with column names c('min', 'max'), and at least two rows, with the first row as eastings (*x*-axis) and the second as northings (*y*-axis). Most often the bbox is generated automatically from the data in subclasses of spatial. The second is a coordinate reference system (CRS) class object defining the coordinate reference system, and may be set to 'missing', represented by NA in R, by CRS (as.character(NA)), its default value. We can use getClass code in R as mentioned below to return the complete definition of a class, including its slot names and the types of their contents:

> library(sp)

> getClass("Spatial")

Slots:

Name: bbox proj4string

Class: matrix CRS

Known Subclasses:

Class "SpatialPoints", directly

Class "SpatialLines", directly

Class "SpatialPolygons", directly

Class "SpatialPointsDataFrame", by class "SpatialPoints", distance 2

Class "SpatialPixels", by class "SpatialPoints", distance 2

Class "SpatialGrid", by class "SpatialPoints", distance 3

Class "SpatialPixelsDataFrame", by class "SpatialPoints", distance 3

Class "SpatialGridDataFrame", by class "SpatialPoints", distance 4

Class "SpatialLinesDataFrame", by class "SpatialLines", distance 2

Class "SpatialPolygonsDataFrame", by class "SpatialPolygons",distance 2

The coordinate reference system (CRS) is defined by projection argument (projargs), which is a character type of data as its only slot value, which may be a missing value NA.

> getClass("CRS")

Slots:

Name: projargs

Class: character

If it is not missing, it should be a PROJ.4-format string describing the projection. For geographical coordinates, the simplest such string is "+proj=longlat", using "longlat", which also shows that eastings always go before northings in sp classes. Even the PROJ.4-format for geographical coordinate system may be written using EPSG code e.g. "+init=epsg:4326". The complete definition of geographical coordinate system may be checked by typing the code proj4string(x) of a spatial object x and may be found as follows.

"+proj=longlat +datum=WGS84 +no_defs +ellps=WGS84 +towgs84=0,0,0"

Strings for other PROJ.4-formats may be found from any open source GIS environment or from internet.

Spatial Point Data Frame in R

The contents of objects in a SpatialPointsDataFrame class are presented in Fig. 6.1. Since the SpatialPointsDataFrame class extends SpatialPoints as an object, it also inherits the information contained in the Spatial class object. The data slot of the SpatialPointsDataFrame class is the slot where the information about the object is kept, which is inherited from a data.frame object. In few SpatialPointsDataFrame object, data slot may be missing containing only the coordinate information.

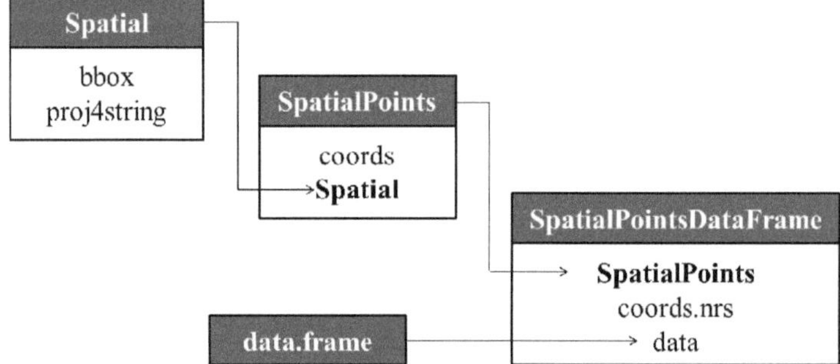

Fig. 6.1: Spatial point classes and their slots
Source: Bivand R.S., Pebesma, E. and Gomez-Rubio, V. 2013.

Spatial Lines in R

Lines have been represented in a simple form as a sequence of points. A Line object in R is a matrix of 2D coordinates, without NA values. A list of Line objects forms the Lines slot of a Lines object as depicted in Fig. 6.2. An identifying character tag is used for constructing SpatialLines objects using the same approach as was used above for matching ID values for spatial points. The slots available with Line and Lines classes are given below.

> *getClass("Line")*

Slots:

Name: coords

Class: matrix

Known Subclasses: "Polygon"

> *getClass("Lines")*

Slots:

Name: Lines ID

Class: list character

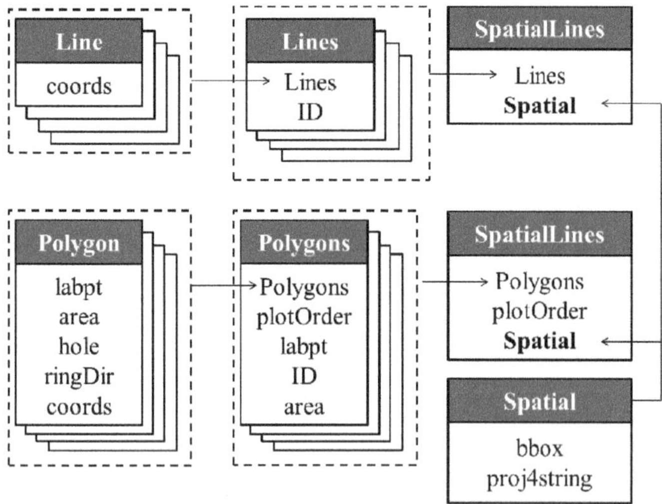

Fig. 6.2: Spatial lines and spatial polygon classes and slots; thin arrows show subclass extensions, thick arrows the inclusion of lists of objects
Source: Bivand R.S., Pebesma, E. and Gomez-Rubio, V. 2013.

It is the SpatialLines object that contains the bounding box and projection information for the list of Lines objects stored in its lines slot.

> *getClass("SpatialLines")*

Slots:

Name: lines bbox proj4string

Class: list matrix CRS

Extends: "Spatial"

Known Subclasses: "SpatialLinesDataFrame"

Spatial Polygons

The basic representation of a polygon is a closed line, a sequence of point coordinates where the first point is the same as the last point. A set of polygons is made of closed lines separated by NA points. The nesting of classes for polygons is the same as that for lines, but the successive objects have more slots.

> *getClass("Polygon")*

Slots:

Name: labpt area hole ringDir coords

Class: numeric numeric logical integer matrix

Extends: "Line"

The Polygon class extends the Line class by adding slots needed for polygons and checking that the first and last coordinates are identical. The extra slots are a label point, taken as the centroid of the polygon, the area of the polygon in the metric of the coordinates, whether the polygon is declared as a hole or not – the default value is a logical NA, and the ring direction of the polygon.

> *getClass("Polygons")*

Slots:

Name: Polygons plotOrder labpt ID area

Class: list integer numeric character numeric

The Polygons class contains a list of valid Polygon objects, an identifying character string, a label point taken as the label point of the constituent polygon with the largest area, and two slots used as helpers in plotting using R graphics functions, given this representation of sets of polygons. These set the order in which the polygons belonging to this object should be plotted, and the gross area of the polygon, equal to the sum of all the constituent polygons.

> *getClass("SpatialPolygons")*

Slots:

Name: polygons plotOrder bbox proj4string

Class: list integer matrix CRS

Extends: "Spatial"

Known Subclasses: "SpatialPolygonsDataFrame"

The top level representation of polygons is as a SpatialPolygons object, a set of Polygons objects with the additional slots of a Spatial object to contain the bounding box and projection information of the set as a whole.

Spatial Polygons Data Frame Objects

As with other spatial data objects, SpatialPolygonsDataFrame objects bring together the spatial representations of the polygons with data (Fig. 6.2). The identifying tags of the Polygons in the polygon slot of a Spatial Polygons object are matched with the row names of the data frame to make sure that the correct data rows are associated with the correct spatial objects.

Spatial Grid and Spatial Pixel Objects

Grids are regular objects requiring much less information to define their structure. Once the single point of origin is known, the extent of the grid can be given by the cell resolution and the numbers of rows and columns present in the full grid. This representation is typical for remote sensing and raster GIS, and is used widely for storing data in regular rectangular cells, such as digital elevation models, satellite imagery, and interpolated data from point measurements, as well as image processing.

> *getClass("GridTopology")*

Slots:

Name: cellcentre.offset cellsize cells.dim

Class: numeric numeric integer

A SpatialGrid object contains GridTopology and Spatial objects, together with two helper slots, grid.index and coords. These are set to zero and to the bounding box of the cell centres of the grid, respectively. The slots available in a SpatialGrid object are mentioned below.

> *getClass("SpatialGrid")*

Class "SpatialGrid" [package "sp"]

Slots:

Name: grid bbox proj4string

Class: GridTopology matrix CRS

Extends: "Spatial"

Known Subclasses: "SpatialGridDataFrame"

Similarly, the slots available in SpatialPixels object is given below and the difference between a SpatialGrid and SpatialPixels object can be noted.

> *getClass("SpatialPixels")*

Class "SpatialPixels" [package "sp"]

Slots:

Name: grid grid.index coords bbox proj4string

Class: GridTopology integer matrix matrix CRS

Extends:

Class "SpatialPoints", directly

Class "Spatial", by class "SpatialPoints", distance 2

Class "SpatialVector", by class "SpatialPoints", distance 2

Class "SpatialPointsNULL", by class "SpatialPoints", distance 2

Known Subclasses: "SpatialPixelsDataFrame"

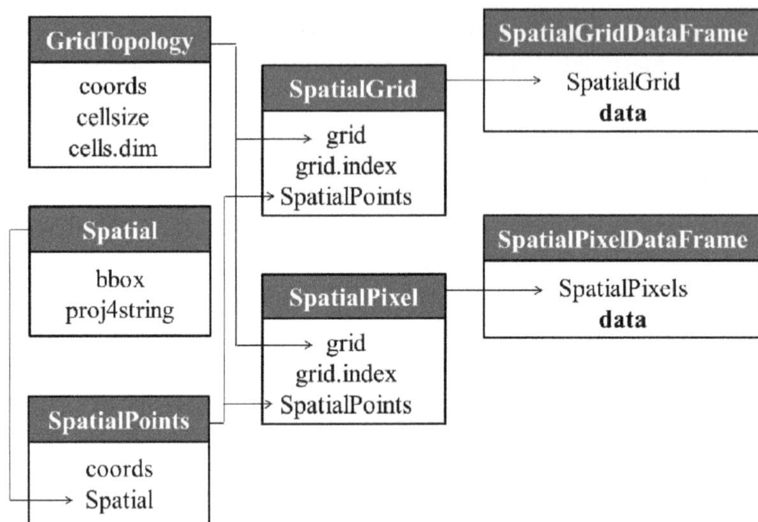

Fig. 6.3: SpatialGrid and spatialPixel classes and their slots; arrows show sub-class extensions *Source*: Bivand R.S., Pebesma, E. and Gomez-Rubio, V. 2013.

When the SpatialGrid or SpatialPixels object are associated with data inherited from a data.frame object, it is converted into a SpatialGridDataFrame or SpatialPixelsDataFrame object, respectively as presented in Fig. 6.3.

Further details on spatial data structure, its handling and plotting is available in Bivand et al. (2013).

Plotting spatial data in R

R has two plotting systems: the 'traditional' plotting system and the Trellis Graphics system, provided by package 'lattice', which is present in default R installations (Sarkar, 2008). Package 'sp' provides plot methods that build on the traditional R plotting system (plot, image, lines, points, etc.), as well as a 'new' generic method called spplot that uses the Trellis system (notably xyplot or levelplot from the 'lattice' package) and can be used for conditioning plots. In the following, few examples of creating and plotting points, polygons, and grid object are presented.

Plotting spatial point data

For example, a spatial data 'sawi_legacy_data.txt' containing soil profile information along with location information of each soil profile from arid western India has been prepared in text format, which may otherwise be prepared in CSV format. Then the data is loaded in R using read.table() code followed by defining the coordinates to make it a spatial object as given below. The projection system of the created spatial data has been further defined as geographic longlat coordinate system as EPSG code *"+init=epsg:4326"*. The coordinates of the profile locations may be viewed by calling data@coords.

```
>data <- read.table("sawi_legacy_data.txt", header=TRUE)
>coordinates(data) <- ~long+lat
>proj4string(data) <- CRS("+init=epsg:4326")
> data@coords
        long          lat
1       72.28230      26.34378
2       70.24269      27.91453
3       71.57806      27.04759
4       72.50969      26.99002
5       74.01729      28.02181
6       71.45142      26.34171
```

The created spatial data may be plotted either using plot() or spplot() function. The plots of spatial data using both the function are given in Fig. 6.4. It is to be noted here that in spplot() function sand content distribution is plotted.

```
>plot(data, pch=6, main="Soil profile locations")
>spplot(data, z="sand", pch=20, main="Soil profile locations",
scales = list(draw = TRUE), cex=2,
key.space = list(x = 1, y = 0.5, corner = c(0,1)),
xlab = "Long (degree)", ylab = "Lat (degree)")
```

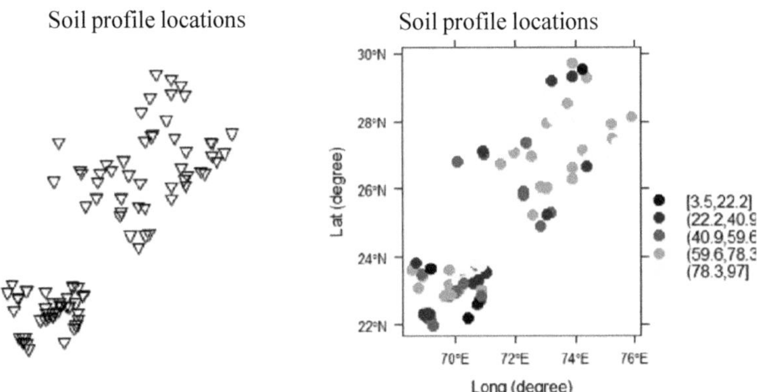

Fig. 6.4: Plot of spatial point data using *plot*() function at left hand side and using *spplot*() function at right hand side

The projection of spatial point data created above is in geographic coordinate system and hence coordinate values are available in decimal degree. However for spatial analysis, it is always better to have the coordinates in metre unit. Therefore, the projections need to be transformed to a system where coordinates are mentioned in metre unit e.g. Universal Transverse Mercator (UTM) or Lambert Conformal Conic (LCC) etc. In the following example, projection of the spatial data has been transformed to UTM system with zone no 42. The transformed projection system of the spatial data may be checked by proj4string() code. It is also to be noted here that the coordinates of the data are now converted into metre unit which was earlier in decimal degrees.

```
>data_trans <- spTransform(data, CRS="+init=epsg:32642")
>proj4string(data_trans)
[1] "+init=epsg:32642 +proj=utm +zone=42 +datum=WGS84 +units=m +no_defs
+ellps=WGS84 +towgs84=0,0,0"
> data_trans@coords
      long        lat
1  827441.4   2916390
2  623010.6   3087861
3  755958.4   2992378
4  848436.0   2989066
5  992903.7   3108447
6  744546.7   2914550
```

Plotting spatial polygon data

In the following example, the polygon boundary of arid western India as shapefile has been loaded in R using *readShapePoly*() function. The projection of the created polygon object needs to be defined if missing. Otherwise, the associated projection information of the shapefile will be loaded. The polygon boundary of the studyarea may be plotted either using *plot*() function or *spplot*() function as follows. The plots of polygon boundary using both the functions are presented in Fig. 6.5.

```
>studyarea <- readShapePoly("arid_western_india.shp")
>proj4string(studyarea) <- CRS("+init=epsg:4326")
>plot(studyarea, main="Arid western India", col="blue")
>spplot(studyarea, main="Arid western India", col="blue",
 scales = list(draw = TRUE),
 xlab = "Long (degree)", ylab = "Lat (degree)")
```

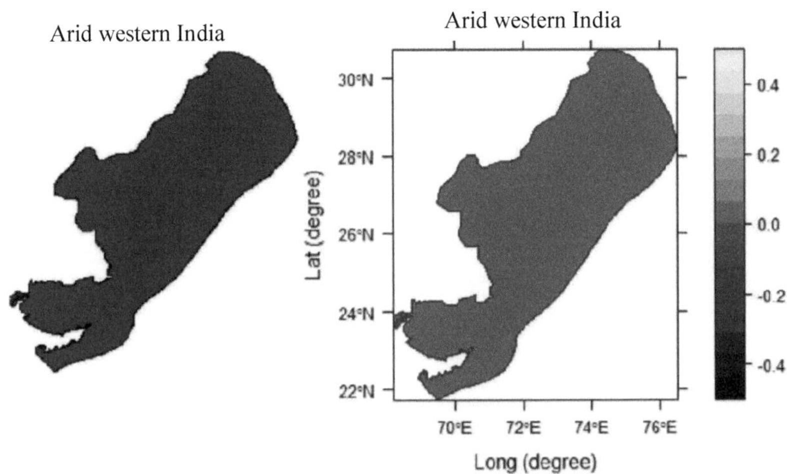

Fig. 6.5: Plot of spatial point data using plot() function at left hand side and using spplot () function at right hand side

Plotting and handling spatial grid data

In the following example, a grid object containing digital elevation model of arid western India with a resolution of 1000×1000 m is loaded in R using *raster*() code available in 'raster' package.. The DEM of arid western India in 1 km grid was extracted from the DEM of whole world in 1 km grid. The projection of the imported DEM is in geographic coordinate system as shown below.

```
>dem <- raster("dem1km_awi.tif")
>proj4string (dem)
> proj4string (dem)
[1] "+proj=longlat +datum=WGS84 +no_defs +ellps=WGS84 +towgs84=0,0,0"
```

The full description of the DEM can be viewed as follows. It may be noted here that the dimension of the grid is 1081 rows and 998 columns with a resolution of about 0.0083 degree. The minimum and maximum elevation of the study area is -11 m and 934 m, respectively.

```
> dem
class : RasterLayer
dimensions : 1081, 998, 1078838 (nrow, ncol, ncell)
resolution : 0.00833193, 0.008335477 (x, y)
extent : 68.18819, 76.50346, 21.71854, 30.72919 (xmin, xmax, ymin, ymax)
coord. ref. : +proj=longlat +datum=WGS84 +no_defs +ellps=WGS84
+towgs84=0,0,0
```

data source : E:\Manuscript_digital soil mapping_arid western India\dem1km_awi.tif

names : dem1km_awi

values : -11, 934 (min, max)

The DEM may be projected in metre coordinate system using projectRaster code function as follows. After the DEM is reprojected in UTM projection system, the cell resolution changes to 833×924 m. It may also be noted that cell values representing elevation is resampled while reprojecting and thus minimum and maximum elevation changes to -11.98206 m and 1572.762 m respectively.

> dem_utm <- projectRaster(dem, crs = "+init=epsg:32642")

> dem_utm

class : RasterLayer

dimensions : 1116, 1046, 1167336 (nrow, ncol, ncell)

resolution : 833, 924 (x, y)

extent : 411040.4, 1282358, 2397181, 3428365 (xmin, xmax, ymin, ymax)

coord. ref. : +init=epsg:32642 +proj=utm +zone=42 +datum=WGS84 +units=m +no_defs +ellps=WGS84 +towgs84=0,0,0

data source : in memory

names : dem1km_awi

values : -11.98206, 1572.762 (min, max)

The DEM may be plotted in R using both *plot*() and *spplot*() functions. In Fig. 6.6, the DEM as plotted by both the functions are shown.

>plot(dem, col=terrain.colors(10),xlab="Longitude", ylab="Latitude",

main="Digital Elevation Model, Arid western India")

>spplot(dem, scales = list(draw = TRUE))

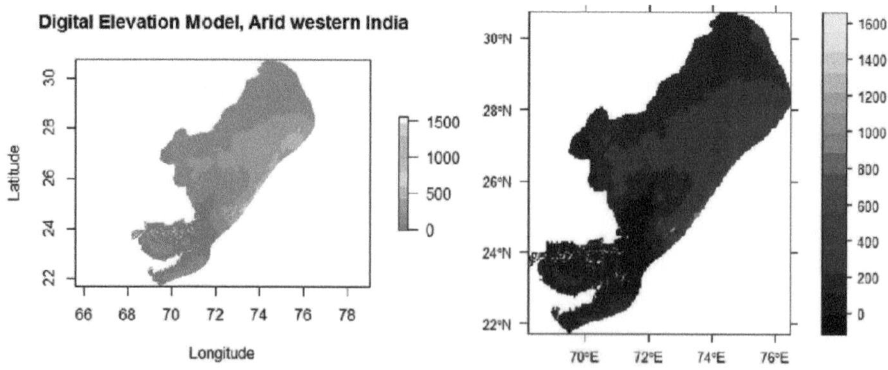

Fig. 6.6: Plot of spatial grid data on digital elevation model using *plot*() function at left hand side and using *spplot*() function at right hand side

Even the DEM may also be plotted using image function as follows

```
>image(dem, col = terrain.colors(10))
```

Cropping the DEM for a region of interest may also be done using following codes. The region of interest can be drawn on the screen by *drawExtent*() code. Otherwise the extent of the region of interest can also be specified by writing the lower left and upper right corner. In Fig. 6.7, an example of cropping the DEM using *drawExtent*() function is presented.

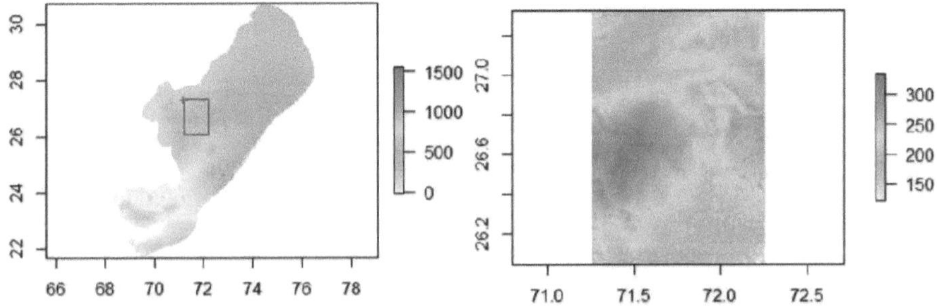

Fig. 6.7: Cropping of DEM using *drawExtent*() function in R

```
#plot the DEM
>plot(dem)
#Define the extent of the crop by clicking on the plot
>cropbox1 <- drawExtent()
#crop the raster, then plot the new cropped raster
>demcrop1 <- crop(dem, cropbox1)
#plot the DEM
>plot(demcrop1)
#define the cropping extent
cropbox2 <-c(71.5,72.0,26.5,27)
demcrop2 <- crop(dem, cropbox2)
#plot the DEM
plot(demcrop2)
```

The histogram of elevation values in the DEM may be created using following codes and is plotted in Fig. 6.8. It shows that major pixels of the region are below 500 m.

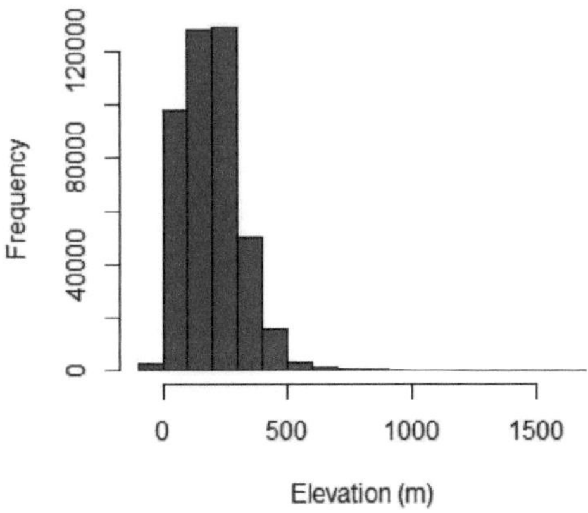

Fig. 6.8: Histogram of elevation values in the DEM of arid western India

hist(dem, main="Distribution of elevation values",

col= "purple",

maxpixels=1167336)

Further details of different function for handling spatial data may be found in http://rspatial.r-forge.r-project.org/gallery/

Plotting spatial data in R using other packages of R

Apart from spatial data plotting using *plot*() and *spplot*() function of sp package, there are several packages available in R by which spatial data can be handled and plotted in R e.g. rasterVis, RgoogleMaps, rworldmap, plotKML, GSIF etc. Detail descriptions of these packages are given in Table 6.3.

Table 6.3: Different packages of R for plotting and handling spatial data

R package	Description
rasterVis: Visualization Methods for Raster Data	Methods for enhanced visualization and interaction with raster data. It implements visualization methods for quantitative data and categorical data, both for univariate and multivariate rasters. It also provides methods to display spatiotemporal rasters, and vector fields. Further detailed description
RgoogleMaps: Overlays on Static Maps	Serves two purposes: (i) Provide a comfortable R interface to query the Google server for static maps, and (ii) Use the map as a background image to overlay plots within R. This requires proper coordinate scaling.
Rworldmap: Mapping Global Data	Enables mapping of country level and gridded user datasets
plotKML: Visualization of Spatial and Spatio-Temporal Objects in Google Earth	Writes sp-class, spacetime-class, raster-class and similar spatial and spatio-temporal objects to KML following some basic cartographic rules. (http://plotkml.r-forge.r-project.org/)
GSIF: Global Soil Information Facilities	Global Soil Information Facilities - tools (standards and functions) and sample datasets for global soil mapping (http://gsif.r-forge.r-project.org/)

Here, few examples from rasterVis, rworldmap and plotKML packages are shown. Further details of these packages may be found in http://pakillo.github.io/R-GIS-tutorial/, http://gsif.isric.org/doku.php/wiki:tutorial_plotkml, http://plotkml.r-forge.r-project.org/ and http://gsif.isric.org/doku.php.

In the following example, *gmap*() function of rasterVis package is used to get google maps of India and Jodhpur in R environment. The plots of these extracted maps are plotted in Fig. 6.9. It is to be noted here that to run the rasterVis package, other associated packages e.g. 'sp', 'raster', 'maptools', 'rgeos' and their dependencies are to be loaded in R environment.

```
>library(sp) # classes for spatial data
>library(raster) # grids, rasters
>library(rasterVis) # raster visualisation
>library(maptools)
>library(rgeos)
# and their dependencies
>library(dismo)
>library(XML)
>mymap <- gmap("India") # choose whatever country
```

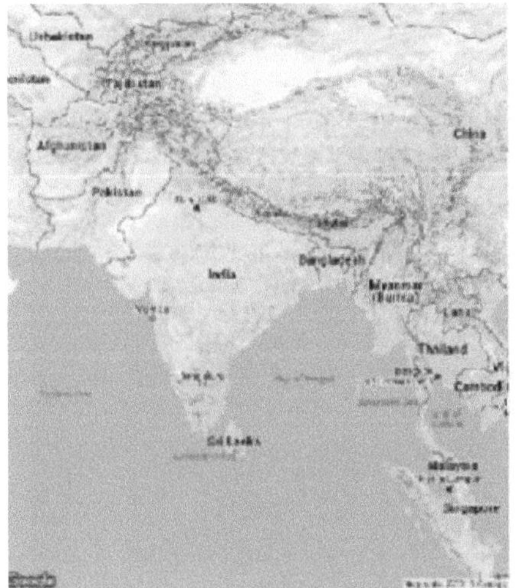

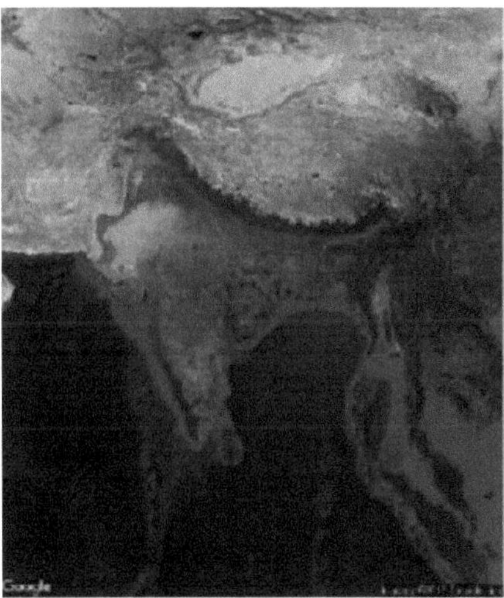

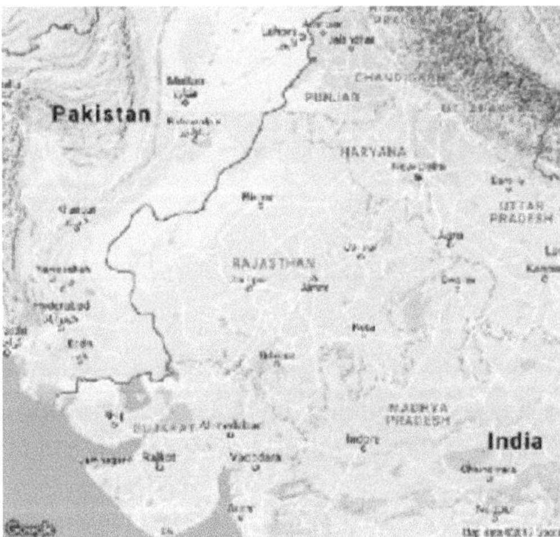

Fig. 6.9: Google map of India and Jodhpur as plotted through 'rasterVis' package of R

```
>plot(mymap)
# Choose map type
>mymap <- gmap("India", type = "satellite")
>plot(mymap)
# Choose zoom level
>mymap <- gmap("India", type = "satellite", exp = 2)
>plot(mymap)
# Save the map as a file in your working directory for future use
mymap <- gmap("India", type = "satellite", filename = "India.gmap")
# get a map for a region drawn at hand
>mymap <- gmap("Jodhpur")
>plot(mymap)
```

Maps of the whole world at different resolution can also be accessed through 'rworldmap' package and an example is show below in Fig. 6.10.

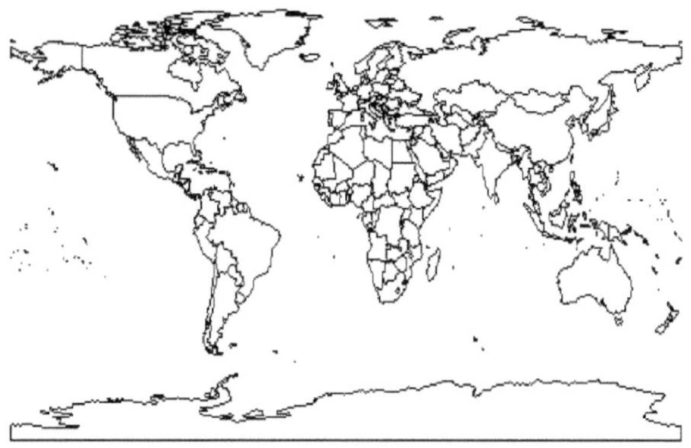

Fig. 6.10: Shapefile of the whole world as plotted through 'rworldmap' package of R

```
library(rworldmap)
>newmap <- getMap(resolution = "coarse") # different resolutions available
>plot(newmap)
```

Spatial data may also be plotted on google earth by creating kml file of spatial data. In the following example, spatial point data on soil profiles from arid western India is loaded and converted to a kml file 'plotKml' project (Fig. 6.11).

```
>library (plotKml)
>data_location <- read.table("sawi_legacy_data_location_sorted.txt", header=TRUE)
```

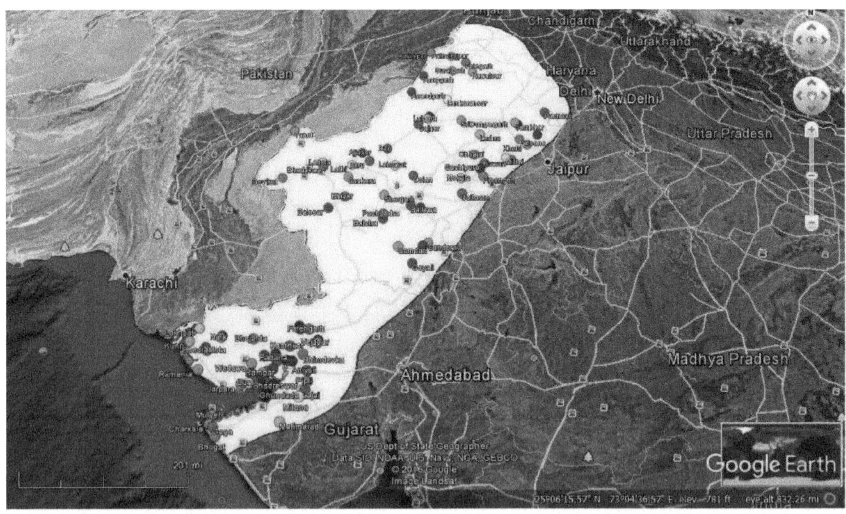

Fig. 6.11: Spatial point locations of soil profiles in arid western India on google Earth plotted through plotKml package

```
>coordinates(data_location) <- ~long+lat
>proj4string(data_location) <- CRS("+init=epsg:4326")
>plotKML(data_location)
```

Similarly, the spatial grid data e.g. DEM of arid western India may also be plotted using plotKml package of R as follows (Fig. 6.12).

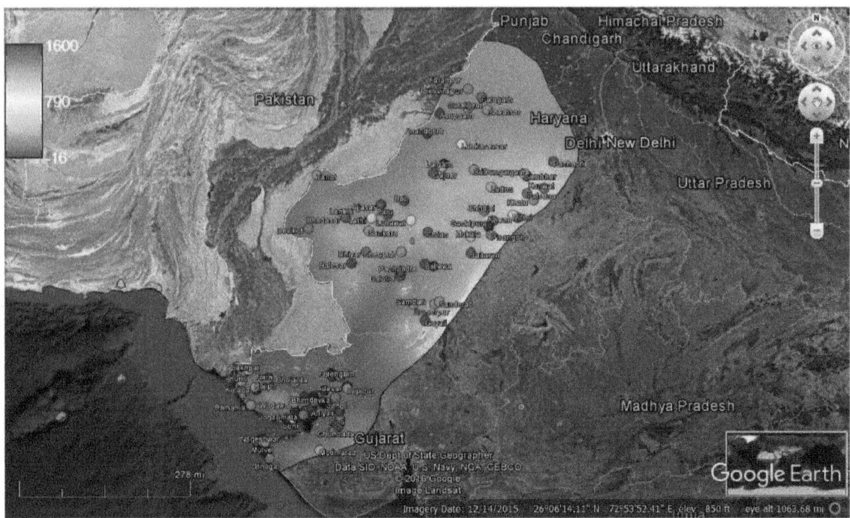

Fig. 6.12: Spatial grid data on DEM of arid western India on google Earth plotted through plotKml package

```
>dem <- raster("dem1km_awi.tif")
>proj4string (dem)
[1] "+proj=longlat +datum=WGS84 +no_defs +ellps=WGS84 +towgs84=0,0,0"
>plotKML(dem)
```

Conclusion

Different types of spatial data objects e.g. point, lines, polygons and grids and how they are handled and plotted using different packages of R are discussed in this chapter. Classes and methods of spatial data objects as handled by 'sp' package is further discussed to get an understanding on the spatial data structure as stored, retrieved and analysed in R. Several methods for analyzing spatial data and its plotting are demonstrated through example data set. Even, the capability of R to plot spatial data on Google Earth environment and to extract the Google maps for further analysis have also been demonstrated. Further, a number of packages are available in R for spatial data handling, which are mentioned in the chapter and may be further explored. Overall, it has been shown that R has the capability to handle and plot spatial data as parallel to commonly available GIS environment.

References

Pebesma, E.J. and Bivand, R.S. 2005. Classes and Methods for Spatial Data in R. R News 5 (2), http://cran.r-project.org/doc/Rnews/.

Bivand R.S., Pebesma, E. and Gomez-Rubio, V. 2013. Applied Spatial Data Analysis with R, Second edition. Springer, NY. http://www.asdar-book.org/

Sarkar, D. 2008. Lattice: Multivariate Data Visualization with R. Springer, New York. ISBN 978-0-387-75968-5.

7

Data Type, Data Management and Basic Plotting in R

***Priyabrata Santra*[1] *and Uttam Kumar Mandal*[2]**

[1]*ICAR-Central Arid Zone Research Institute, Jodhpur, Rajasthan, India*
[2]*ICAR-Central Soil Salinity Research Institute, Regional Station Canning Town, West Bengal, India*

Introduction

The first step of geostatistical analysis in R is the creation of dataset which is acceptable in R environment by using the data collected from field survey or laboratory experiments. Different type of data is collected during experiments e.g. sand content, silt content, organic carbon content, geographical location of data points, soil type, land use type etc. Here, we will discuss how these data are converted into a acceptable dataset format for R environment and how we manage these data. In R this task is performed in two steps. First, a data structure is created to hold the data and second is the entering or exporting the data into data structure.

Understanding datasets

A dataset is usually a rectangular array of data with rows representing observation sand columns representing variables. An example of a hypothetical soil dataset is provided in Table 7.1.

Table 7.1: An example data from a soil survey (*adapted redrawn from Kabacoff, R.I. 2011.*)

Soil ID	Date of survey	Organic carbon (%)	Soil texture	Degradation status
1	12/2/2015	0.13	Loam	Moderate
2	8/5/2015	0.6	Sandy loam	High
3	4/20/2015	0.02	Sandy	Low
4	1/24/2015	1.2	Clay loam	Very High

The columns and rows in the dataset are called as variables and observations, respectively in R. The content of the data type may be different in a single dataset, whereas the structure of the dataset may be different. Here, in the example the dataset structure is a rectangular array. Soil ID in the above example dataset is a row or case identifier. Date of survey is the date variable, organic carbon content is a continuous variable, soil texture is a nominal variable, and degradation status is an ordinal variable. In R, an *object* is anything that can be assigned to a variable. This includes constants, data structures, functions, and even graphs. Objects have a mode (which describes how the object is stored) and a class (which tells generic functions like print how to handle it). All objects have two intrinsic attributes: mode and length. The mode is the basic type of the elements of the object; there are four main modes: numeric, character, complex, and logical (FALSE or TRUE). Other modes exist but they do not represent data, for instance function or expression. The length is the number of elements of the object. In the above example, soil ID, date of survey, and organic carbon content would be numeric variables, whereas soil texture and degradation status would be character variables. R refers to case identifiers as row names and categorical variables (nominal, ordinal) as factors.

Data structures

R has a wide variety of objects for holding data, including scalars, vectors, matrices, arrays, data frames, and lists. They differ in terms of the type of data they can hold, how they're created, their structural complexity, and the notation used to identify and access individual elements. Fig. 7.1 shows a diagram of these data structures.

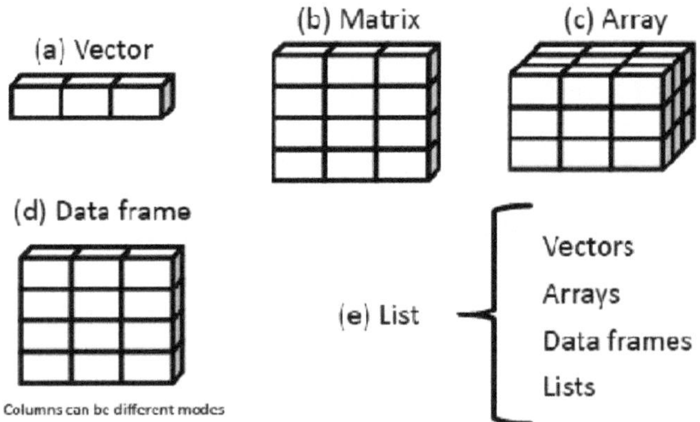

Fig. 7.1: R data structures (*adapted redrawn from Kabacoff, R.I. 2011.*

Vectors

Vectors are one-dimensional arrays that can hold numeric data, character data, or logical data. The vector may be column vector or row vector. In case of column vector data, multiple observations for a single variable is listed, whereas in case of row vector single observation for multiple variables are recorded. The combine function *c*() is used to form the vector. Here are examples of each type of vector:

a <- c(1, 2, 5, 3, 6, -2, 4)

b <- c("one", "two", "three")

c <- c(TRUE, TRUE, TRUE, FALSE, TRUE, FALSE)

Here, a is numeric vector, b is a character vector, and c is a logical vector. It is noted here that the data in a vector must only be one type or mode (numeric, character, or logical) and mix modes of data cannot be added in a single vector.

The elements of a vector can be referred using a numeric vector of positions within brackets. For example, a[c(2, 4)] refers to the 2nd and 4th element of vector a. Here are additional examples:

> *a <- c(1, 2, 5, 3, 6, -2, 4)*

> *a[3]*

[1] 5

> *a[c(1, 3, 5)]*

[1] 1 5 6

> *a[2:6]*

[1] 2 5 3 6 -2

The colon operator used in the last statement is used to generate a sequence of numbers. For example, a <- c(2:6) is equivalent to a <- c(2, 3, 4, 5, 6).

Matrix

A matrix is a two-dimensional array where each element has the same mode (numeric, character, or logical). Matrices are created with the matrix function. The general format is

*myymatrix <- matrix(vector, nrow=number_of_rows, ncol=number_of_columns, byrow=logical_value, dimnames=*list(*char_vector_rownames, char_vector_colnames*))

where *vector* contains the elements for the matrix, nrow and ncol specify the row and column dimensions, and dimnames contains optional row and column labels stored in character vectors. The option byrow indicates whether the matrix should be filled in by row (byrow=TRUE) or by column (byrow=FALSE). The default is by column. The following listing demonstrates the matrix function .

> *y <- matrix(1:20, nrow=5, ncol=4)*

> *y*

[,1] [,2] [,3] [,4]

[1,] 1 6 11 16

[2,] 2 7 12 17

```
[3,] 3 8 13 18
[4,] 4 9 14 19
[5,] 5 10 15 20
> cells <- c(1,26,24,68)
> rnames <- c("R1", "R2")
> cnames <- c("C1", "C2")
> mymatrix <- matrix(cells, nrow=2, ncol=2, byrow=TRUE,
dimnames=list(rnames, cnames))
> mymatrix
C1 C2
R1 1 26
R2 24 68
> mymatrix <- matrix(cells, nrow=2, ncol=2, byrow=FALSE,
dimnames=list(rnames, cnames))
> mymatrix
C1 C2
R1 1 24
R2 26 68
```

In the above example, a 5×4 matrix is first created. Then a 2×2 matrix is created with labels and matrix is filled by row. Finally, a 2×2 matrix is created and the matrix is filled by columns. User can identify rows, columns, or elements of a matrix by using subscripts and brackets. X[*i*] refers to the ith row of matrix X, X[,*j*] refers to *j* th column, and X[*i, j*]refers to the ij^{th} element, respectively. The subscripts *i* and *j* can be numeric vectors in order to select multiple rows or columns, as shown in the following listing.

```
> matrix(1:10, nrow=2, ncol = 5)
> x[2,]
[1] 2 4 6 8 10
> x[,2]
[1] 3 4
> x[1,4]
[1] 7
> x[1, c(4,5)]
[1] 7 9
```

Matrices are two-dimensional and, like vectors, can contain only one data type. When there are more than two dimensions, arrays are used to create dataset. When there are multiple modes of data, data frames are used to create a dataset.

Arrays

Arrays are similar to matrices but can have more than two dimensions. They're created with an array function of the following form:

myarray <- array(vector, dimensions, dimnames)

where *vector* contains the data for the array, *dimensions* is a numeric vector giving the maximal index for each dimension, and *dimnames* is an optional list of dimension labels. The following listing gives an example of creating a three-dimensional (2×3×4) array of numbers.

```
> dim1 <- c("A1", "A2")
> dim2 <- c("B1", "B2", "B3")
> dim3 <- c("C1", "C2", "C3", "C4")
> z <- array(1:24, c(2, 3, 4), dimnames=list(dim1, dim2, dim3))
> z
, , C1
B1 B2 B3
A1 1 3 5
A2 2 4 6
, , C2
B1 B2 B3
A1 7 9 11
A2 8 10 12
, , C3
B1 B2 B3
A1 13 15 17
A2 14 16 18
, , C4
B1 B2 B3
A1 19 21 23
A2 20 22 24
```

Arrays are a natural extension of matrices. They can be useful in programming new statistical methods. Like matrices, they must be a single mode.

Data frames

A *data frame* is a structure in R that holds data and is similar to the datasets found in standard statistical packages (for example, SAS, SPSS, and Stata). The columns are variables and the rows are observations. User can have variables of different types (for example, numeric, character) in the same data frame. Data frames are the main structures to store datasets.

In R, a data frame is more general than a matrix in that different columns can contain different modes of data (numeric, character, etc.). It's similar to the datasets typically seen in SAS, SPSS, and Stata. Data frames are the most common data structure deal with in R. The example soil dataset in Table 1 consists of numeric and character data. Because there are multiple modes of data, user can't contain this data in a matrix. In this case, a data frame would be the structure of choice. A data frame is created with the data.frame() function :

mydata <- data.frame(*col1*, *col2*, *col3*,…)

where *col1, col2, col3,* … are column vectors of any type (such as character, numeric,or logical). Names for each column can be provided with the names function. The following listing makes this clear.

```
>soilID<- c(1, 2, 3, 4)
>oc<- c(0.13, 0.6, 0.02, 1.2 )
>soiltexture<- c("Loam", "Sandy loam", "Sandy", "Clay loam")
>degstatus<- c("Moderate", "High", "Low", "Very High")
>soil<- data.frame(soilID, oc, soiltexture, degstatus)
>soil
```

	soilID	oc	soiltexture	degstatus
1	1	0.13	Loam	Moderate
2	2	0.6	Sandy loam	High
3	3	0.02	Sandy	Low
4	4	1.2	Clay loam	Very High

Each column must have only one mode, but can put columns of different modes together to form the data frame. It is to be noted here that when defining column names care should be taken that it should not contain spaces between letters. There are several ways to identify the elements of a data frame. User can use the subscript notation to use before (for example, with matrices) or can specify column names. Using the soil data frame created earlier, the following listing demonstrates these approaches.

```
>soil[1:2]
  soilID  oc
1   1     0.13
2   2     0.6
3   3     0.02
4   4     1.2
```

```
>Soil[c("soiltexture", "degstatus")]
  soiltexture  degstatus
1 Loam         Moderate
2 Sandy loam   High
3 Sandy        Low
4 Clay loam    Very High
```

```
>soil$oc
[1] 0.13 0.6 0.02 1.2
```

The $ notation in the third example is new. It's used to indicate a particular variable from a given data frame. For example, if user want to cross tabulate soil texture by degstatus, he could use the following code:

```
> table(soil$soiltexture, soil$degstatus)
```

	High	Low	Moderate	Very high
Clay loam	0	0	0	1
Loam	0	0	1	0
Sandy	0	1	0	0
Sandy loam	1	0	0	0

It can get tiresome typing soil $ at the beginning of every variable name, so shortcuts are available. User can use either the *attach*() and *detach*() or *with*() functions to simplify code.

Attach, detach and with: The *attach*() function adds the data frame to the R search path. When a variable name is encountered, data frames in the search path are checked in order to locate the variable. Using the mtcars data frame from R as an example, user could use the following code to obtain summary statistics for automobile mileage (mpg), and plot this variable against engine displacement (disp), and weight (wt):

```
>summary(mtcars$mpg)
>plot(mtcars$mpg, mtcars$disp)
>plot(mtcars$mpg, mtcars$wt)
```

This could also be written as

```
>attach(mtcars)
```

```
>summary(mpg)
>plot(mpg, disp)
>plot(mpg, wt)
>detach(mtcars)
```

The *detach*() function removes the data frame from the search path. Note that *detach*() does nothing to the data frame itself. The statement is optional but is good programming practice and should be included routinely. The limitations with this approach are evident when more than one object can have the same name. Consider the following code:

```
> mpg <- c(25, 36, 47)
> attach(mtcars)
The following object(s) are masked by '. GlobalEnv'; mpg
> plot(mpg, wt)
Error in xy.coords(x, y, xlabel, ylabel, log) :
'x' and 'y' lengths differ
> mpg
[1] 25 36 47
```

Here we already have an object named mpg in our environment when the mtcars data frame is attached. In such cases, the original object takes precedence, which isn't what user want. The plot statement fails because mpg has 3 elements and disp has 32 elements. The *attach*() and *detach*() functions are best used when analyzing a single data frame and unlikely to have multiple objects with the same name. In any case, be vigilant for warnings that say that objects are being masked. An alternative approach is to use the *with*() function. User could write the previous example as

```
>with(mtcars, {
summary(mpg, disp, wt)
plot(mpg, disp)
plot(mpg, wt)
})
```

In this case, the statements within the {} brackets are evaluated with reference to the mtcars data frame. If there's only one statement (for example, *summary*(mpg)), the {} brackets are optional. The limitation of the *with*() function is that assignments will only exist within the function brackets. Consider the following:

```
> with(mtcars, {
stats <- summary(mpg)
stats
})
Min. 1st Qu. Median Mean 3rd Qu. Max.
10.40 15.43 19.20 20.09 22.80 33.90
> stats
Error: object 'stats' not found
```

To create objects that will exist outside of the *with*() construct, use the special assignment operator <<- instead of the standard one (<-). It will save the object to the global environment outside of the *with*() call. This can be demonstrated with the following code:

```
> with(mtcars, {
nokeepstats <- summary(mpg)
keepstats <<- summary(mpg)
})
> nokeepstats
Error: object 'nokeepstats' not found
> keepstats
Min. 1st Qu. Median Mean 3rd Qu. Max.
10.40 15.43 19.20 20.09 22.80 33.90
```

Most books on R recommend using *with*() over *attach*().

Case identifiers: In the soil data example, soilID is used to identify individuals in the dataset. In R, case identifiers can be specified with a row name option in the data frame function.

For example, the statement

```
>soil <-data.frame (soilID, oc, soil texture, degstatus,
                  row.names=soilID)
```

specifies soilID as the variable to use in labeling cases on various printouts and graphs produced by R.

Factors: As we have seen, variables can be described as nominal, ordinal, or continuous. Nominal variables are categorical, without an implied order. Soil texture (Clay Loam, Loam, Sand, Sandy loam) is an example of a nominal variable. Even if Clay Loam is coded as 1 and Loam is coded as 2 in the data, no order is implied. Ordinal variables imply order but not amount. Status (Low, Moderate,

High, Very High) is a good example of an ordinal variable. A soil with a Low degradation status is not same with a soil with Very High status, but not by how much. Continuous variables can take on any value within some range, and both order and amount are implied. Organic carbon data is a continuous variable and can take on values such as 0.02 or 1.2 and any value in between. Categorical (nominal) and ordered categorical (ordinal) variables in R are called factors. Factors are crucial in R because they determine how data will be analyzed and presented visually. The function factor() stores the categorical values as a vector of integers in the range [1... *k*] (where *k* is the number of unique values in the nominal variable), and an internal vector of character strings (the original values) mapped to these integers. For example, assume that you have the vector

soiltexture ("Loam", "Sandy Loam", "Loam", "Loam")

The statement *soiltexture <- factor(soil texture)* stores this vector as (1, 2, 1, 1) and associates it with 1=Loam and 2=Sandy Loam internally (the assignment is alphabetical). Any analyses performed on the vector soiltexture will treat the variable as nominal and select the statistical methods appropriate for this level of measurement. For vectors representing ordinal variables, add the parameter ordered=TRUE to the factor() function. Given the vector

>degstatus ("Moderate", "High", "Low", "Very High")

the statement *status <- factor(degstatus, ordered=TRUE)* will encode the vector as (3, 2, 1, 3) and associate these values internally as 1=High, 2= Low, 3=Moderate, 4= Very High. Additionally, any analyses performed on this vector will treat the variable as ordinal and select the statistical methods appropriately.

By default, factor levels for character vectors are created in alphabetical order. Therefore, this does not work for the degstates factor, because the order "High", "Low", "Moderate", and "Very High" does not make same. For ordered factors, the alphabetical default is rarely sufficient. Override the default by specifying a levels option. For example,

>degstatus <- factor(status, order=TRUE,

levels=c("Low", "Moderate", "High", "Very high")

would assign the levels as 1=Low, 2=Moderate, 3= High and 4=Very High. Be sure that the specified levels match your actual data values. Any data values not in the list will be set to missing. The following listing demonstrates how specifying factors and ordered factors impact data analyses.

> soilID <- c(1, 2, 3, 4)

> oc <- c(0.13, 0.6, 0.02, 1.2)

> soiltexture <- c("Loam", "Sandy Loam", "Sandy", "Clay Loam")

> status <- c ("Moderate", "High", "Low" "Very high")

> soiltexture <- factor (soiltexture)

```
> degstatus <- factor(degstatus, order=TRUE)
> soil <- data.frame(soilID, oc, soiltexture, degstatus)
> str(soil)
'data.frame': 4 obs. of 4 variables:
$ soilID: num 1 2 3 4
$ oc : num 0.13 0.6 0.02 1.2
$ soiltexture : Factor w/ 4 levels ("Clay loam", "Loam",.. : 2 4 3 1
$ degstatus : Ord.factor w/ 4 levels "High"<"Low"<..: 3 1 2 4
> summary(soil)
```

soilID		oc		soiltexture		degstatus	
Min	: 100	Min	: 0.0200	Clay loam	: 1	High	: 1
Ist Qu	: 1.75	Ist Qu	: 0.1025	Loam	: 1	Low	: 1
Median	: 2.50	Median	: 0.3650	Sandy	: 1	Moderate	: 1
Mean	: 2.50	Mean	: 0.4875	Sandy Loam	: 1	Very high	: 1
3rd Qu	: 3.25	3rd Qu	: 0.7500				
Max	: 4.00	Max	: 1.2000				

First, enter the data as vectors. Then specify that soiltexture is a factor and degstatus is an ordered factor. Finally, combine the data into a data frame. The function *str(object)* provides information on an object in R (the data frame in this case). It clearly shows that soiltexture is a factor and degstatus is an ordered factor, along with how it's coded internally. Note that the *summary()* function treats the variables differently. It provides the minimum, maximum, mean, and quartiles for the continuous variable oc, and frequency counts for the categorical variables soiltexture and degstation.

Lists

Lists are the most complex of the R data types. Basically, a list is an ordered of collection of objects (components). A list allows to gather a variety of (possibly unrelated) objects under one name. For example, a list may contain a combination of vectors, matrices, data frames, and even other lists. To create a list using the *list()* function :

```
>mylist <- list(object1, object2, ...)
```

where the objects are any of the structures seen so far. Optionally, you can name the objects in a list:

```
>mylist <- list(name1=object1, name2=object2, ...)
```

The following listing shows an example.

```
> g <- "My First List"
```

```
> h <- c(25, 26, 18, 39)
> j <- matrix(1:10, nrow=5)
> k <- c("one", "two", "three")
> mylist <- list(title=g, ages=h, j, k)
> mylist
$title
[1] "My First List"
$ages
[1] 25 26 18 39
[[3]]
[,1] [,2]
[1,] 1 6
[2,] 2 7
[3,] 3 8
[4,] 4 9
[5,] 5 10
[[4]]
[1] "one" "two" "three"
> mylist[[2]]
[1] 25 26 18 39
> mylist[["ages"]]
[[1] 25 26 18 39
```

In this example, create a list with four components: a string, a numeric vector, a matrix, and a character vector. combine any number of objects and save them as a list.

User can also specify elements of the list by indicating a component number or a name within double brackets. In this example, mylist[[2]] and mylist[["ages"]] both refer to the same four-element numeric vector. Lists are important R structures for two reasons. First, they allow you to organize and recall disparate information in a simple way. Second, the results of many R functions return lists. It's up to the analyst to pull out the components that are needed. You'll see numerous examples of functions that return lists in later chapters.

Data input

Now that you have data structures, you need to put some data in them! As a data analyst, you're typically faced with data that comes to you from a variety of sources and in a variety of formats (Fig. 7.2). Your task is to import the data into your tools, analyze the data, and report on the results. R provides a wide range of tools for importing data. The definitive guide for importing data in R is the *R Data Import/Export* manual available at http://cran.r-project.org/doc/manuals/R-data.pdf.

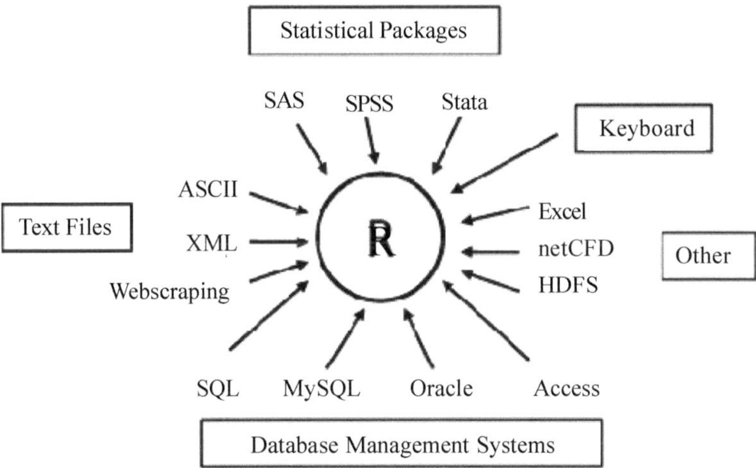

Fig. 7.2: Sources of data that can be implemented into R
(*adapted redrawn from Kabacoff, R.I. 2011.*)

Entering data from the keyboard

Perhaps the simplest method of data entry is from the keyboard. The *edit*() function in R will invoke a text editor that will allow to enter data manually. Here are the steps involved:

(i) Create an empty data frame (or matrix) with the variable names and modes you want to have in the final dataset.

(ii) Invoke the text editor on this data object, enter data, and save the results back to the data object.

In the following example, a data frame named mydata with three variables: age (numeric), gender (character), and weight (numeric) is created. Then invoke the text editor, add data, and save the results.

```
>mydata <- data.frame(age=numeric(0),
gender=character(0), weight=numeric(0))
>mydata <- edit(mydata)
```

Assignments like age=numeric(0) create a variable of a specific mode, but without actual data. Note that the result of the editing is assigned back to the object itself. The *edit*() function operates on a copy of the object. If you don't assign it a destination, all of edits will be lost!

Importing data from a delimited text file

You can import data from delimited text files using read table(), a function that reads a file in table format and saves it as a data frame. Here's the syntax:

```
>mydataframe <- read.table(file, header=logical_value,
                 sep= "delimiter")
```

where *file* is a delimited ASCII file, header is a logical value indicating whether the first row contains variable names (TRUE or FALSE), sep specifies the delimiter. Note that the sep parameter allows you to import files that use a symbol other than a comma to delimit the data values. You could read tab-delimited files with sep="\t". The default is sep=" ", which denotes one or more spaces, tabs, new lines, or carriage returns.

Importing data from Excel

The best way to read an Excel file is to export it to a comma-delimited file from within Excel and import it to R using the method described earlier. On Windows systems you can also use the RODBC package to access Excel files. The first row of the spread sheet should contain variable/column names. First, download and install the RODBC package.

```
>install.packages("RODBC")
```

You can then use the following code to import the data:

```
>library(RODBC)

>channel <- odbcConnectExcel("myfile.xls")

>mydataframe <- sqlFetch(channel, "mysheet")

>odbcClose(channel)
```

Here, myfile.xls is an Excel file, mysheet is the name of the Excel worksheet to read from the workbook, channel is an RODBC connection object returned by odbcConnectExcel(), and mydataframe is the resulting data frame. RODBC can also be used to import data from Microsoft Access.

Excel 2007 uses an XLSX file format, which is essentially a zipped set of XML files. The xlsx package can be used to access spreadsheets in this format. Be sure to download and install it before first use. The *read.xlsx*() function imports a worksheet from an XLSX file into a data frame. The simplest format is *read.xlsx(file, n)* where *file* is the path to an Excel 2007 workbook and *n* is the number of the worksheet to be imported. For example, on a Windows platform, the code

```
>library(xlsx)
>workbook <- "c:/myworkbook.xlsx"
>mydataframe <- read.xlsx(workbook, 1)
```

imports the first worksheet from the workbook myworkbook.xlsx stored on the C:drive and saves it as the data frame mydataframe. The xlsx package can do more than import worksheets. It can create and manipulate Excel XLSX files as well. Programmers who need to develop an interface between R and Excel should check out this relatively new package.

Importing data from SPSS

SPSS datasets can be imported into R via the *read.spss*() function in the foreign package. Alternatively, you can use the *spss.get*() function in the Hmisc package. *spss.get*() is a wrapper function that automatically sets many parameters of read. *spss*() for you, making the transfer easier and more consistent with what data analysts expect as a result. First, download and install the Hmisc package (the foreign package is already installed by default):

```
>install.packages("Hmisc")
```

Then use the following code to import the data:

```
>library(Hmisc)
>mydataframe <- spss.get("mydata.sav", use.value.labels=TRUE)
```

In this code, mydata.sav is the SPSS data file to be imported, use.value.labels=TRUE tells the function to convert variables with value labels into R factors with those same levels, and mydataframe is the resulting R data frame.

Importing data from SAS

A number of functions in R are designed to import SAS datasets, including *read.ssd*() in the foreign package and *sas.get*() in the Hmisc package. Unfortunately, if you're using a recent version of SAS (SAS 9.1 or higher), you're likely to find that these functions don't work for you because R hasn't caught up with changes in SAS file structures. There are two solutions that is recommended. You can save the SAS dataset as a comma-delimited text file from within SAS using PROC EXPORT, and read the resulting file into R using the method described in earlier section . Here's an example:

SAS program

```
proc export data=mydata
outfile="mydata.csv"
dbms=csv;
run;
```

R program:

```
>mydata <- read.table("mydata.csv", header=TRUE, sep= ",")
```

Alternatively, a commercial product called Stat Transfer does an excellent job of saving SAS datasets (including any existing variable formats) as R data frames.

Creating a data frame in R Console

A new data frame may also be created in R console for example a data frame leadership is created below in R console with 10 variables; Manager, 'date', 'country', 'gender', 'age' ,'q1', 'q2', 'q3', 'q4' and 'q5'

```
>Manager c(1,2,3,4,5)
>date <- ("10/24/08", "10/28/08", "10/1/08", "10/12/08", "5/1/08")
>country <- ("us", "us", "uk", "uk", "uk")
>gender <- c("M", "F", "F", "M", "F")
>age <- c(32, 45, 25, 39, 99)
>q1 <- c(5, 3, 3, 2)
>q2 <- c(4, 5, 5, 3, 2)
>q3 <- c(5, 2, 5, 4, 1)
>q4 <- c(5, 5, 5, NA, 2)
>q5 <- c(5, 5, 2, NA, 1)
>Leadership <- data.frame (Manager, date country, gender, age, q1, q2, q3, q4, q5 stringasFactor=False)
```

Data Management

Creating new variables

Typical research project 1 needs to create new variables and transform existing ones. This is accomplished with statements of the form:

```
>variable <- expression
```

A wide array of operators and functions can be included in the *expression* portion of the statement. Table 7.2 lists R's arithmetic operators. Arithmetic operators are used when developing formulas.

Table 7.2: Arithmetic operators (*adapted redrawn from Kabacoff, R.I. 2011.*

Operator	Description
+	Addition
-	Subtraction
*	Multiplication
/	Division
^ or **	Exponeniation
x%%y	Modulus (x mod y) 5%%2 is 1
x%/%y	Integer division 7%/%3 is 2

Recoding involves creating new values of a variable conditional on the existing values of the same and/or other variables. R's logical operators are used to recode data and these are listed in Table 7.3. Logical operators are expressions that return TRUE or FALSE.

Table 7.3: Logical operators in R (*adapted redrawn from Kabacoff, R.I. 2011.*

Operator	Description
<	Less than
<=	Less than or equal to
>	Greater than
>=	Greater than or equal to
= =	Exactly equal to
!=	Not equal to
!x	Not x
x | y	x or y
x & y	x and y
isTRUE (x)	Test if x is TRUE

Renaming variables

If the user is not happy with variable names, he can change them interactively or programmatically. Let's say that user wants to change the variables Manager to Manager ID and date to test Date. The statement is *fix*(leadership)

Programmatically, the reshape package has a *rename*() function that's useful for altering the names of variables. The format of the *rename*() function is

>rename(*dataframe*, *c*(*oldname*="*newname*", *oldname*="*newname*",...))

Finally, you can rename variables via the *names*() function . For example:

>*names*(*leadership*)[2] <- *"testDate"*

would rename date to test Date. In a similar fashion,

>*names(leadership)*[6:10] <- *c("item1", "item2", "item3", "item4", "item5")*

would rename q1 through q5 to item1 through item5.

Recoding values to missing

One can use assignments to recode values to missing. For example in a data leadership, missing age values were coded as 99. Before analyzing this dataset, you must let R know that the value 99 means missing in this case can accomplish this by recoding the variable:

>*leadership$age[leadership$age == 99]* <- *NA*

Any value of age that's equal to 99 is changed to NA

Missing values

In a project of any size, data is likely to be incomplete because of missed questions, faulty equipment, or improperly coded data. In R, missing values are represented by the symbol NA (not available). Impossible values (for example, dividing by 0) are represented by the symbol NaN (not a number). Unlike programs such as SAS, R uses the same missing values symbol for character and numeric data. R provides a number of functions for identifying observations that contain missing values. The function is.na() allows you to test for the presence of missing values.

Assume that you have a vector:

>*y*<- *c(1, 2, 3, NA)*

then the function

>*is.na(y)*

returns c(FALSE, FALSE, FALSE, TRUE).

Excluding missing values from analyses

Once you've identified the missing values, you need to eliminate them in some way before analyzing your data further. The reason is that arithmetic expressions and functions that contain missing values yield missing values. Luckily, most numeric functions have a na.rm=TRUE option that removes missing values prior to calculations and applies the function to the remaining values:

>*x* <- *c(1, 2, NA, 3)*

>*y* <- *sum(x, na.rm=TRUE)*

Here, y is equal to 6.You can remove *any* observation with missing data by using the *na.omit*() function. *na.omit*() deletes any rows with missing data.

Date values

Dates are typically entered into R as character strings and then translated into date variables that are stored numerically. The function *as.Date*() is used to make this translation. The syntax is *as.Date*(*x*, "*input_format*"), where *x* is the character data and *input_format* gives the appropriate format for reading the date (see Table 7.4).

Table 7.4: Date formats in R (*adapted redrawn from Kabacoff, R.I. 2011.*

Symbol	Meaning	Example
%d	Day as a number (0–31)	01–31
%a	Abbreviated weekday	Mon
%A	Unabbreviated weekday	Monday
%m	Month (00–12)	00–12
%b	Abbreviated month	Jan
%B	Unabbreviated month	January
%y	2-digit year	16
%Y	4-digit year	2016

The default format for inputting dates is yyyy-mm-dd. The statement

```
>mydates <- as.Date(c("2007-06-22", "2004-02-13"))
```

converts the character data to dates using this default format. In contrast,

```
>strDates <- c("01/05/1965", "08/16/1975")
>dates <- as.Date(strDates, "%m/%d/%Y")
```

reads the data using a mm/dd/yyyy format.

Two functions are especially useful for time-stamping data. *Sys.Date*() returns today's date and *date*() returns the current date and time. As I write this, it's July 17, 2016 at 5:25 pm. So executing those functions produces

```
>Sys.Date()
[1] "2016-07-16"
> date()
[1] "SunJul 16 17:25:21 2016"
```

You can use the format(*x*, format= "*output_format*") function to output dates in as pecified format, and to extract portions of dates:

```
> today <- Sys.Date()
> format(today, format= "%B %d %Y")
[1] "July 17 2016"
> format (today, format= "%A")
[1] "Sunday"
```

When R stores dates internally, they're represented as the number of days since January 1, 1970, with negative values for earlier dates. That means you can perform arithmetic operations on them. Finally, you can also use the function difftime() to calculate a time interval and express it as seconds, minutes, hours, days, or weeks. Let's assume that I was born on October 1, 1977. How old am I?

```
> today <- Sys.Date()
> dob <- as.Date("1977-10-01")
> difftime (today, dob, units= "days")
Time difference of 14169 days
```

This indicates that, I am 14169 days old.

Although less commonly used, you can also convert date variables to character variables. Date values can be converted to character values using the *as.character*() function:

```
>strDates <- as.character(dates)
```

Type conversions

In the previous section, we discussed how to convert character data to date values, and vice versa. R provides a set of functions to identify an object's data type and convert it to a different data type. Type conversions in R work in a similar fashion to those in other statistical programming languages. For example, adding a character string to a numeric vector converts all the elements in the vector to character values. User can use the functions listed in Table 7.5 to test for a data type and to convert it to a given type.

Table 7.5: Type conversion functions (*adapted redrawn from Kabacoff, R.I. 2011.*

Test	Convert
is.numeric()	as.numeric()
is.character()	as.character()
is.vector()	as.vector()
is.matrix()	as.matrix()
is.data.frame()	as.data.frame()
is.factor()	as.factor()
is.logical()	as.logical()

Functions of the form *is.datatype*()return TRUE or FALSE, whereas *as. datatype*() converts the argument to that type.

Sorting data

Sometimes, viewing a dataset in a sorted order can tell quite a bit about the data. For example, which managers are most deferential? To sort a data frame in R, use the *order*() function. By default, the sorting order is ascending. Prepend the sorting variable with a minus sign to indicate a descending order. The following examples illustrate sorting with the leadership data frame. The statement

>newdata <- leadership[order(leadership$age),]

creates a new dataset containing rows sorted from youngest manager to oldest manager. The statement

>attach(leadership)

>newdata <- leadership[order(gender, age),]

>detach(leadership)

sorts the rows into female followed by male, and youngest to oldest within each gender. Finally,

>attach(leadership)

>newdata <-leadership[order(gender, -age),]

>detach(leadership)

sorts the rows by gender, and then from oldest to youngest manager within each gender.

Merging datasets

If your data exist in multiple locations, you'll need to combine them before moving forward. This section shows you how to add columns (variables) and rows (observations) to a data frame.

To merge two data frames (datasets) horizontally, use the *merge*() function. In most cases, two data frames are joined by one or more common key variables (that is an inner join). For example:

>total <- merge(dataframeA, dataframeB, by="ID")

merges dataframeA and dataframeB by ID. Similarly,

>total <- merge(dataframeA, dataframeB, by=c("ID","Country"))

merges the two data frames by ID and Country. Horizontal joins like this are typically used to add variables to a data frame. For joining two matrices or data frames horizontally, don't need to specify a common key, use the *cbind*() function:

>total <- cbind(A, B)

To join two data frames (datasets) vertically, use the *rbind*() function :

Subsetting datasets

R has powerful indexing features for accessing the elements of an object. These features can be used to select and exclude variables, observations, or both. The following sections demonstrate several methods for keeping or deleting variables and observations.

Selecting (keeping) variables

It's a common practice to create a new dataset from a limited number of variables chosen from a larger dataset. The elements of a data frame are accessed using the notation *dataframe*[*row indices*, *column indices*]. User can use this to select variables. For example:

>*newdata* <- *leadership*[, *c*(*6:10*)]

selects variables q1, q2, q3, q4, and q5 from the leadership data frame and saves them to the data frame newdata. Leaving the row indices blank (,) selects all the rows by default. The statements

>*myvars* <- *c*(*"q1"*, *"q2"*, *"q3"*, *"q4"*, *"q5"*)

>*newdata* <-*leadership*[*myvars*]

also select variables q1, q2, q3, q4, and q5 from the leadership data frame and saves them to the data frame newdata.

Excluding (dropping) variables

There are many reasons to exclude variables. For example, if a variable has several missing values, you may want to drop it prior to further analyses. Let's look at some methods of excluding variables. You could exclude variables q3 and q4 with the statements

>*myvars* <-*names*(*leadership*) *%in%* c(*"q3"*, *"q4"*)

>*newdata* <- *leadership*[*!myvars*]

Selecting observations

Selecting or excluding observations (rows) is typically a key aspect of successful data preparation and analysis. Several examples are given in the following listing.

>*newdata*<- *leadership*[1:3,]

>*newdata*<- *leadership*[*which*(*leadership$gender*=="M" &

>*leadership$age*> 30),]

The examples in the previous two sections are important because they help to describe the ways in which logical vectors and comparison operators are interpreted within R. Understanding how these examples work will help you to interpret R code in general. Now that you've done things the hard way, let's look at a shortcut. The subset function is probably the easiest way to select variables and observations. Here are two examples:

```
>newdata<- subset(leadership, age>= 35 | age< 24,
select=c(q1, q2, q3, q4))
>newdata<- subset(leadership, gender=="M" & age> 25,
select=gender:q4)
```

In the first example, you select all rows that have a value of age greater than or equal to 35 or age less than 24. You keep the variables q1 through q4. In the second example, you select all rows that have gender M and age greater than 25. Here, you keep the variabtes gender through q4.

Random samples

Sampling from larger datasets is a common practice in data mining and machine learning. For example, you may want to select two random samples, creating a predictive model from one and validating its effectiveness on the other. The *sample()* function enables you to take a random sample (with or without replacement) of size *n* from a dataset.

You could take a random sample of size 3 from the leadership dataset using the statement *mysample<- leadership[sample(1:nrow(leadership), 3, replace=FALSE),]* The first argument to the *sample()* function is a vector of elements to choose from. Here, the vector is 1 to the number of observations in the data frame. The second argument is the number of elements to be selected, and the third argument indicates sampling without replacement. The *sample()* function returns the randomly sampled elements, which are then used to select rows from the data frame.

Working with graphs

R is an amazing platform for building graphs. *pdf("mygraph.pdf")* will save the graph as a PDF document named mygraph.pdf in the current working directory:

```
>pdf("mygraph.pdf")
>attach(mtcars)
>plot(wt, mpg)
>abline(lm(mpg~wt))
>title("Regression of MPG on Weight")
>detach(mtcars)
>dev.off()
```

In addition to *pdf()*, you can use the functions *win.metafile()*, *png()*, *jpeg()*, *bmp()*, *tiff()*, *xfig()*, and *postscript()* to save graphs in other formats. Creating a new graph by issuing a high-level plotting command such as *plot()*, *hist()* (for histograms), or *boxplot()* will typically overwrite a previous graph.

Graphical parameters

User can customize many features of a graph (fonts, colors, axes, titles) through options called *graphical parameters*. One way is to specify these options through the *par*() function. Values set in this manner will be in effect for the rest of the session or until they are changed. The format is *par*(*optionname=value*, *optionname=value*, ...). Specifying *par*() without parameters produces a list of the current graphical settings. Adding the no.readonly=TRUE option produces a list of current graphical settings that can be modified.

Symbols and lines

As you've seen, you can use graphical parameters to specify the plotting symbols and lines used in your graphs. The relevant parameters are shown in Table 7.6.

Table 7.6: Parameters for specifying symbols and lines (*adapted redrawn from Kabacoff, R.I. 2011.*

Parameter	Description
pch	Specifies the symbol to use when plotting points
cex	Specifying the symbol size. cex is a number indicating the amount by which plotting symbols should be scaled relative to the default. 1 = default, 1.5 is 50% larger, 0.5 is 50% smaller, and so forth
lty	Specifies the line type
lwd	Specifies the line width. lwd is expressed relative to the default (default = 1). For example, lwd = 2 generates a line twice as wide as default

Colors

There are several color-related parameters in R. Table 7.7 shows some of the common ones.

Table 7.7: Parameters for specifying color in plots of R (*adapted redrawn from Kabacoff, R.I. 2011.*

Parameter	Description
col	Default plotting color. Some functions (such as lines and pie) accept a vector of values that are recycled. For example, if col = c("red", "blue") and three lines are plotted, the first line will be red, the second blue and third red.
col.axis	Color for axis text
col.lab	Color for axis labels
col.main	Color for titles
col.sub	Color for subtitles
fg	The plot's foreground color
bg	The plot's background color

Text characteristics

Graphic parameters are also used to specify text size, font, and style. Parameters controlling text size are explained in Table 7.4. Font family and style can be controlled with font options (Table 7.8).

Table 7.8: Parameters specifying text size, font family, font size, font style in plots of R (*adapted redrawn from Kabacoff, R.I. 2011.*

Parameter	Description
cex	Number indicating the amount by which plotted text should be scaled relative to the default. 1 = default, 1.5 is 50% larger, 0.5 is 50% smaller etc.
cex.axis	Magnification of axis text relative to cex
cex.lab	Magnification of axis labels relative to cex
cex.main	Magnification of titles relative to cex
cex.sub	Magnification of subtitles relative to cex
font	Integer specifying font to use for plotting text, 1 = plain, 2 = bold, 3 = italic, 4 = bold italic, 5 = symbol (in Adobe symbol coding)
font.axis	Font for axis text
font.lab	Font for axis labels
font.main	Font for titles
font.sub	Font for subtitles
ps	Font point size (roughly 1/72 inch). The text size = ps*cex
family	Font family for drawing text. Standard values are *serif, sans and mono*

For example, all graphs created after the statement

>*par(font.lab*=3, *cex.lab*=1.5, *font.main*=4, *cex.main*=2)

will have italic axis labels that are 1.5 times the default text size, and bold italic titles that are twice the default text size.

Graph and margin dimensions

Finally, you can control the plot dimensions and margin sizes using the parameters listed in Table 7.9.

Table 7.9: Parameters for defining plot dimensions and margin in R (*adapted redrawn from Kabacoff, R.I. 2011.*

Parameter	Description
pin	Plot dimensions (width, height) in inches
mai	Numerical vector indicating margin size where c(bottom, left, top, right) is expressed in inches
mar	Numerical vector indicating margin size where c(bottom, left, top, right) is expressed in lines. The default is c(5,4,4,2)+0.1

The code

>*par(pin*=c(4,3), *mai*=c(1,.5, 1, .2))

produces graphs that are 4 inches wide by 3 inches tall, with a 1-inch margin on the bottom and top, a 0.5-inch margin on the left, and a 0.2-inch margin on the right.

Adding text, customized axes, and legends

Many high-level plotting functions (for example, plot, hist, boxplot) allow you to include axis and text options, as well as graphical parameters. For example, the following adds a title (main), subtitle (sub), axis labels (xlab, ylab), and axis ranges (xlim, ylim).

Titles

Use the *title*() function to add title and axis labels to a plot. The format is

>*title*(*main*=*"main title"*, *sub*=*"sub-title"*,

xlab=*"x-axis label"*, *ylab*=*"y-axis label"*)

Graphical parameters (such as text size, font, rotation, and color) can also be specified in the *title*() function.

Axes

Rather than using R's default axes, you can create custom axes with the *axis*() function. The format is

>*axis*(*side*, at=, labels=, pos=, lty=, col=, las=, tck=, ...)

where each parameter is described in Table 7.10.

Table 7.10: Axis option in R (*adapted redrawn from Kabacoff, R.I. 2011.*

Option	Description
side	An integer indicating the side of the graph to draw the axis (1 = bottom, 2 = left, 3 = top, 4 = right)
at	A numeric vector indicating where tick marks should be drawn
labels	A character vector of labels to be placed at the tick marks (if NULL, the at values will be used)
pos	The coordinate at which the axis line is to be drawn (that is, the value on the other axis where it crosses)
lty	Line type
col	The line and tick mark color
las	Labels are parallel (=0) and perpendicular (=2) to the axis
tck	Length of tick mark as a function of the plotting region (a negative number is outside the graph, a positive number is inside, 0 suppresses tick, 1 creates gridlines); the default is -0.01

Reference lines

The *abline*() function is used to add reference lines to our graph. The format is

>*abline*(h=*yvalues*, v=*xvalues*)

It is specifically used to draw 1:1 line when observed versus predicted values are plotted as scatter plot. For example, if the observed and predicted values ranges from 0 to 1 then abline should be plotted by abline (h=1, v=0)

Legend

When more than one set of data or group is incorporated into a graph, a legend can help to identify what's being represented by each bar, pie slice, or line. A legend can be added (not surprisingly) with the *legend*() function. The format is

```
>legend(location, title, legend, ...)
```

Text annotations

Text can be added to graphs using the *text*() and *mtext*() functions. The function *text*() places text within the graph whereas *mtext*() places text in one of the four margins. The formats are

```
text(location, "text to place", pos, ...)
mtext("text to place", side, line=n, ...)
```

Combining graphs

R makes it easy to combine several graphs into one overall graph, using either the *par*() or *layout*() function. Our focus here is on the general methods used to combine them. With the *par*() function, user can include the graphical parameter mfrow=c(*nrows,ncols*) to create a matrix of *nrows* × *ncols* plots that are filled in by row. Alternatively, use mfcol=c(*nrows, ncols*) to fill the matrix by columns.

The *layout*() function has the form *layout*(*mat*) where *mat* is a matrix object specifying the location of the multiple plots to combine. In the following code, one figure is placed in row 1 and two figures are placed in row 2:

```
>attach(mtcars)
>layout(matrix(c(1,1,2,3), 2, 2, byrow = TRUE))
>hist(wt)
>hist(mpg)
>hist(disp)
>detach(mtcars)
```

Summary

In this chapter, different types of data handled by R are discussed e.g. data frame, matrix, array, list, nominal data, ordinal data, factor data, date type data etc. Different options to create a dataset and to import them in R environment are also elaborated. For example, steps to import text data, ASCII data, excel spreadsheet data etc. are mentioned in the chapter with examples. Moreover, the handling of date type of data is also discussed. Finally, few basics of plotting in R environment are given. All the examples codes given in the chapter are either dealt with some common dataset available in R environment or was created as a new dataset and can be tried your own dataset to understand them fully. For detailed description of all R codes discussed here may be accessed from R open source environment or from Kabacoff (2011) and Paradis (2005).

References

Kabacoff, R.I. 2011. R in Action: Data Analysis and Graphics with R. Manning Publications Co. 20 Baldwin Road PO Box 261 Shelter Island, NY 11964.

Paradis, E. 2005. R for Beginners. Institut des Sciences de l'Evolution Universite Montpellier II. F-34095 Montpellier cedex 05 France.

8

Semivariogram Modeling in R

***Priyabrata Santra*[1] *and Debashish Chakraborty*[2]**

[1]*ICAR-Central Arid Zone Research Institute, Jodhpur, Rajasthan, India*
[2]*ICAR-Indian Agricultural Research Institute, New Delhi, India*

Introduction

Spatial variation structure of soil properties is a key input for digital soil mapping and generally determined through geostatistical techniques (Webster and Oliver 2007). In geostatistics, spatial variation is expressed by semivariogram $\hat{\gamma}(h)$, which measures the average dissimilarity between data separated by a vector *h*. It is generally computed as half of the average squared difference between the components of data pairs:

$$\hat{\gamma}(h) = \frac{1}{2N(h)} \sum_{i=1}^{N(h)} \left[z(x_i) - z(x_i + h)\right]^2 \quad (1)$$

where, *N*(*h*) is the number of data pairs within a given class of distance and direction, $z\,(x_i)$ is the value of the variable at the location x_i, $z\,(x_i+h)$ is the value of the variable at a lag of *h* from the location x_i. Experimental semivariograms [$\hat{\gamma}(h)$] as obtained from Eq (1) needs to be fitted in standard model to obtain three standard spatial variations parameters: nugget (C_0), sill ($C + C_0$) and range (*a*). Weighted least square technique are commonly used for fitting purpose. Weights are generally assigned as inversely proportional to the number of pairs for a particular lag or the lag distance. During semivariogram calculation, maximum lag distance is generally taken as half of the minimum extent of sampling area to minimize the border effect. Best-fit model with the lowest value of residual sum of square is selected to characterize the spatial variation of that particular soil property.

Semivariogram models

Following semivariogram models are commonly used to characterize the spatial variation e.g., spherical, exponential, Gaussian, and linear models, which are referred by short name as 'Sph', Exp', 'Gau', 'Lin', respectively in gstat package of R.

Spherical model: $\hat{\gamma}(h)\gamma(h) = C_0 + C\left[1.5\frac{h}{a} - 0.5\left(\frac{h}{a}\right)^3\right]$ if $0 \le h \le a$ otherwise $C_0 + C$ (2)

Exponential model: $\hat{\gamma}(h)\gamma(h) = C_0 + C\left[1.5\frac{h}{a} - 0.5\left(\frac{h}{a}\right)^3\right]$ for $h \ge 0$ (3)

Gaussian model: $\gamma(h) = C_0 + C\left[1 - \exp\left\{\frac{-h^2}{a^2}\right\}\right]$ for $h \ge 0$ (4)

Matern model: $\gamma(h) = C_0 + C_1\left[\frac{h}{a}\right]$ if h<a otherwise = $C_0 + C_1$ (5)

In all these semivariogram models, nugget, sill and range were expressed by C_0, $(C+C_0)$ and a, respectively. In case of exponential and Gaussian models, a represents the theoretical range. Practical range for these two semivariogram models is considered as the lag distance for which semivariogram value is ~95% of sill. Nugget (C_0) defines the micro-scale variability and measurement error for the respective soil property, whereas partial sill (C) indicates the amount of variation which can be defined by spatial correlation structure.

Other than these four models, several other semivariogram models are available in 'gstat' package of R, which can be seen by the command *vgm*() in R console. Among these, following seven unit semivariogram models are described below in Table 8.1. It is to be noted here that Exp(), Gau() and Bes() models reach their sill asymptotically (as h →∞). The effective range is the distance where the variogram reaches 95% of its maximum, and this is 3a for Exp(a), for Gau(a) and 4a for Bes(a).

Other than these unit models, there are few complex semivariogram models are also available in R. These are Exponential class (Exc), Matern (Mat), M. Stein's parameterization (Ste), wave (Wav), hole (Hol), spline (Spl), Legendre (Leg), Measurement error (Err) and Intercept (Int) model.

Table 8.1: Unit semivariogram model available in R

Model	Short form	γ(h)	h range
Nugget	Nug	0 1	$h = 0$ $h > 0$
Circular	Cir	$\frac{2h}{\pi a}\sqrt{1-\left(\frac{h}{a}\right)^2}+\frac{2}{\pi}\arcsin\frac{h}{a}$ 1	$0 \le h \le a$ $h > a$
Pentaspherical	Pen	$\frac{15h}{8a}-\frac{5}{4}\left(\frac{h}{a}\right)^3+\frac{3}{8}\left(\frac{h}{a}\right)^5$ 1	$0 \le h \le a$ $h > a$
Bessel	Bes	$1-\frac{h}{a}K_1\left(\frac{h}{a}\right)$	$h \ge 0$
Logarithmic	Log	0 $\log(h+a)$	$h = 0$ $h \ge 0$
Power	Pow	h^a	$h \ge 0, 0 < a \le 2$
Periodic	Per	$1-\cos\left(\frac{2\pi h}{a}\right)$	$h \ge 0$

Creating gstat object

The first step in developing semivariogram model is to create a gstat object. The *gstat* function available in 'gstat' package of R is used to create the gstat object. The details of the 'gstat' package are available in Pebesma and Wesseling (1998), Pebesma (2003) and Pebesma and Benedikt (2015). Different arguments of *gstat* function are given below. It is not necessary to use always all the arguments as mentioned below to create a gstat object.

```
>gstat(g, id, formula, locations, data, model = NULL, beta,
nmax = Inf, nmin = 0, omax = 0, maxdist = Inf, force = FALSE,
dummy = FALSE, set, fill.all = FALSE,
fill.cross = TRUE, variance = "identity", weights = NULL, merge,
degree = 0, vdist = FALSE, lambda = 1.0)
```

In the above function, argument 'g' is a gstat object to which the present gstat object will be appended. If it is missing, a new gstat object is created. The argument 'id' is the identifier of a new variable which is generally created using formula argument and if 'id' argument is missing, *varn* is used with *n* the number for this variable. If a cross variogram is entered, 'id' should be a vector with the two id values, e.g. c("log.SAND", "log.CLAY"). The 'formula' argument defines the dependent variable as a linear model of independent variables. For example, the dependent variable has name *z*, for ordinary and simple kriging the formula *z*~1 is used whereas for universal kriging, if *z* is linearly dependent on *x* and *y*, the formula *z*~*x*+*y* is used. The 'locations' argument indicates the formula with only independent variables that define the spatial data locations (coordinates), e.g. ~*x*+*y*. If data has a coordinates method to extract its coordinates this argument can be ignored. The 'data' argument defines the data frame containing the dependent variable, independent variables, and locations. The 'model' argument indicates the variogram model for the 'id', which is defined by a call to *vgm*() function. The 'beta' argument is used for simple kriging and simulation based on simple kriging. It is a vector with the trend coefficients including intercept. If no independent variables are defined the model only contains an intercept and this should be the expected value or mean. For local kriging, the arguments 'nmax', 'nmin', 'omax', 'maxdist' and 'force' are used. The argument 'nmax' defines number of nearest observations that should be used for a kriging prediction or simulation, where nearest is defined in terms of the space of the spatial locations. The argument, 'nmin' defines a number and if the number of nearest observations within distance 'maxdist' is less than the defined 'nmin', a missing value will be generated, unless force=TRUE is declared. The function 'omax' defines maximum number of observations to select per octant (3D) or quadrant (2D) and it is only relevant if 'maxdist' has been defined as well. The 'maxdist' argument defines a distance and only observations within the distance of 'maxdist' from the prediction location are used for prediction or simulation. If 'maxdist' is combined with 'nmax', both criteria are applied. The argument 'force' is used for force neighbourhood selection. In case 'nmin' is given, search beyond 'maxdist' until 'nmin' neighbours are found. A missing value is returned if this is not possible. 'dummy' is a logical argument and if it is TRUE the data is considered as a dummy variable and it is only necessary for unconditional simulation. The 'set' argument passes the list of optional parameters to gstat. The argument 'x' defines the gstat object to print. 'fill.all' and 'fill.cross' arguments are logical to print the direct and cross variograms. If 'fill.all' is TRUE, all of the direct variogram and, depending on the value of 'fill.cross' also all cross variogram model slots in 'g' with the given variogram model will be filled. If 'fill.cross' is TRUE, all of the cross variograms are filled and if it is FALSE only all direct variogram model slots in 'g' with the given variogram model is filled and only if 'fill.all' is defined. Other arguments of gstat functions are not commonly used; however details of them can be seen from http://www.gstat.org/ (Pebesma, 2004).

Calculating experimental variogram

The next step to creating gstat object is the calculation of experimental semivariogram values from spatial data. Basically it calculates the distances for all possible pairs among spatially distributed data. Simultaneously, squared difference of values of the target variable for each pair and semivariance cloud is prepared. From semivariance cloud, average semivariance for a bin of lag distance was calculated as per eq (1) and is represented by g(h) where h is the average lag distance for that particular bin. In gstat package of R, variogram are calculated using *variogram*() function as follows.

```
>variogram(object, locations, X, cutoff, width = cutoff/15,
alpha = 0, beta = 0, tol.hor = 90/length(alpha), tol.ver =
90/length(beta), cressie = FALSE, dX = numeric(0), boundaries =
numeric(0), cloud = FALSE, trend.beta = NULL, debug.level = 1,
cross = TRUE, grid, map = FALSE, g = NULL, ..., projected = TRUE,
lambda = 1.0, verbose = FALSE, covariogram = FALSE, PR = FALSE,
pseudo = -1)
```

In the above function, several arguments need to be provided. It is not necessary that all arguments are to be defined for executing the *variogram*() function because several arguments are optional. In the following few essential arguments are discussed. Description of rest arguments can be found in gstat manual. The argument 'object' indicates an object of class gstat and direct and cross (residual) variograms are calculated for all variables and variable pairs defined in the object. The argument 'locations' defines spatial data locations. If the data has been already transformed to a spatial data using *coordinates*(*data*) <- ~x+y function, then 'locations' argument may be ignored in *variogram*() function. Otherwise, a formula with only the coordinate variables in the right hand (explanatory variable) side e.g. ~x+y needs to be defined. The argument 'X' is optional, which is a list indicating a matrix with regressors/covariates for each variable. The 'cutoff' argument indicates spatial separation distance up to which point pairs are included in semivariance estimates. By default, the cutoff value is defined as the diagonal of the box spanning the data is divided by three. The 'width' argument defines the width of subsequent distance intervals into which data point pairs are grouped for semivariance estimates. The 'alpha' argument defines the direction in plane (x,y), in positive degrees clockwise from positive y (North): alpha=0 for direction North (increasing y), alpha=90 for direction East (increasing x). It is an optional argument and if not defined omnidirectional variogram will be calculated. Similarly, the 'beta' argument defines the direction in z direction, in positive degrees up from the (x,y) plane. It is specifically required for calculating 3D semivariogram. The 'boundaries' argument defines the numerical vector with distance interval upper boundaries and values mentioned in the vector should be strictly increasing. The argument 'cloud' is logical and if defined as TRUE, semivariogram cloud will be calculated.

Fitting variogram

The next step is to fit the calculated semivariogram values to a model. Weighted least square technique (WLS) is followed to fit the semivariogram values in model. If no weight is used ordinary least square (OLS) is followed. The function *fit.variogram*() available in gstat package is used to fit the variogram.

```
>fit.variogram(object, model, fit.sills = TRUE, fit.ranges = TRUE,
fit.method = 7, debug.level = 1, warn.if.neg = FALSE)
```

In the above function, 'object' argument defines a sample variogram which is an output from *variogram*() function. The argument 'model' is the variogram model, which needs to be defined e.g. model = *vgm*(1, "Sph", 10, 1). The argument 'fit.sills' is a logical value which determines whether the partial sill coefficients (including nugget variance) should be fitted; or logical vector, which determines for each partial sill parameter whether it should be fitted or fixed. The argument 'fit.ranges' is also a logical argument which determines whether the range coefficients (excluding that of the nugget component) should be fitted; or a logical vector, which determines for each range parameter whether it should be fitted or fixed. The argument 'fit.method' defines the method of fitting semivariogram values in defined model using weighted least square technique. The default method uses weights $N(h)/h^2$ with $N(h)$ is the number of point pairs and h is the distance. Other values of fit.method, are given in Table 8.2 along with the weight to be used in fitting process.

Table 8.2: Fitting methods along with weight criteria available in gstat package

Fit.method	Fit by	weight
0	gstat	- (no fit)
1	*gstat*	N_j
2	gnuplot	$\frac{N_j}{\{\lambda(h_j)\}^2}$
3	gnuplot	N_j
4	gstat	$\frac{N_j}{\{\lambda(h_j)\}^2}$
5	gstat	REML
6	gstat	No weight (OLS)
7	gstat	$\frac{N_j}{h_j^2}$

Example of semivariogram fitting in R

In the following, we discuss few basic steps to calculate and fit semivariogram using example soil data of arid western India, which is also given in Appendix-1. Here, the spatial structure estimation of sand content is demonstrated step by step.

Creating gstat object on sand content

At first, a gstat object named 'gsawi', is created from the spatial data 'sawi'. It is to be noted here that 'sawi' had already been converted to a spatial data frame, otherwise, 'locations' argument need to be specified in the following command.

>gsawi <- gstat(formula = SAND~1, data = sawi)

In the 'sawi' database, coordinates are referred by 'Easting' and 'Northing' value and therefore it was converted to spatial data frame by the command *coordinates(sawi) <- ~Easting + Northing*. It may be argued that longitude and latitude value in decimal format could also be used while converting the data to a spatial data frame. Semivariogram values would have been the same for both coordinate systems however the lag distance unit would be in degree unit in later case, which is difficult to interpret for field situation specifically the 'range' parameter. Therefore, meter coordinate systems are most commonly used.

Calculating semivariogram values of gstat object

In the next step, variogram values are calculated from the gstat object, e.g. 'gsawi' in this example, using following command and the output is saved in a data frame, 'vggsawi'. Further, calculated semivariogram values are plotted in Fig. 8.1.

>vggsawi <- variogram(gsawi)

>plot(vggsawi, plot.nu=TRUE)

It is to be noted here that 15 lag classes or bins are created by default with width of each class equal to cutoff distance divided by 15. Again, the cutoff distance for semivariogram calculation is taken as one third of the diagonal distance of study area as a default option. In this example, the diagonal distance of the study area is about 1116 km, and thus cutoff was set as about 372 km or 3.72×10^5 m, which is clearly found in Fig. 8.1. In the figure, average semivariogram values against centre point of each lag class are plotted.

After visualizing basic spatial variation structure of the target variable as depicted in Fig. 8.1, necessary customization may be done by changing the argument value of *varigram*() function to capture required spatial variation structure for specific application. For example, it is observed in Fig. 8.1, that semivariogram reaches to sill just after three points. Therefore, semivariogram values can be calculated for a maximum distance of about 200 km or 2×10^5 m to capture the variation within this lag distance.

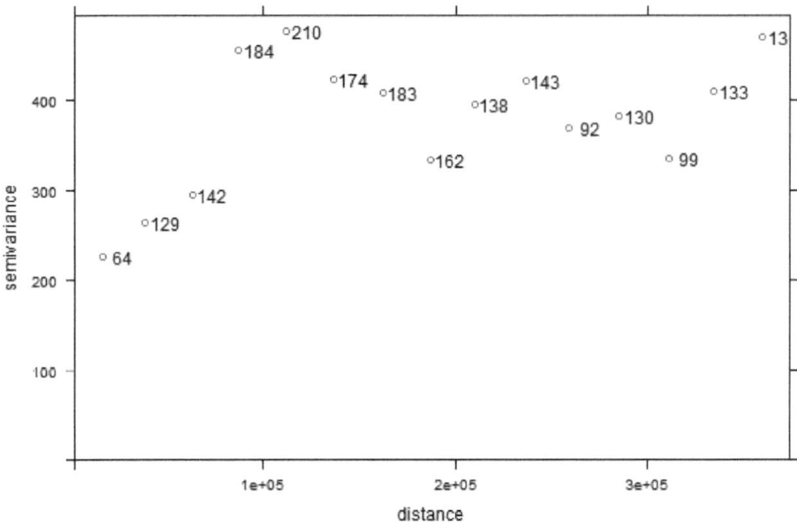

Fig. 8.1: Semivariogram plot of sand content obtained from default calculation using *variogram*() function

Semivariogram values were further calculated within a lag distance of 200 km using boundaries argument. In this example, width of each lag classes are maintained 10 km for first 9 classes and then increased to 25 km and 50 km, respectively for last two classes. The resulting semivariogram values are plotted in Fig. 8.2.

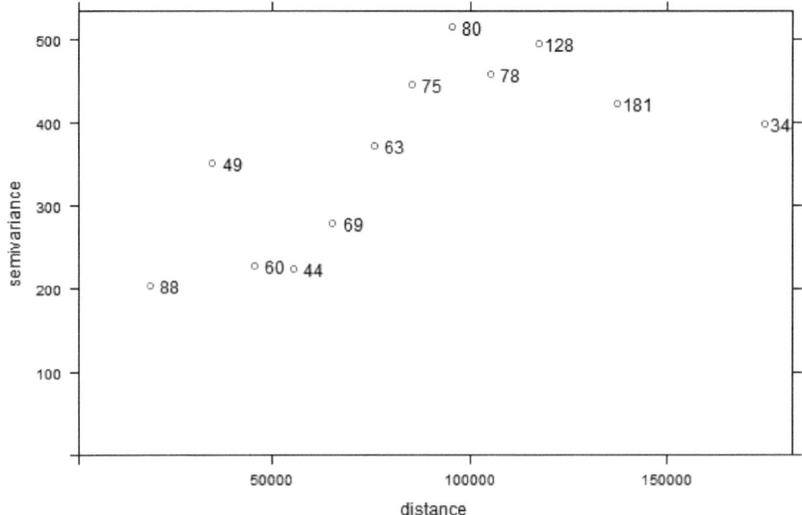

Fig. 8.2: Semivariogram plot of sand content obtained by defining the lag width using ‘boundaries’ argument of *variogram*() function

```
>vggsawi <- variogram(gsawi, boundaries = c(0.3e5, 0.4e5, 0.5e5, 0.6e5, 0.7e5, 0.8e5, 0.9e5, 1e5, 1.1e5, 1.25e5, 1.5e5, 2e5))
>plot(vggsawi, plot.nu=TRUE)
```

Calculation of semivariogram values for a customized lag width can also be done using 'cutoff' and 'width' arguments of *variogram*() function. Only difference of this method with 'boundaries' argument is that lag width for each lag class will be always equal using 'cutoff' and 'width' arguments whereas using 'boundaries' argument, one can define unequal width for each lag class. Smaller width can be specified at shorter lag distances to capture details of small scale variation. However, care should be taken on the number of data pairs using which average semivariogram value for each lag class is calculated while reducing the width. Although no clear recommendation on minimum number of data pairs to be used for each lag class is available, however, a thumb rule of $N(h) \geq 30$ in a lag class in followed considering the central limit theorem. This also gives an idea on minimum number of spatial data points required for geostatistical analysis.

From the scatter plots of semivariogram as depicted in Fig. 8.1 or Fig. 8.2, knowledge on inherent spatial varioation structure can be judged and accordingly specific model can be chosen. However, it is always best to apply few standard models *e.g.* exponential, spherical, Gaussian etc. on experimental semivariogram values and calculate fitting errors. The model with least fitting error quantified through sum of squared errors can be judged as the best model.

The model to be fitted on experimental semivariogram values can be defined using *vgm*() function as given below. Name of variogram model is to be specified by its short name as discussed above. Along with the model, initial value of semivariogram parameters e.g. nugget, partial sill and range are also to be mentioned while defining the model structure. How good the approximation of model structure is to be checked by plotting experimental semivariogram and model structure together.

```
>vgm <- vgm(nugget = 100, psill = 400, range = 1e+5, model="Sph")
>plot(vggsawi, vgm)
```

In this example, spherical model with nugget (C_0) = 100, partial sill (C) = 400 and range (a) = 100 km is defined. Other semivariogram models can also be defined by only replacing 'Sph' with short name of chosen model e.g. 'Exp' for expeonential model, 'Gau' for Gaussian model and 'Lin' is for Linear model etc.

In the next step, selected variogram model is fitted on calculated semivariogram values, which are already available in 'vggsawi' using *fit.variogram*() function as follows. The fitted semivariogram structure is plotted along with their calculated values are plotted in Fig. 8.3.

```
>vgmgsawi <- fit.variogram(vggsawi, vgm, fit.method=7)
>plot(vggsawi, vgmgsawi)
```

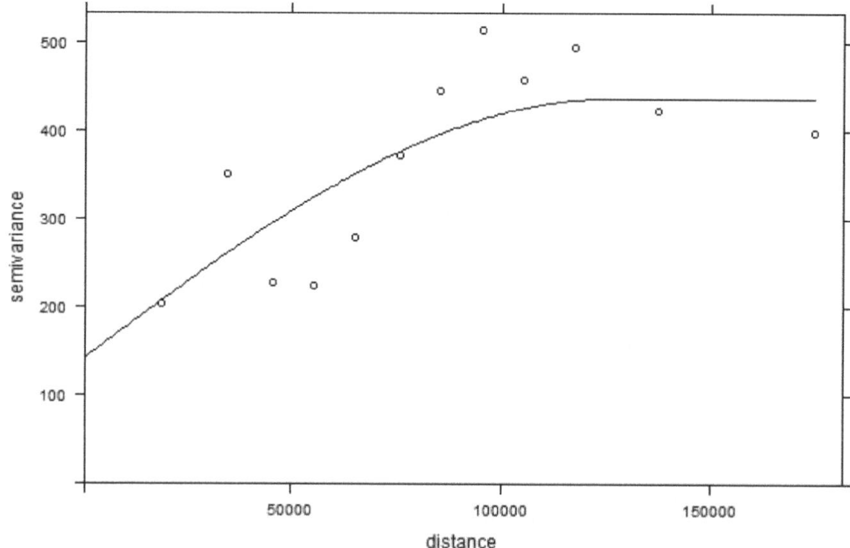

Fig. 8.3: Fitted semivariogram structure of sand content using spherical model

In the above example, fit.method=7 is used for which weights are assigned as $N(h)/h^2$, as given in Table 8.2. The fitting performance can be evaluated by checking sum of squared error (SSErr), which can be accessed from fitted model structure 'vgmgsawi' as follows.

>attr(vgmgsawi, "SSErr")

>attr(vgmgsawi, "SSErr")

[1] 0.0008506435

From above, it can be seen that SSErr is 8.50×10^{-4} for spherical model. Repeating the above fitting steps with 'Exp' and 'Gau' model, SSErr was found 9.16×10^{-4} and 9.10×10^{-4}, respectively. It indicates that spherical model is the best to fit the calculated semivariogram values of sand content in arid western India. Model parameters can be accessed for its physical interpretation as follows.

>vgmgsawi

model psill range

1 Nug 142.8092 0.0

2 Sph 293.9945 125420.5

The model structure shows two components, one is nugget component and another is model component. From the above example, it is observed that partial sill of the nugget model is 142.8092, which refers to actual nugget (C_0) whereas partial sill of spherical model is 293.9945, which refers to partial sill (C), and thus the total sill (C+C_0) is 436.8037. The range parameters is about 125 km, which signifies that sand content varies as per spatial variation structure obtained above surrounding a point in the field up to a distance of about 125 km, beyond which variation can be considered random.

Plotting the semivariogram plot in R

Fitted semivariogram model can be saved as image file in a defined computer directory using following commands in R script.

>tiff (filename = "E:/variogram_sand.tiff", res = 300, pointsize = 8, width = 1500, height = 1500)

>par(mfrow=c(1,1))

>plot(vggsawi, vgmgsawi, xlab="Lag distance (m)", ylab="Semivariance", main="Variogram of Sand (%)")

>dev.off()

In the above example, variogram model is saved as 'variogram_sand.tiff in E:/ directory of the computer with a resolution of 300 dpi and size of 1500×1500 pixels.

Summary

Knowledge of soil property at each location of a field is often required to manage the resources. However, it is not feasible to measure soil property at multiple locations in field, which necessitates its approximation or estimation at unsampled locations. Classical interpolation method is often followed for estimation purpose however soil varies a particular spatial variation structure as per Jenny soil formation theory and thus these methods do not justify the estimation process. Therefore, calculating spatial variation structure becomes an essential requirement for estimation or prediction purpose. Semivariogram defines the spatial variation structure of a soil property in field condition. In this chapter, basic semivariogram structure, different semivariogram models and fitting methods are discussed. Further these methods are demonstrated with 'gstat' package of R using an example data from arid western India. The semivariogram structure fitted through this process will be further used in kriging process, by which estimation at unsample locations will be made.

References

Pebesma, E.J. and Wesseling, C.G. 1998. Gstat, a program for geostatistical modelling, prediction and simulation, Computers & Geosciences 24 (1): 17-31.

Pebesma, Z. E. 2003. Gstat user's manual. Dept. of Physical Geography, Utrecht University P.O. Box 80.115, 3508 TC, Utrecht, The Netherlands, http://www.gstat.org/gstat.pdf

Pebesma, Z.E. and Benedikt, G. 2015. 'gstat' package. https://r-forge.r-project.org/projects/gstat/

Rossiter, D.G. 2012. Technical Note: Co-kriging with the gstat package of the R environment for statistical computing. http://www.css.cornell.edu/faculty/dgr2/teach/R/R_ck.pdf

Webster, R. and Oliver, M. A. 2007. Geostatistics for environmental scientists. John Wiley & Sons. pp. 315.

9

Kriging in R for Digital Soil Mapping

Priyabrata Santra[1], Mahesh Kumar[1], N.R. Panwar[1] and Uttam Kumar Mandal[2]

[1]ICAR-Central Arid Zone Research Institute, Jodhpur, Rajasthan, India
[2]ICAR-Central Soil Salinity Research Institute, Regional Station Canning Town, West Bengal, India

Introduction

In geographical location, the observations taken in proximity tend to be more alike than observations made at points that are far apart-that is, the values tend to be correlated. An area of study that takes this sort of spatial consideration into account is called geostatistics. A primary motivation for sampling is to make meaningful estimates of values at nearby positions in space and time. When values are assumed independent, the best estimate for an unmeasured point will be the mean. However, for correlated values nearby values are estimated by interpolation. The simplest interpolation scheme is a nearest neighbor estimate for which the unknown values are based on the closest measured location. Inverse distance is another sort of interpolation technique where weights are estimated based on the inverse distance squared. Kriging is a geostatistical interpolation technique that considers both the distance and the degree of variation between known data points when estimating values in unknown areas.

It makes the best use of existing knowledge by taking account of the way that a property varies in space through the variogram model. In its original formulation a kriged estimate at a place was simply a linear sum or weighted average of the data in its neighbourhood. Weights are based on the distance between the measured points, the prediction locations, and the overall spatial arrangement among the measured points. Since then kriging has been elaborated to tackle increasingly complex problems in mining, petroleum engineering, pollution control and abatement, and public health. The term is now generic, embracing several distinct kinds of kriging, both linear and non-linear.

Principles of kriging

Surface maps of soil properties are prepared using the semivariogram parameters through kriging. In the kriging process, estimates of soil attributes at unsampled locations, $z(u)$, are made using weighted linear combinations of known soil attributes $z(u_a)$ located within a neighborhood $W(u)$ centered around u.

$$z^*(u) = \sum_{\alpha=1}^{n(u)} \lambda_\alpha z(u_\alpha) \tag{1}$$

where λ_α is the weight assigned to datum $z(u_\alpha)$ located within a given neighborhood, $W(u)$ centered on u. Weights for n number of neighbourhood points were chosen as such so as to minimize the estimation or error variance, $\sigma_E^2(u) = Var\{z^*(u) - z(u)\}$ under the constraint of no-bias of the estimator. The kriged map for each soil property was prepared using 'gstat' package of R (Pebesma *et al.* 2003).

Kriging weights

When the kriging equations are solved to obtain the weights, λ_α, in general the only large weights are those of the points near to the point or block to be kriged. The nearest four or five might contribute 80% of the total weight, and the next nearest ten almost all of the remainder. The weights also depend on the configuration of the sampling. We can summarize the factors affecting the weights as follows.

i). Near points carry more weight than more distant ones. Their relative proportions depend on the positions of the sampling points and on the variogram: the larger is the nugget variance, the smaller are the weights of the points that are nearest to target point or block.

ii). The relative weights of points also depend on the block size: as the block size increases, the weights of the nearest points decrease and those of the more distant points increase, until the weights become nearly equal.

iii). Clustered points carry less weight individually than isolated ones at the same distance.

Kriging and its variation

Once the spatial variation parameters are identified, estimates at unknown locations within the study area can be done through kriging technique. Ordinary kriging (OK) technique is the mostly used technique for spatial estimation of soil properties. However, different variants of OK can be used as per the data availability and their characteristics. In the following sections, different types of kriging techniques are discussed briefly.

Ordinary point kriging, block kriging and simple kriging

Ordinary kriging estimates the value of soil attributes following the equation given in Eq. (1). When the kriging predictions are made for a block instead of a point, then it is called ordinary block kriging. When the mean of the random variable to be kriged is known to us, we can incorporate this in the kriging process, which is called simple kriging. If the random variable to be kriged is highly skewed then logarithmic transformation is done and ordinary kriging is performed over transformed variable, which is called as lognormal kriging. Here it is to be noted that proper care should be taken to back-transform the kriged output and its error. Detailed theory of these kriging techniques along with examples may be found in Webster and Oliver (2007). In all these kriging techniques, when the kriging equations are solved to obtain the weights, large weights are assigned to those sampled points near to the point or block to be kriged. The larger is the nugget variance, the smaller are the weights of the points that are nearest to target point or block. The relative weights of points also depend on the block size: as the block size increases, the weights of the nearest points decrease and those of the more distant points increase, until the weights become nearly equal. Clustered points carry less weight individually than isolated ones at the same distance.

Universal kriging, regression kriging or kriging with external drift

Universal kriging (UK), regression kriging (RK) or kriging with external drift (KED) are sometimes considered as synonymous. All these three kriging techniques assume that spatial process of a random variable is divided in two components, one is deterministic and other is random. The first part is predicted by a deterministic trend model and second part is by ordinary kriging technique. The trend component may be predicted by auxiliary variable, data of which is available at all predicted grids. These auxiliary variables may be any landscape features *e.g.* elevation, slope, land use type etc. or environmental covariates e.g. remote sensing signatures, rainfall pattern etc. or even be the coordinates. When the trend part of the random variable is modeled using spatial coordinates and residual part is modelled through ordinary kriging then it is called universal kriging. Otherwise, if auxiliary environmental or landscape covariates are used to model the trend part, it is called regression kriging or kriging with external drift. The difference between RK and KED lies in the computation process. In case of RK, the trend and residual are modeled separately and output from these two steps are joined together to get the output. Whereas, in case of KED, modeling of trend and residual is done simultaneously while assigning the weights in the kriging process.

Co-kriging

Co-kriging is the extension of ordinary kriging of a single variable to two or more variables. There must be some co-regionalization among the variables for it to be profitable. It is particularly useful if some property that can be measured cheaply at many sites is spatially correlated with one or more others that are expensive to measure and are measured at many fewer sites. It enables us to estimate the more sparsely sampled property with more precision by co-kriging using the spatial information from the more intensely measured one.

Indicator kriging and probability kriging

Indicator kriging is a non-linear, non-parametric form of kriging in which continuous variables are converted to binary ones (indicators). It is becoming popular because it can handle distributions of almost any kind, and empirical cumulative distributions of estimates can be computed and thereby provides confidence limits on them. It can also accommodate 'soft' qualitative information to improve prediction. Probability kriging was proposed by Sullivan (1984) because indicator kriging does not take into account the proximity of a value to the threshold, but only its position. It uses the rank order for each value, z(x), normalized to 1 as the secondary variable to estimate the indicator by co-rkiging.

Examples of kriging in R

In the following sections, examples of kriging in R has been presented using dataset on soil of arid western India ('sawi'), which is also available as Appendix-1. From this database, examples of preparing map of sand content through different geostatistical methods are presented. All geostatistical analyses presented here were carried out using 'gstat', 'sp' and other required auxiliary packages of R (Pebesma and Wesseling 1998, Pebesma 2003, Pebesma and Bivand 2005, Bivand et al. 2013).

Reading prediction grid

The basic requirement to start kriging is to prepare a grid file where predictions are to be made. This prediction grid is the GIS grid with a desired pixel resolution. Generally this prediction grid is to be prepared in GIS environment and then exported as *.txt file with some header information including the lower left coordinates, cell size, nodata value etc. For those pixels, where data is filled with nodata value, prediction will not be made. Therefore for an irregular shaped study area there may be large number of pixels with nodata value and -99 is most commonly used as nodata value. For a small and regular shaped study area, this grid may be prepared in excel sheet also. Each cell in excel spreadsheet may be considered as a pixel and a predefined value (e.g. 1) may be filled in each cell. The number of rows and column defines the length and width of the grid file. The excel sheet may be saved as text file and then necessary header file may be added, which contains the following information: ncols, nrows, xllcorner, yllcorner, cellsize, NODATA_value. The prepared text file for the prediction grid needs to be imported in R environment using following *readGDAL* command.

>mask<- readGDAL("mask1_awi.txt")

>names(mask)

The command reads the prediction grid, mask1_awi.txt as mask in the R environment. Associated files in mask can be checked by *names* command.

Inverse distance weighting interpolation in R

Inverse distance weighting (IDW) interpolation is followed in those cases where defining the spatial variation structure through variogram modeling is difficult. In such cases, IDW is the most appropriate method to predict the target variable at unsampled location. This can be achieved in R through *krige* command as follows:

>sawi.krig.sand<- krige(formula = SAND~1, locations = sawi, newdata = mask)

In the above example, *krige* function performs the inverse distance weighting method of interpolation on measured data on sand content available in the file 'sawi' and prediction are made in the grid 'mask'. The prediction grid as obtained from IDW method is saved by default as 'var1.pred' along with the prediction variance as 'var1.var' in the spatial data frame 'sawi.krig.sand'. the prediction grid of sand content can be plotted through spplot function as follows and is depicted in Fig. 9.1.

>spplot(sawi.krig.sand["var1.pred"], main = "Inverse distance weighting")

It can be seen from the Fig. 9.1 that most locations within the arid western India have sand content >60%.

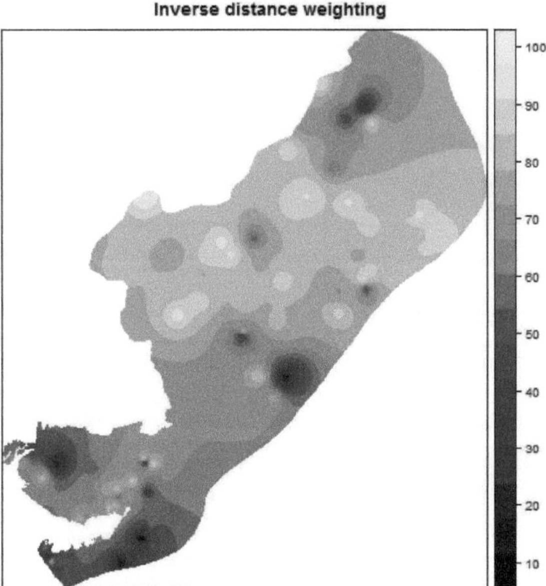

Fig. 9.1: Predicted sand content through inverse distance weighting

Ordinary point kriging in R

Ordinary kriging (OK) in R can be done in R through *krige* function by incorporating the variogram model structure as given below.

>sawi.krig.sand<- krige(formula = SAND~1, locations = sawi, newdata = mask, model =vgmgsawi)

In this example, OK is performed using 'vgmgsawi' model, which was previously developed from the spatial data on sand content in 'sawi' dataset. Predicted sand content through OK is presented in Fig. 9.2. The difference in prediction through IDW and OK can be noted from Fig. 9.1 and Fig. 9.2. The prediction through OK is found much smoother than the IDW prediction.

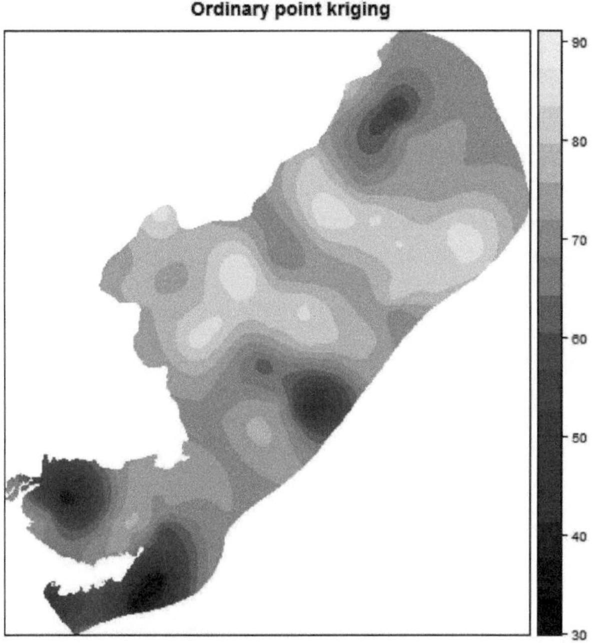

Fig. 9.2: Predicted sand content through ordinary point kriging

Ordinary block kriging in R

Ordinary block kriging can be done in R by defining the size of blocks in *krige* fuction. In case of point kriging, predictions are made on each point or pixel of the prediction grid whereas in case of block kriging predictions are made in each block which generally consists of number of pixels. In the following example, the pixel size or the cell size of the predction grid, mask, was 1000 m whereas predictions were made in each block of 5000 m.

>sawi.krig.sand<- krige(formula = SAND~1, locations = sawi, newdata = mask, model =vgmgsawi, block=c(5000,5000))

Predicted sand content through block kriging is presented in Fig. 9.3. It may be noticed here that there is not much difference in predicted maps through point kriging and block kriging. However, the difference may be visualized if we increase the block size.

It may be questioned here that through resampling method in GIS, we can prepare the map of desired cell size from a map with finer cell resolution and therefore what is the need for ordinary block kriging. Here it is to be noted that ordinary block kriging generally averages the estimates and always gives the prediction variance as an additional output, which we cannot obtain from resampling method. Larger will be block size, the lesser will be the prediction variance for each block which means low uncertainty of prediction. Therefore, as per need of the application of the predicted map, block size may be chosen which have a direct effect on uncertainty. For example, if one is interested to prepare a soil map for application in farmer's field the map should be of finer resolution whereas if he is interested to prepare the map for its application in a regional level policy decision, the block size may be increased up to a desired level.

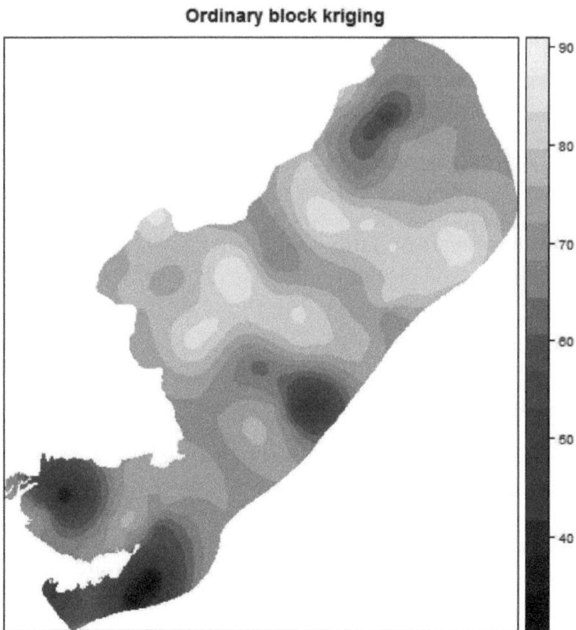

Fig. 9.3: Predicted sand content through ordinary block kriging

Simple kriging in R

When the mean of random variable to be kriged is known, then it is incorporated in the kriging process by defining the '*beta*' attribute of *krige* function as the sample mean in the measured data as follows:

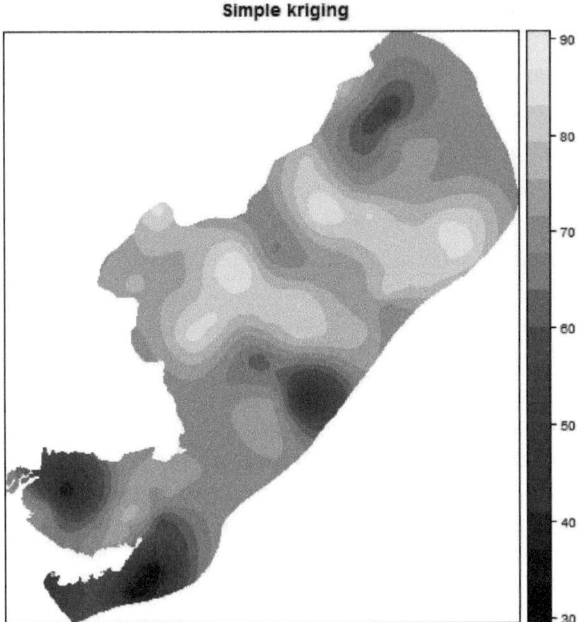

Fig. 9.4: Predicted sand content through simple kriging

>sawi.krig.sand<- krige(formula = SAND~1, locations = sawi, newdata = mask, model = vgmgsawi, beta=mean(sawi$SAND))

In the above example, mean of sand content in 'sawi' database was given as an additional attribute in kriging process, which leads to simple kriging. The predicted map through simple kriging is presented in Fig. 9.4. It can be seen from the figure that the predicted map is smoother than ordinary point kriging or block kriging.

Universal kriging in R

Often the target variable to be kriged has been found related with direction, for example, the data has a specific trend towards a specific direction. In such cases, universal kriging (UK) is followed as follows:

>sawi.krig.sand<- krige(formula = SAND~x+y, locations = sawi, newdata = mask, model = vgmgsawi, block=c(5000,5000))

In the above example, it is similar with ordinary block kriging except the formula definition where sand was shown related with *x* and *y* direction and is expressed as *SAND~x+y*. Other than direction, data may have trend with other auxiliary variables e.g. environmental variable landscape features etc. and in those cases UK can be applied, which is discussed in this chapter later as universal kriging with trend. Here it is to be noted that name of direction variable (*e.g. x* and *y* here) need to be same in both measured data variable as well as in prediction grid otherwise error will appear. The predicted sand content through UK is presented in Fig. 9.5. The difference in prediction from OK can be observed specifically from the limit of prediction and its distribution within the study area.

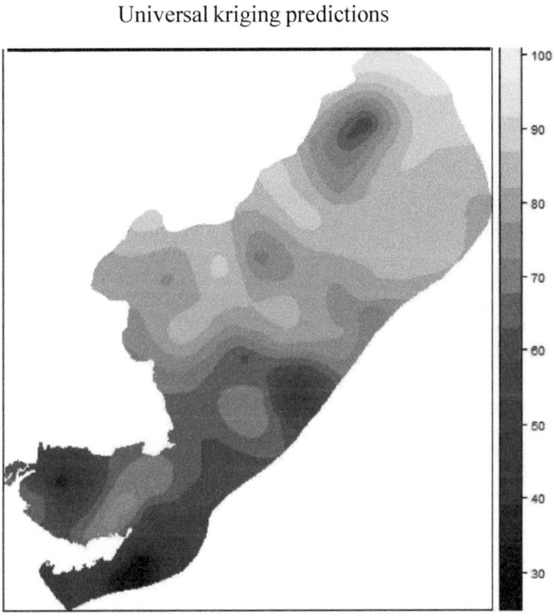

Universal kriging predictions

100
90
80
70
60
50
40
30

Fig. 9.5: Predicted sand content through universal kriging

Universal kriging with trend and regression kriging in R

In recent times, data on environmental variables or landscape features are available in plenty in various open source platforms. These data can be used in the kriging process to improve the accuracy of predicted maps. These auxiliary data are called covariates and the advanced kriging processes which can handle these covariates are known as universal kriging with trend and regression kriging (RK) or kriging with external drift (KED). In the following, we discuss how to process covariates in R environment for applying universal kriging with trend and regression kriging.

Processing covariates

Covariate grids can be imported in R environment by *readGDAL* function. In the following example, soil is a covariate grid has been imported in R environment as follows:

```
>mask$soil<- readGDAL("soil.txt")$band1
```

In the above example, 'soil.txt' is the covariate grid which has been read in 'mask' grid as the variable 'soil'. It is to be noted here that whenever *readGDAL* function is applied it reads and saves the files in band1 by default. Here, an extra space for the variable 'soil' is created in mask where the soil.txt grid is copied from its by default position 'band1'. Available grids in mask can be checked by the names command as follows, which shows two spaces: 'band1' by default and newly copied grid 'soil'.

>names(mask)

[1] "band1" "soil"

The boundary saved in mask can also be saved separately as a grid file and band1 may be deleted as follows:

>mask$boundary<- mask$band1

> mask$band1 <- NULL

>names(mask)

[1] "soil" "boundary"

The data type in soil grid can be checked by class function, which shows the data in soil grid as integer data. However, the soil grid imported here is the soil suborder association map and the values represents the map unit and thus needs to be converted to factor type for its further analysis.

>class(mask$soil)

[1] "integer"

The attribute table of soil grid can be found from the table function as follows. This table shows that map unit 27 covers 26519 pixels of the soil grid.

>table(mask$soil)

27	28	29	31	32	35	36	37	40	48	58
26519	18007	51237	17076	33031	67606	11044	13229	60116	6291	14355

Soil grid can be converted from integer data type to factor data type as follows. The attribute table of soil grid can further be checked.

>mask$soil<- as.factor(mask$soil)

>class(mask$soil)

[1] "factor"

Plotting covariates in R

Once the covariate data is prepared in R, it can be plotted in R through spplot function as follows. Details of plotting attributes may be found from Kabacoff (2011) or from CRAN. The plot of sub-order association in arid western India is presented in Fig. 9.6.

>spplot(mask, zcol = "soil", col.regions = brewer.pal(11,"Set3"), xlim=c(2758619,3703301),

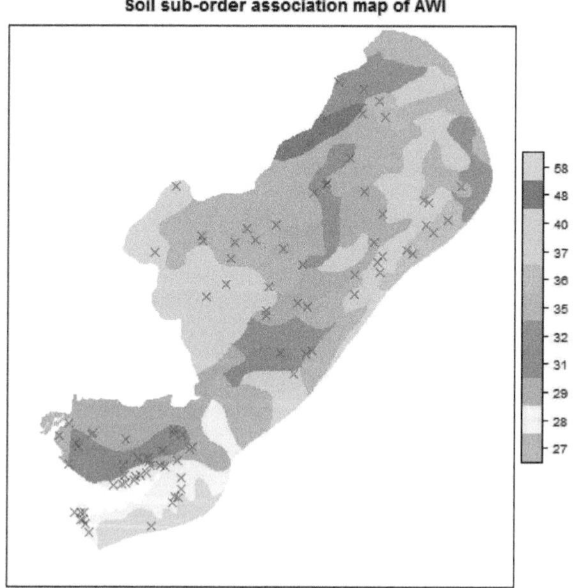

Fig. 9.6: Soil sub-order association map of arid western India

+ ylim=c(3724304,4751924), main="Soil sub-order association map of AWI", sp.layout=list("sp.points",sawi, pch=4, cex=1))

Extracting the soil factor data in original spatial dataset

The soil factor data from soil grid can be extracted to spatial data 'sawi', in which data on sand content is available. It can be done by over function as follows.

>sawi$soil<- over(sawi, mask)[[1]]

Extracted soil data in 'sawi' dataset can be checked by table function. It can be seen that 5 sampling points lies in map unit 27 and similarly as on. It is very important to check that at least one sampling lies in each map unit of the soil grid. Otherwise, it will create error in the process of universal kriging or regression kriging. In this example, 11 map units were present in the soil grid and the sampling points in 'sawi' dataset represent all map unit.

>table(sawi$soil)

27	28	29	31	32	35	36	37	40	48	58
5	17	21	12	7	7	6	2	10	2	3

Universal kriging with trend

To perform universal kriging with trend, variogram on residual needs to be developed first. In the following example, kriging of sand content with a trend of soil sub-order association on it was targeted. Hence, the residual variogram of sand content after de-trending was done as follows. It is to be noted here that when gstat object is created for computing and developing variogram model of residual, formula in gstat function should be written as *SAND~soil* or accordingly for other applications.

```
> gsawi.res <- gstat(formula = SAND~soil, data = sawi)
> vggsawi.res <- variogram(gsawi.res)
> vgmgsawi.res <- vgm(nugget = 100, psill = 250, range = 1e+05, model = "Sph")
> vgmgsawi.res <- fit.variogram(vggsawi.res, vgmgsawi.res, fit.method=7)
```

Once the variogram on residual is created universal kriging with trend can be performed as follows. One advantage of this process is that trend model development and kriging functions are operated simulataneously and therefore no need to fit the trend and do the kriging sperately.

```
> sawi.uk <- krige(SAND~soil, locations = sawi, model = vgmgsawi.res, newdata=mask)
```

Once the kriging has been done, the predicted map can be plotted using spplot function as follows.

```
>spplot(sawi.uk["var1.pred"], main = "universal kriging with trend")
```

The predicted map of sand content through universal kriging with trend is presented in Fig. 9.7. The map of sand content shown here is different from the nap prepared through OK or simple kriging. The trend effect of soil sub-order association on sand content is clearly visible in the map.

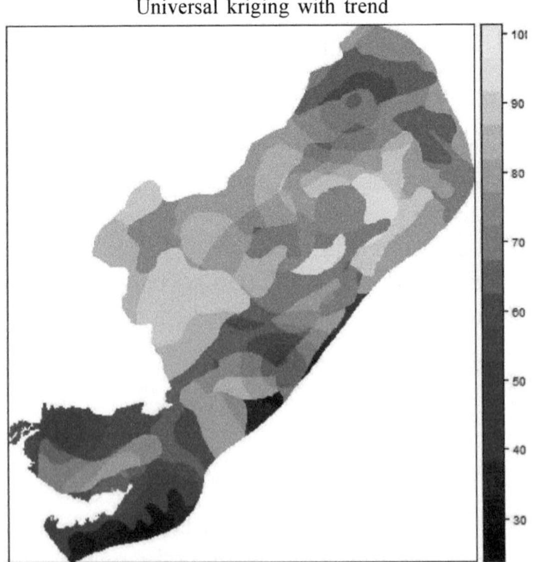

Fig. 9.7: Predicted sand content through universal kriging with trend

Regression kriging or kriging with external drift

For regression kriging, trend model is to be fitted first. In R, trend between sand content and soil sub-order association can be done through 'lm' function. Once the trend has been fitted, results of the fitted model can be seen by 'summary' function. Here, the adjusted R^2 value of the fitted model was found 0.2257.

```
>trend<- lm(SAND~soil, data = sawi)
>summary(trend)
```

Call:

lm(formula = SAND ~ soil, data = sawi)

Residuals:

Min 1Q Median 3Q Max

-49.314 -10.180 4.584 11.663 28.735

Coefficients:

Estimate	Std. Error	t value	Pr(>|t|)		
(Intercept)	78.620	8.588	9.155		3.93e-14 ***
soil28	-22.755	9.770		-2.329	0.02234 *
soil29	-14.896	9.556		-1.559	0.12293
soil31	-4.382	10.222	-0.429	0.66931	
soil32	-12.206	11.244	-1.085	0.28092	
soil35	2.309	11.244	0.205		0.83785
soil36	-1.953	11.628	-0.168	0.86702	
soil37	-8.720	16.067	-0.543	0.58880	
soil40	9.240	10.518	0.878		0.38228
soil48	-0.070	16.067	-0.004	0.99653	
soil58	-46.253	14.024	-3.298	0.00145 **	—

Signif.codes: 0 '***' 0.001 '**' 0.01 '*' 0.05 '.' 0.1 ' ' 1

Residual standard error: 19.2 on 81 degrees of freedom

Multiple R-squared: 0.3108, Adjusted R-squared: 0.2257
F-statistic: 3.653 on 10 and 81 DF, p-value: 0.0004711

By using the fitted trend model, prediction of sand content can be made at all possible locations in mask, grid by using predict function. However, for this purpose, spatial grid dataframe was first converted to spatial pixel dataframe. After prediction the grid was reconverted to spatial grid dataframe.

```
>class(mask)
  [1] "SpatialGridDataFrame"
  attr(,"package")
  [1] "sp"
>mask.pix<- as(mask, "SpatialPixelsDataFrame")
>mask.pix$predtrend<- predict(trend, new = mask.pix)
>class(mask.pix)
  [1] "SpatialPixelsDataFrame"
  attr(,"package")
  [1] "sp"
>mask$predtrend<- as(mask.pix, "SpatialGridDataFrame")$predtrend
```

The residual of the fitted trend model is to be saved in original data set because there is need to prepare the residual map from point data on residuals. The residual values of the fitted trend model 'trend' can be saved in 'sawi' dataset as follows. It can be checked by exploring the variables in 'sawi' dataset by *names* function. The variogram of residual sand content after de-trending the effect of soil-sub order association is required and for this purpose '*vgmgsawi.res*' prepared earlier may be used.

```
>sawi$res<- trend$residuals
>names(sawi)
[1] "RAINFALL" "SOIL  DEPTH" "SAND" "SILT" "CLAY" "OC"
[7] "CACO3" "PH" "EC" "BD" "FC" "PWP"
[13] "soil" "res"
```

Prediction of residual in 'mask' grid can be done through ordinary kriging. It is to be noted here that '*vgmgsawi.res*' variogram model is used to predict the residual variable, 'res' in 'sawi' dataset.

```
>sawi.res.krig<- krige(formula = res~1, locations = sawi, newdata = mask, model = vgmgsawi.res)
```

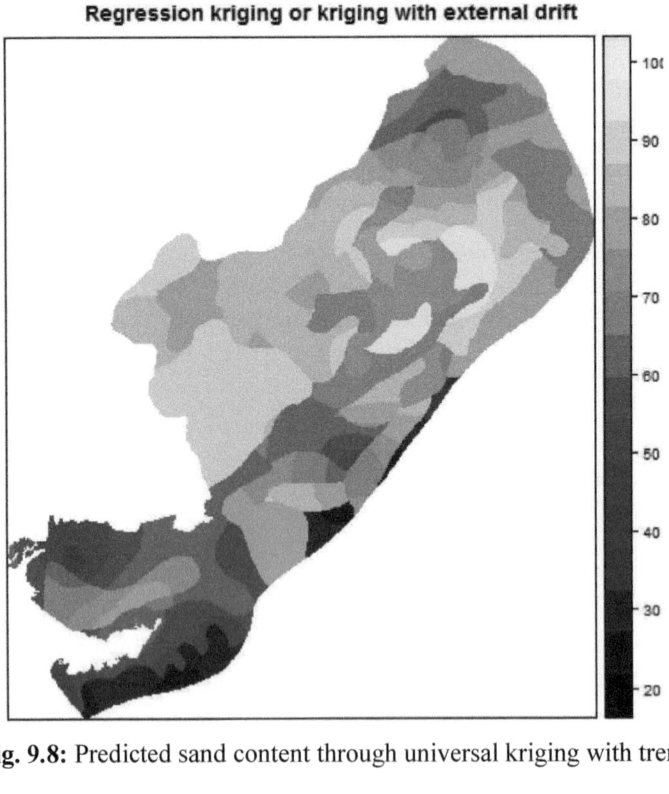

Regression kriging or kriging with external drift

Fig. 9.8: Predicted sand content through universal kriging with trend

The final RK prediction grid can be prepared by adding the trend predicted grid and residual grid as follows. Finally the RK predicted grid is plotted through *spplot* function, which is presented in Fig. 9.8.

```
>mask$residualpred<- sawi.res.krig$var1.pred
>mask$rkprediction<- mask$predtrend + mask$residualpred
>spplot(mask["rkprediction"], main = "Regression kriging or kriging with external drift")
```

Spatial stochastic simulation

Other than kriging, stochastic simulation is another advanced geostatistical technique by which several possible realizations of the target variable can be created based on its spatial variation structure from measured data. Additional attribute on number of simulation to be incorporated in the *krige* function other than normal kriging as follows.

```
>sawi.krig.sand.sim<- krige(SAND~1, sawi, newdata = mask, vgmgsawi, nsim = 9)
```

In the above example, 9 realizations of sand content has been created, which are plotted using spplot function and presented in Fig. 9.9.

```
>spplot(sawi.krig.sand.sim, col.regions = bpy.colors(), xlim=c(2758619,3703301),
ylim=c(3724304,4751924), main ="Sand Content [%]")
```

Average estimates of all these realizations can be prepared to develop sand content map. Standard deviation of these maps can also be calculated to represent the uncertainty of map.

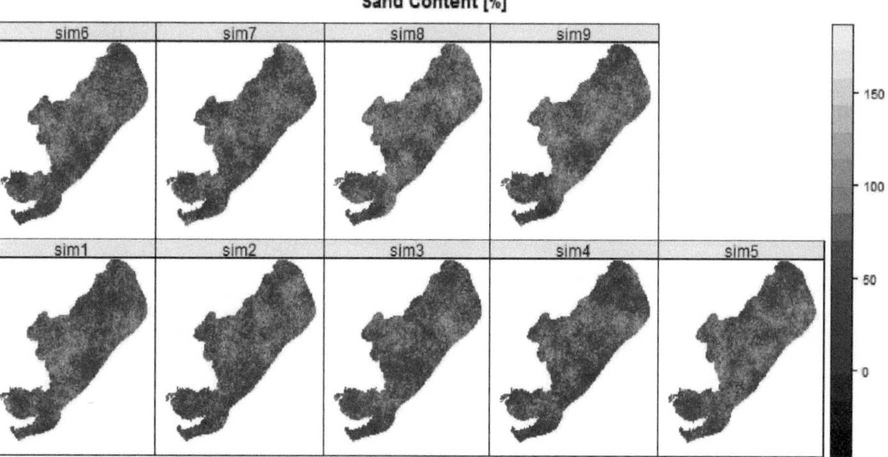

Fig. 9.9: Simulated maps of sand content prepared through stochastic simulation

Extracting the predicted values on observed point and plotting

Observed and predicted values of target variable can be compared to evaluate the performance of geostatistical technique and to further calculate different evaluation indicaes e.g. root mean squared error, mean squared deviation ratio etc. predicted values at sampled locations can be extracted from the predicted map by over function, an example of which is given below. In this example, predicted sand content at sampling coordinates of 'sawi' dataset has been extracted as an additional variable 'SAND_pred' from the predicted sand content map, 'sawi.krig.sand'. Observed and predicted values are plotted through plot function to see their distribution along 1:1 line.

```
>sawi$SAND_pred<- over(sawi, sawi.krig.sand)[[1]]

>plot(sawi$SAND, sawi$SAND_pred,

main = “(a) Observed vs predicted values of sand content (%)”, xlab = “Observed sand content (%)”, ylab = “Predicted sand content (%)”, xlim=c(0,100), ylim=c(0,100), xaxs=“i”, yaxs=“i”, pch=21, col=“black”, cex=1, font=1, font.lab=2, abline(a = 0, b = 1, col= “black”))
```

Digital soil mapping

Digital soil mapping (DSM) uses quantitative models that relate field observations on soil type or a soil property to spatially exhaustive environmental data. Environmental data should represent important soil-forming factors that explain the spatial variation of the target soil attribute. For example, derivatives from DEMs represent the influence of topography on soil formation, while satellite imagery represents the effects of vegetation, parent material, and climate. Once a model is fitted to the data, it can be used to spatially predict the soil attribute at unobserved locations given the observed environmental data at these locations. Conventional soil maps (CSM) are generally created using free survey. In free survey, the soil surveyor uses a conceptual (mental) soil-landscape model to select suitable observation locations. The surveyor uses landscape features, aerial photographs, topographical maps, digital elevation models (DEMs), and past experiences in similar landscapes. Conventional soil mapping results in a soil type map. Conceptually, DSM and CSM are very similar. Both approaches use a soil-landscape model to predict soil at unobserved locations. The main difference is that in CSM the soil-landscape model is a qualitative model based on soil surveyor expert knowledge, while in DSM the soil-landscape model is a quantitative model. Because of its qualitative nature, CSM is often considered an art. The main criticisms of CSM include irreproducibility, soil bodies being represented as discrete, homogeneous entities, and the lack of quantified measures of uncertainty (Kempen et al. 2011). Digital soil mapping does not suffer from these shortcomings. It is reproducible, and maps are easy to update because prediction models can be stored and run again when new data are available. Despite these advantages, DSM methods have some drawbacks. Complex soil forming processes might be difficult to quantify and represent by environmental explanatory variables, while these can be more easily taken into account in CSM. The success of DSM depends on the availability of (up-to-date) soil data and environmental data layers.

Digital soil mapping for organic C stock using regression kriging

The regression kriging approach was applied to map the soil organic carbon (SOC) stock in Warangal district of Telengana representing semiarid tropical region of India. Soil samples were collected at every 20 cm depth interval up to a depth of 60 cm from 51 profiles, one from each Mandal in Warangal district. A surface map of soil organic C stock was prepared using ordinary kriging as well as using regression kriging (Fig. 9.10). As soil organic C stock is strongly affected by environmental variables, these variables have been widely explored to estimate the SOC stock across the region. The environmental variables (covariates) like digital elevation model (DEM), slope, curvature, normalized difference vegetation index, soil texture and Mandal-wise annual normal rainfall data were explored and included as independent variables to establish the linear regression model for estimating the SOC stock. The 90-m SRTM-DEM (Shuttle Radar Topography Mission-DEM) was downloaded from srtm.csi.cgiar.org website and imported to ARC-GIS and converted to slope and curvature map of the district. MODIS (Moderate Resolution Imaging Spectro-radiometer) data (http://ladsweb.nascom.nasa.gov/data/) of 16 days composite vegetation index during the study period (2009-2010) and soil texture map collected from NBSS&LUP for Warangal district was used as input data. The linear regression model gave R^2 value of 0.46 for predicting the SOC stock of the district and the predicted value varied from 20-80 Mg ha^{-1}. The

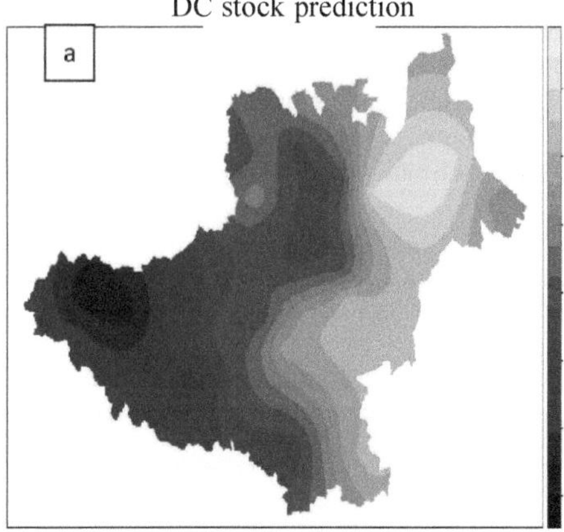

Fig. 9.10: Organic C stock Mg/ha/60 cm depth of soil of Warangal district using ordinary kriging (a) and regression kriging (b)

residual values (or errors) interpolated with ordinary kriging predicted SOC stock between -2.0 to 1.5 Mg ha^{-1}. The residual and regression predicted maps were added to obtain the final regression kriging map. The regression kriging interpolated spatial distribution map of SOC stock showed clear influence of the environmental variables and enhanced the precision of estimating the SOC stock compared to the ordinary kriging.

Summary

This chapter presents an overview of the most recent developments in the field of geostatistics and describes their application to soil science. Various interpolation (kriging) techniques capitalize on the spatial correlation between observations to predict attribute values at unsampled locations using information related on one or several attributes. An important contribution of geostatistics is the assessment of the uncertainty about unsampled values.

References

Bivand, R.S., Pebesma, E. and Gomez-Rubio, V. 2013. Applied spatial data analysis with R, Second edition. Springer, NY, http://www.asdar-book.org/

Kabacoff, R.I. 2011. R in Action: Data analysis and graphics with R. Manning Publications Co. 20 Baldwin Road PO Box 261 Shelter Island, NY 11964.

Kempen, B., Brus, D.J., Stoorvogel, J.J., Gerard, B.M.H. and de Vries, F. 2011. Efficiency comparison of conventional and digital sol mapping for updating soil maps. Soil Science Society of America Journal 76, 2097-2115.

Pebesma, E.J. and Bivand, R.S. 2005. Classes and methods for spatial data in R. R News 5 (2), http://cran.r-project.org/doc/Rnews/.

Pebesma, E.J. and Wesseling, C.G. 1998. Gstat, a program for geostatistical modelling, prediction and simulation, Computers & Geosciences 24 (1), 17-31.

Pebesma, Z.E. 2003. Gstat user's manual. Dept. of Physical Geography, Utrecht University P.O. Box 80.115, 3508 TC, Utrecht, The Netherlands, http://www.gstat.org/gstat.pdf

Sullivan, J. 1984. Conditional Recovery Estimation Through Probability Kriging-Theory and Practice. In: Geostatistics for Natural Resource Characterization (Part-1) (Eds. G. Verly), M. David, A.G. Journel and A. Marechal). NATO ASI Series. Dordrecht, Reidel Publishing Co., P. 365-384.

Webster, R. and Oliver, M. A. 2007. Geostatistics for environmental scientists. John Wiley & Sons.

10

Indicator Kriging and Natural Resource Management

Partha Pratim Adhikary and Ch. Jyotiprava Dash

ICAR-Indian Institute of Soil and Water Conservation, Koraput, Odisha, India

Introduction

Spatial datasets always exhibit two common features (i) the occurrence of a few very large concentrations (hot-spots) and (ii) the presence of data below the detection limit (black spots). Extreme values can strongly affect the characterization of spatial patterns, and subsequently the prediction. Several approaches exist to handle strongly positively skewed histograms (Saito and Goovaerts 2000). One common approach is to first transform the data (e.g. normal, Box Cox or lognormal), perform the analysis in the transformed space, and back-transform the resulting estimates. Such transform, however, does not solve problems created by the presence of numerous black spot data, since either it yields a spike of similar transformed values or, in the case of the normal-score transform, it requires a necessarily subjective ordering of all equally-valued observations. Moreover, except for the normal score transform (Deutsch and Journel 1998), it does not guarantee the normality of the transformed histogram, which is required to compute confidence intervals for the estimates. Last, the back-transform of estimated moments is not straightforward and can introduce bias if not done properly (Saito and Goovaerts 2000); for example, lognormal kriging estimates cannot simply be exponentiated. Another way to attenuate the impact of extreme values is to use more robust statistics and estimators. The non-parametric approach of indicator kriging (IK) falls within that category (Journel 1983, Goovaerts 2001). The basic idea is to discretize the range of variation of the environmental attribute by a set of thresholds (*e.g.* deciles of sample histogram, detection limit, regulatory threshold) and to transform each observation into a vector of indicators of non-exceedence of each threshold. Kriging is then applied to the set of indicators and estimated values are assembled to form a conditional cumulative distribution function (ccdf). The mean or median of the probability distribution can be used as an estimate of the concentration of the material in question (Barabas et al. 2001, Cattle et al. 2002, Goovaerts et al. 2005).

Indicator Kriging (IK) as a technique in natural resource management is over thirty years old. Since its introduction in the geostatistical sphere by Journel in 1983, many authors have worked on the IK algorithm or its derivatives. The original intention of Journel was the estimation of local uncertainty by the process of derivation of a local cumulative distribution function (cdf). The beauty of IK is that it is non-parametric, therefore did not rely upon the assumption of a particular distribution model for its results. From slow beginnings in the early eighties as a technique in natural resource estimation, IK has grown to become one of the most widely-used algorithms, despite the relative difficulty in its application. It is one of the prime non-linear geostatistical techniques used today.

Indicator Kriging: Theory

The concept of the indicator approach is the binomial coding of data into either 1 or 0 depending upon its relationship to a cut-off value, z_k. During the indicator kriging procedure, first the indicator codes are generated by the indicator function, which is below a desired threshold value z_{th}. It is written as follows:

$$I(x;z_{th}) = \begin{cases} 1, if\ z(x) \geq z_{th} \\ 0, otherwise \end{cases} \tag{1}$$

Then, the semivariogram $\gamma_I(h)$ is used to quantify the spatial correlation of the indicator codes, I (x_i, z_k), and it is written as follows:

$$\gamma_I(h) = \frac{1}{2N(h)} \sum_{i=1}^{N(h)} [I(x_i;z_{th}) - I(x_i + h;z_{th})]^2 \tag{2}$$

The indicator kriging estimator, $I^\wedge (x_o; z_{th})$, at the location x_o can be calculated by

$$I^\wedge (x_o; z_{th}) = \sum_{i=1}^{n} \lambda_i I(x_i;z_{th}) \tag{3}$$

and the indicator kriging system given $\sum\lambda_i = 1$ is

$$\sum_{j=1}^{n} \lambda_j \gamma_I (x_j - x_i) = \gamma_I (x_o - x_i) - \mu \tag{4}$$

where λ_j is the weighted coefficient, γ_I is the semivariance of the indicator codes at the respective lag distance, and μ is the Lagrange multiplier.

This is a non-linear transformation of the data value, into either 1 or 0. Values which are much greater than a given cut-off, z_k, will receive the same indicator value as those values which are only slightly greater than that cut-off. Similarly, the values which are much lower than a given threshold; will receive the same indicator value as those values which are only slightly lower than that cut-off. Thus indicator transformation of data is an effective way of limiting the effect of very high and very low values. Simple or Ordinary kriging of a set of indicator-transformed values will provide a resultant value between 0 and 1 for each point estimate. This is in effect an estimate of the proportion of the values in the neighbourhood which are greater or lower than the indicator or threshold value.

The outcome of IK is a conditional cumulative distribution function (ccdf). A distribution of local uncertainty or possible values conditional to data in the neighbourhood of the sampling site need to be estimated. This distribution of grades can be used for many purposes, in addition to the deriving the average value. Any relevant criteria may be used to derive the estimate required, not simply the arithmetic mean of the local distribution.

The practice of IK involves calculating and modelling indicator variograms (that is, variograms of indicator-transformed data) at a range of cut-offs or thresholds which should cover the range of the input data. This approach is termed Multiple Indicator Kriging (MIK). One approximation is to simply infer the variogram for the median of the input data and to use this for all cut-offs. This so-called Median IK approach is very fast, since the kriging weights do not depend on the cut-off being considered.

Criticisms and Extols

A frequent criticism of the indicator approach is that the binary coding amounts to discarding some of the information in the data. In theory, this loss of information can be compensated by accounting for indicator values defined at different thresholds, which is using indicator cokriging instead of kriging. Practice has shown, however, that indicator cokriging improves little over indicator kriging (Goovaerts 1994, Pardo-Igúzquiza and Dowd 2005) because cumulative indicator data carry substantial information from one threshold to the next one, and all indicator values are available at each sampled location (isotopic or equally-sampled case). Another way to increase the resolution of the discrete ccdf is to conduct a fine discretization of the continuous sample distribution using a large number of thresholds. For example, 15 indicator cutoffs were used by Lark and Fergusson (2004) to map the risk of soil nutrient deficiency in a field of Nebraska. Goovaerts et al. (2005) used indicator kriging with 22 thresholds to model probabilistically the spatial distribution of arsenic concentrations in groundwater of Southeast Michigan. Cattle et al. (2002) used 100 threshold values to characterize the spatial distribution of urban soil lead contamination. The extreme situation is to identify the set of thresholds with the sample dataset, which is to use as many thresholds as observations. In few cases, typically only the observations closest to the interpolated location were used as thresholds. Such tailoring of thresholds to the local information available leads to a better resolution of the discrete ccdf by selecting low thresholds in the low-valued parts of the study area and high thresholds in the high-valued parts (Saito and Goovaerts 2000, Lloyd and Atkinson 2001, Cattle et al. 2002).

The above mentioned Median IK has its own assumptions and drawbacks. Since IK generates at each point a cumulative distribution, this should be non-decreasing and valued between zero and one. These two requirements are sometimes not met, leading to so-called order relations violations. The trade-off costs for the finer resolution of the ccdf are the tedious inference and modeling of multiple indicator semivariograms, as well as the increasing likelihood that the estimated probabilities won't honor the axioms of a cumulative distribution function: all probabilities must be valued between 0 and 1 and form a non-decreasing function of the threshold value. Failure to honor such constraints, referred to as order relation deviations, requires a posteriori correction of the set of estimated probabilities (Deutsch and Journel 1998).

Many methods have been proposed to counteract the order relations issue – the most commonly-used involves direct correction of the indicator values (Deutsch and Journel 1998). Another approach, proposed by, Dimitrakopoulos and Dagbert (1992), involves the use of nested indicator variables - in other words, indicator variables which are defined by successively halving the data set to define the thresholds. This nested indicator kriging approach eliminates any problems associated with order relations, but suffers from data deficiency problems, especially at high thresholds.

To keep these deviations within reasonable limits, Deutsch and Lewis (1992) recommend using no more than 9–15 thresholds. Several authors have proposed alternate implementations of the indicator approach that reduce the proportion and magnitude of order relation deviations, while maintaining a reasonable resolution for the ccdf. For example, Pardo-Igúzquiza and Dowd (2005) developed a procedure that requires solving a single indicator cokriging system at each location, leading to far fewer order relation problems than the traditional indicator kriging. Two other implementation tips (Goovaerts 1997) are to avoid sudden changes in indicator semivariogram parameters from one threshold to the next, and to select thresholds z_k so that within each search neighborhood there is at least one datum from each class $(z_{k-1}, z_k]$. This is ensured by using locally adaptive thresholds and the same semivariogram model for all thresholds (Saito and Goovaerts 2000, Lloyd and Atkinson 2001). For large datasets Cattle et al. (2002) developed a program where indicator semivariograms are computed and modeled locally whereas the same 100 global thresholds are used across the entire study area.

The IK algorithm has been extended to not only include the indicator transform of the data, but also the data itself. This approach, first postulated by Sullivan (1984), and also covered by Isaaks (1984), is termed probability kriging, and is essentially indicator co-kriging between the indicator-transformed data and the uniform $(0\rightarrow1)$ transform of the sample data. As with most co-kriging, the downside is the calculation and modelling of the cross-variograms between the two data types in addition to the univariate indicator variograms.

Rivoirard (1993) used the relationships between the cross-variograms of indicators at adjacent cut-offs to draw conclusions about the nature of the processes influencing the distribution of data values. This work led to the definition of the so-called mosaic and diffusion models, among others, which lead to a particular style of IK or other non-linear estimation algorithm. Despite much theoretical development, in practice it is the straightforward implementation of MIK, using non-nested indicator transforms of data at multiple cut-offs leading to the definition of local distributions of natural resource concentration.

Merits of using Indicator Kriging

The primary motivation behind the use of IK in most earth science applications is that it is non-parametric. Moreover, it can also address mixed data populations. Since IK actually partitions the overall sample distribution by a number of thresholds, there is no need to fit or assume a particular analytically-derived distribution model for the data. MIK requires the inference of a variogram model at each cut-off, and can handle different anisotropies at different cut-offs. But if the anisotropy changes too much between adjacent thresholds, the order relations violations become prohibitively large, but if the changes are gradual then the situation depicted can easily be handled.

IK is a favoured option for highly-skewed data sets, as it offers a practical way of treating the upper tail of the distribution which does not depend entirely on an arbitrary upper cut value, but IK allows the practitioner to use features of the actual data for defining any upper tail treatment. While unconstrained estimation is not advocated wherever there is the opportunity of defining constraining domains, if it is a necessity then MIK is an approach which can minimize the smoothing under certain conditions.

One of the great benefits of the indicator transformation is that it allows common coding of diverse data types and their integration into the single process. Since all data is transformed to $0 \rightarrow 1$ space, other, secondary data types can easily be accommodated by the same coding scheme.

The users can also estimate the uncertainty at unsampled locations, via the inference of a distribution of values. This data can be used to derive an expected value, but also to yield risk-qualified outcomes, such as the probability of exceeding a given grade – the cut-off grade, or to map a given percentile of the distribution.

Practical corrections to use indicator Kriging

Treatment of Upper and Lower Tails

An indicator kriging will provide an estimate of the proportion of resource above each of the indicator cut-offs or thresholds assessed. To reduce this data into an estimate of mean quantity or quantity above a cut-off, it is a requirement that each indicator class interval be assigned a threshold. A number of sensitivities must be considered when undertaking the task of class interval threshold assignment. If indicator thresholds have been carefully selected with adequate regard to the input distribution, then the distribution of cut-offs within many classes will be nearly linear. The average value of the input data will normally suffice for the assignment of threshold in these classes.

The distribution of thresholds in the uppermost and lowermost classes of the distribution will not normally be linear. In the case of a positively-skewed distribution, the greatest estimation sensitivities relate to the threshold assigned to the uppermost class. Distribution skew and outliers both influence the threshold distribution in this class, which requires a more sophisticated method of mean threshold selection if parameter over-estimation or underestimation is to be avoided. Deutsch and Journel (1998) proposed a model based on a hyperbolic distribution for representing the parameter distribution above the uppermost threshold. Both of these variables may be judged from the sample parameter distribution.

The Data Dilemma

When applying multiple indicator kriging to resource estimation, it is necessary to settle on a finite number of thresholds that adequately represent the input data distribution shape. There is always a competition between the number of thresholds selected and the time available for the required analysis. Additional indicators may be included to discriminate between components of mixed population distributions. The disadvantage of these deciles is that many of the indicator thresholds will be concentrated at the lower end of the positively skewed distribution. Fewer indicator thresholds will represent the higher values. The answer to this dilemma is to collect more data. Enough close-spaced data needs to be collected from representative areas to allow the better definition of the high-value continuity and the short-range continuity of lower-value indicators.

Initially Use the Median Indicator Kriging

Median indicator kriging (Median IK) uses the median indicator variogram to define the continuity conditions for all indicators. This method is a simplified form of MIK that might be considered in the early stages of a resource project, when sample data is sparse and it is difficult or impossible to define threshold continuity for a full range of indicators. The median indicator variogram is typically the most robust of all indicators, it tends to have the greatest range of continuity, and it is the easiest to define with some confidence from sparse data.

The application of variograms from a single indicator to all thresholds reveals the main assumption associated with the median indicator method. This is that the direction and range of continuity does not vary with changing thresholds. Experience from full indicator variography studies shows that continuity almost always varies with indicator threshold, and invariably declines with increasing indicator threshold. This will tend to result in an overestimation of the quantiles of the upper grade classes with Median IK, resulting in a higher-than-normal expected grade. In practical terms, Median IK is not a recommended technique where the data permits full estimation of a set of indicator variograms.

Change of Support

Change of support is the generation of a distribution of thresholds for non-point data (Dowd 1992, Matheron 1982). Unlike ordinary kriging, inverse distance, and other linear estimation methods, the indicator transform is non-linear, as is the logarithmic transform, the uniform transform, and the normal scores transform. The consequence of this is that one cannot average indicators linearly, and thus cannot obtain a spatial distribution by averaging a series of point distributions derived at a smaller scale. The Ordinary Kriging (or inverse distance) corollary is to discretise the space by a series of points, estimate the value at each point, and carry out the arithmetic mean to derive the spatial pattern. However, if the statistics required from the IK distribution at each point is known and fixed, such as the median or the mean value, then one can subdivide the space into points, estimate the ccdf by MIK at each point, deriving the desired statistic, and then averaging the result.

One downside of this approach is the extra computational effort. A short cut would be to use the same data configuration for each point, thus yielding only one matrix inversion on the left-hand side of the kriging system; the difference between the points would be in the different weights obtained due to the differing positions of each point in space.

The traditional approach to the change of support in MIK has been to apply a variance correction factor on a global basis to the point statistics, exactly as one might do with point kriged data. There are a number of common techniques for achieving this. Perhaps the most widely used is the affine correction, a simple factoring of variance from point to (theoretical) block variance.

Another approach is to use the indirect lognormal correction (Isaaks and Srivastava 1989). The most elegant and theoretically correct solution to deriving spatial values is to move beyond the realm of estimation to that of sequential indicator simulation - this approach provides a truly local change of support, conditional only upon values in the neighbourhood.

Variogram Inference

In IK, there is the need to calculate a variogram and develop a model for each threshold value. For multiple (10 to 12) thresholds this was once a major exercise. However, advances in both computer speed and memory capacity and in software technology have produced new generations of fast and efficient variogram generation and modelling software. Such software allows the calculation of a full set of indicator variograms over all thresholds in one pass; this allows the scientists to iterate between the various cut-offs and ensure a smooth variation in the variogram parameters. This is turn will serve to minimize the number of order relations corrections required. The combination of modern software and hardware has all but eliminated the time penalty of variogram inference, and a full set of thresholds may now be generated and modelled in a few hours for a moderately-large data set.

Applications in natural resource management

Geology, Geochemistry and Geomorphology

Indicator kriging can be applied to assess the lake sediment geochemical distribution. An important problem associated with the analysis of the geochemical information is the presence of skewed distributions with high coefficients of variation. Another problem is that values below the detection limit are grouped and unresolved. In these situations, two traditional solutions are proposed (Jimenez-Espinosa et al. 1999): (i) remove the extreme values, based on geological or probabilistic criteria; (ii) transform the data by means of a smoothing function or by taking logarithms. The first approach is not acceptable when the data carry the most valuable structural information, not to mention their economic weight. Log transformations are non-linear, and that calls for non-linear estimation techniques (i.e. disjunctive kriging), which requires a hypothesis about the distribution. Indicator kriging (IK) provides an alternative approach to modeling the spatial distribution of positively skewed populations, and has the potential to overcome both of these limitations (Panahi et al. 2004) and used widely to predict the geochemical distribution of lake sediments.

Indicator kriging can successfully be applied to improve the prediction of geological models. Scientists always wish to keep the generated the geological models as accurate as possible. The focus was mainly put on the statistics of the mechanical parameters of soil and rock masses that go directly into the safety calculations. Besides the statistical features of the mechanical parameters, care must also be given to the statistics associated with the geometrical parameters of the geological layers. One way to statistically model the underground geometry is through the use of geostatistical interpolation techniques generally known as indicator kriging. The smoothing effects around zero value zones can be reduced significantly by the use of indicator kriging (Marinoni 2003).

Nonlinear geostatistics is commonly used in ore grade estimation and seldom used in lithological characterization. Categorization of lithological units is essential in ore grade estimation, and this can be done based on the lithological information obtained from drill-hole data. Geostatistical nonlinear indicator kriging (IK) can be a better approach to delineate different lithological units of an iron ore deposit

(Rao and Narayana 2015). IK can also be used to estimate the slate deposite in the geological layers and the ways to mine those suatainably.

Soil Science

There are many applications of IK in soil science. It has been used successfully to assess the urban soil lead contamination (Cattle et al. 2002). Yu-Pin et al. (2010) used IK to delineate and map the spatial patterns of soil heavy metal (Cr, Cu, Ni and Zn) pollution in Chunghua County, central Taiwan. Mining and mineral-processing activities can modify the environment in a variety of ways. Sulfide mineralization is notorious for producing waters with high metal contents. Arsenic is commonly associated with sulfide mineralization and is considered to be toxic in the environment at low levels. The IK can also be used for the evaluation of arsenic potential contamination in abandoned mining areas (Antunes and Albuquerque 2013). Soil nutrient management is an important issue to raise the crop production and productivity. The deficiency of soil nutrients leads to crop loss and sometimes failure and the presence of excess nutrients in the soil may lead to toxicity and luxary consumption. Risk mapping of soil nutrient deficiency or excess can be done with adequate accuracy using indicator kriging (Lark and Ferguson 2004). IK can also be used as complementary use of bio-indicators at different tropic levels. The use of biological indicators is widespread in environmental monitoring, although it has long been recognised that each bio-indicator is generally associated with a range of potential limitations and shortcomings. To circumvent this problem, indicator kriging can be adopted the complementary use of bio-indicators representing different trophic levels and providing different type of information to integrate knowledge and to estimate the overall health state of ecosystems. Soil salinity risk can be assessed effectively using IK.

Water Science

In the water science applications, IK can be used to determine the water quality parameters influencing sprinkler irrigation performance. The normal irrigation water quality mapping and management can be done using IK. The heavy metals in the ground water can be quantified with high precision using indicator kriging (Adhikary et al. 2011). As groundwater nitrate pollution associated with agricultural activity is an important environmental problem in the management of this natural resource. Therefore, specific measures aimed to control the risk of water pollution by nitrates must be implemented to minimize its impact on the environment and potential risk to human health. The spatial probability distribution of nitrate contents exceeding a threshold or limit value, established within the quality standard, will be helpful to managers and decision-makers. Categorical Indicator Kriging can be used for assessing the risk of groundwater nitrate pollution (Chica-Olmo et al. 2014). Indicator kriging provides a way to use traditional water table data to quantify probability of soil saturation to evaluate predicted spatial distributions of runoff generation risk, especially for the new generation of water quality models incorporating saturation excess runoff theory. When spatial measurements of a variable are transformed to binary indicators (i.e., 1 if above a given threshold value and 0 if below) and the resulting indicator semivariogram is modeled, indicator kriging produces the probability of the measured variable to exceed the threshold value. Indicator kriging gives quantified probability of saturation or, consistent with saturation excess runoff theory, runoff generation risk with depth to water table as the variable and the threshold set near the soil surface (Lyon et al. 2006).

Climate Science

Uses the geostatistical methods like indicator kriging (IK) to address the problem of estimating values of precipitation at locations from which measurements have not been taken is highly demanding. Several problems or issues like (i) lognormality of the data, (ii) non-stationarity of the data and (iii) anisotropy of the spatial continuity an easily be addressed by IK.

Conclusions

Indicator kriging is now widely used in the natural resource management as an estimation technique over a wide range of soil, minerals, geology and environments, because it offers practical solutions to some common estimation problems. In particular the issue of mixed or poorly-domained distributions, and the general trend away from the so-called parametric techniques, has probably enhanced the acceptance of IK. The appeal of being able to generate (at least in theory) 'recoverable' resources have undoubtedly contributed to the popularity of the approach.

However, the use of IK has had its downside. Particular criticisms have been the relative difficulty of deriving true distributions, the nuisance of order relations, and the sheer work involved in inferring variogram models at multiple thresholds. The scientists, aided and abetted by academic research, have come up with solutions to almost all of the perceived problems, some more elegant than others.

References

Adhikary, P.P., Dash, J.Ch., Bej, R. and Chandrasekharan, H. 2011. Indicator and probability kriging methods for delineating Cu, Fe, and Mn contamination in groundwater of Najafgarh Block, Delhi, India. Environmental Monitoring and Assessment 176: 663-676.

Antunes, I.M.H.R. and Albuquerque, M.T.D. 2013. Using indicator kriging for the evaluation of arsenic potential contamination in an abandoned mining area (Portugal). Science of the Total Environment 442: 545-552.

Barabás, N,, Goovaerts, P. and Adriaens, P. 2001. Geostatistical assessment and validation of uncertainty for three-dimensional dioxin data from sediments in an estuarine river. Environmental Science and Technology 35(16): 3294-3301.

Cattle, J.A., McBratney, A.B. and Minasny, B. 2002. Kriging method evaluation for assessing the spatial distribution of urban soil lead contamination. Journal of Environmental Quality 31: 1576-1588.

Chica-Olmo, M., Luque-Espinar, J.A., Rodriguez-Galiano, V., Pardo-Igúzquiza, E. and Chica-Rivas, L. 2014. Categorical indicator kriging for assessing the risk of groundwater nitrate pollution: The case of Vega de Granada aquifer (SE Spain). Science of the Total Environment 470-471: 229–239.

Deutsch, C.V. and Journel, A.G. 1998. GSLIB: Geostatistical Software Library and User's Guide. 2. Oxford University Press; New York, NY: p. 369.

Deutsch, C.V. and Lewis, R. 1992. Advances in the practical implementation of indicator geostatistics. Proceedings of the 23rd International APCOM Symposium; Tucson, AZ, Society of Mining Engineers. pp. 169-179.

Dimitrakopoulos, R. and Dagbert, M. 1992. Sequential modelling of relative indicator variables: dealing with multiple lithology types, in Geostatistics Troia '92, Soares, A Editor, Kluwer 1993.

Dowd, P.A. 1992. A review of recent developments in geostatistics. Computers and Geosciences 17(10): 1481-1500.

Goovaerts, P., AvRuskin, G., Meliker, J., Slotnick, M., Jacquez, G.M. and Nriagu, J. 2005. Geostatistical modeling of the spatial variability of arsenic in groundwater of Southeast Michigan. Water Resources Research 41(7): W07013 10.1029.

Goovaerts, P. 1994. Comparative performance of indicator algorithms for modeling conditional probability distribution functions. Mathematical Geology 26(3): 389-411.

Goovaerts, P. 1997. Geostatistics for Natural Resources Evaluation. Oxford University Press; New York, NY: pp. 483.

Goovaerts, P. 2001. Geostatistical modelling of uncertainty in soil science. Geoderma. 103: 3-26.

Isaaks, E.H. and Srivastava, R.M. 1989. An introduction to applied geostatistics. Oxford University Press, New York.

Isaaks, E.H. 1984. Risk qualified mappings for hazardous waste sites: a case study in distribution-free geostatistics. Master's thesis, Stanford University, Stanford, California, USA.

Jimenez-Espinosa, R., Sousa, A.I. and Chica-olmo, M. 1999. Identification of geochemical anomalies using principal component analysis and factorial kriging analysis. Journal of Geochemical Exploration 46: 245–256.

Journel, A.G. 1983. Nonparametric estimation of spatial distributions. Mathematical Geology. 15(3): 445-468.

Lark, R.M. and Ferguson, R.B. 2004. Mapping risk of soil nutrient deficiency or excess by disjunctive and indicator kriging. Geoderma 118(1): 39-53.

Lloyd, C.D. and Atkinson, P.M. 2001. Assessing uncertainty in estimates with ordinary and indicator kriging. Computers and Geosciences 27(8): 929-937.

Lyon, S.W., Lembo Jr, A.J., Walter, M.T. and Steenhuis, T.S. 2006. Defining probability of saturation with indicator kriging on hard and soft data. Advances in Water Resources 29: 181-193.

Marinoni, O. 2003. Improving geological models using a combined ordinary-indicator kriging approach. Engineering Geology 69: 37-45.

Matheron, G. 1982. La destructuration des hautes teneurs et le krigeage des indicatrices. Centre de Geostatistiqueet de Morphologie Mathematique. Note N-761, 33p.

Panahi, A., Cheng, Q. and Bonham-Carter, G.F. 2004. Modelling lake sediment geochemical distribution using principal component, indicator kriging and multifractal power-spectrum analysis: A case study from Gowganda, Ontario. Geochemistry: Exploration, Environment, Analysis 4: 59-70.

Pardo-Igúzquiza, E. and Dowd, P.A. 2005. Multiple indicator cokriging with application to optimal sampling for environmental monitoring. Computers and Geosciences 31(1): 1-13.

Rao, V.K. and Narayana, A.C. 2015. Application of nonlinear geostatistical indicator kriging in lithological categorization of an iron ore deposit. Current Science 108(3): 413-421.

Rivoirard, J. 1992. Relations between the indicators related to a regionalised variable, in Geostatistics Troia '92, Soares, A Editor, Kluwer.

Saito, H. and Goovaerts, P. 2000. Geostatistical interpolation of positively skewed and censored data in a dioxin contaminated site. Environmental Science and Technology 34(19): 4228-4235.

Sullivan, J. 1984. Conditional recovery estimation through probability kriging: theory and practice, in Geostatistics for natural resources characterisation, G Verly *et.al.*, editors, pp. 365-84, Riedel, Dordrecht, Holland.

Yu-Pina, L., Bai-Youa, C., Guey-Shinb, S. and Tsun-Kuoa, C. 2010. Combining a finite mixture distribution model with indicator kriging to delineate and map the spatial patterns of soil heavy metal pollution in Chunghua County, Central Taiwan. Environmental Pollution 158: 235-244.

11

Introduction of QGIS for Spatial Data Handling

Priyabrata Santra and Mahesh Kumar

ICAR-Central Arid Zone Research Institute, Jodhpur, Rajasthan, India

Introduction

Quantum Geographical Information System (QGIS) is an open source GIS environment. The software package is available freely in web https://www.qgis.org/en/site/forusers/download.html) and can be installed in a variety of computer platform *e.g.* Windows, Linux, Mac OS X, Android etc. Latest information on QGIS is available in http://www.qgis.org/en/site/. Basically, the QGIS GUI is divided into five areas: (i) Menu Bar, (ii) Tool Bar (iii) Map Legend, (iv) Map View and (v) Status Bar. QGIS allows users to define a global and project-wise coordinate reference system (CRS) for layers without a pre-defined CRS. It also allows the user to define custom coordinate reference systems and supports on-the-fly (OTF) projection of vector and raster layers. All of these features allow the user to display layers with different CRSs and have them overlay properly. QGIS has support for approximately 2,700 known CRSs. Definitions for each CRS are stored in a SQLite database that is installed with QGIS. The CRSs available in QGIS are based on those defined by the European Petroleum Search Group (EPSG) and the Institute Geographique National de France (IGNF) and are largely abstracted from the spatial reference tables used in Geospatial Data Abstraction Library (GDAL). For Post GIS layers, QGIS use the spatial reference identifier that was specified when the layer was created. For data supported by OGR, QGIS relies on the presence of a recognized means of specifying the CRS. In the following sections, few basic GIS analysis using example spatial data from arid western India is discussed. However, readers are advised to go through different tools and functions available in QGIS after installation of the system in personal computers.

Georeferencing a map in QGIS

Georeferencing is the first step to work in GIS environment. For example, spatial information in the form of hard copy of a survey map is available; however, the geographical coordinate reference system of it needs to be assigned to further work with it. Once geographic coordinates of the map are defined, then it is ready to apply different GIS tools and technique on the map for further spatial analysis. If you think about the coordinates system of a map, it will be clear to you that even before georeferencing, every map has its own coordinate system, which is actually the Cartesian coordinate system. Through georeferencing, the Cartesian coordinate system is converted to geographical coordinate system. Therefore, georeferencing can be defined as the process of assigning real-world coordinates to each pixel of the raster. Many times these coordinates are obtained by doing field surveys - collecting coordinates with a Global Positioning System (GPS) device for few easily identifiable features in the image or map. To start georefencing a map, known coordinates within the map area should preferably be distributed throughout the map, which is also called as ground control points (GCPs). Geographical coordinates of few permanent features within the map either measured in field or collected from secondary source are often used as GCPs and these permanent features may be crossing of roads, highways, railway line, river bridge etc. Otherwise, the coordinates at crossings of two grid lines printed on a map are also used as GCPs. In the following, the stepwise procedure to georeference the soil map of India is discussed. The soil map of India depicted below in Fig. 11.1 is also available in http://eusoils.jrc.ec.europa.eu/esdb_archive/EuDASM/Asia/maps/IN1000_SOTO.htm.

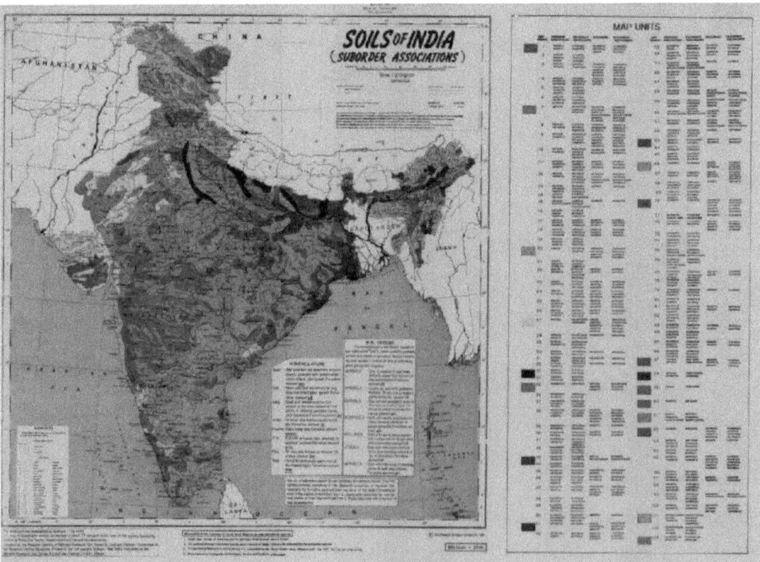

Fig. 11.1: Soil suborder association map of India

Source: (NBSS&LUP published map and freely available at internet https://esdac.jrc.ec.europa.eu/ESDB_Archive/EuDASM/Asia/images/maps/download/IN1000_SOTO.jpg)

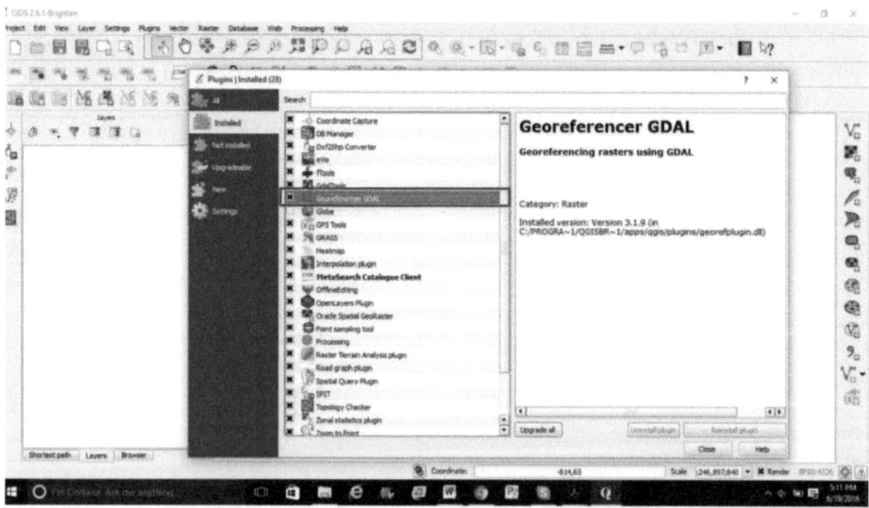

Fig. 11.2: Georeferencer GDAL plugin in QGIS

Georeferencing in QGIS is done via the 'Georeferencer GDAL' plugin (Fig. 11.2). This is a core plugin - meaning it is already part of your QGIS installation. You just need to enable it. Go to Plugins -> Manage and Install Plugins -> and enable the Georeferencer GDAL plugin in the Installed tab.

After enabling the Georeferencer GDAL plugin, the georeferencer window can be opened through Raster → Georeferencer available in menu bar. Georeferencer window has two sub-windows, the top one deals with the raster map and bottom one handles the GCP table. Raster map to be georeferenced needs to be opened using the raster symbol (Fig. 11.3). Once the raster map is selected for open, it will ask for the coordinate reference system. The coordinates reference system

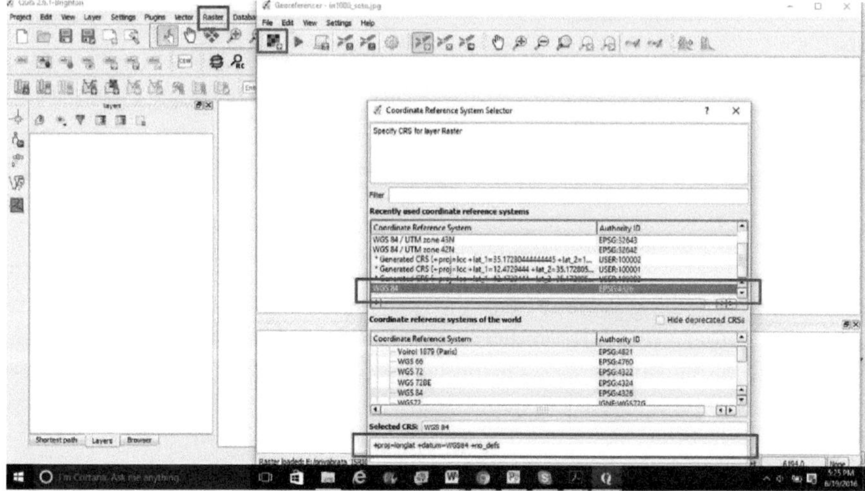

Fig. 11.3: Adding the raster map for georeferencing

printed in the scanned map needs to be used here otherwise if the GCPs were collected through field survey with GPS, Geographic coordinate system (WGS, EPSG: 4326) may be selected. In the example soil map of India presented in Fig. 11.1, grid lines of latitude and longitude were printed on the map, and hence EPSG:4326 coordinate system is selected.

Once the coordinate system is assigned, the GCP points needs to be added by selecting 'Add points' as given in Fig. 11.4. For example, here GCP 1 is added as 68°E and 36°N and once it is added, the coordinates are appeared in the GCP table. Likewise, several GCP points need to be added and the minimum number of GCP points is four to run the georefernecer tool. Here, five GCP points are added. Once, the GCP points are assigned, georeferencer tool needs to be run.

However, before georeferencing process to start, it will ask for transformation settings. Thin plate spline is selected here as the transformation algorithm. Nearest neighbor method was selected as the resampling method. Here, the name and location of the final georeferenced raster map needs to be mentioned. Selection of checkbox on 'Load on QGIS when done' will lead to display of the final map after georeferencing.

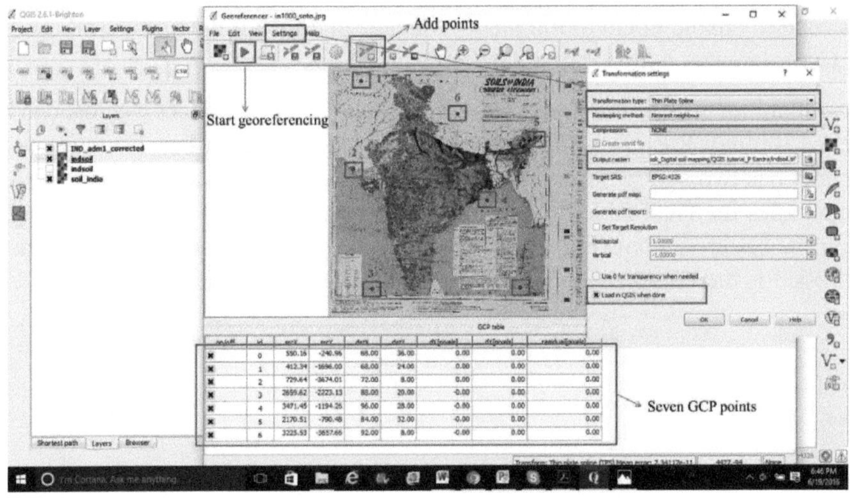

Fig. 11.4: Addition of GCP points assigning transformation settings for georeferencing purpose

Once the gereferencing is completed, the final map is displayed in QGIS window. To know whether the georeferenced has been done correctly or not, a spatial polygon map of the study area with similar CRS may be opened and checked. For example, here state boundary map of India with WGS 84 and lat/long geographic CRS is overlayed over the georeferenced map, which is shown in Fig. 11.5. It is found from overlay of the vector boundary over raster georeferenced map that state as well as national boundary map is georeferenced satisfactorily. Here it is to be noted that more precise GCPs are given as input, higher is the accuracy of the georeferenced map.

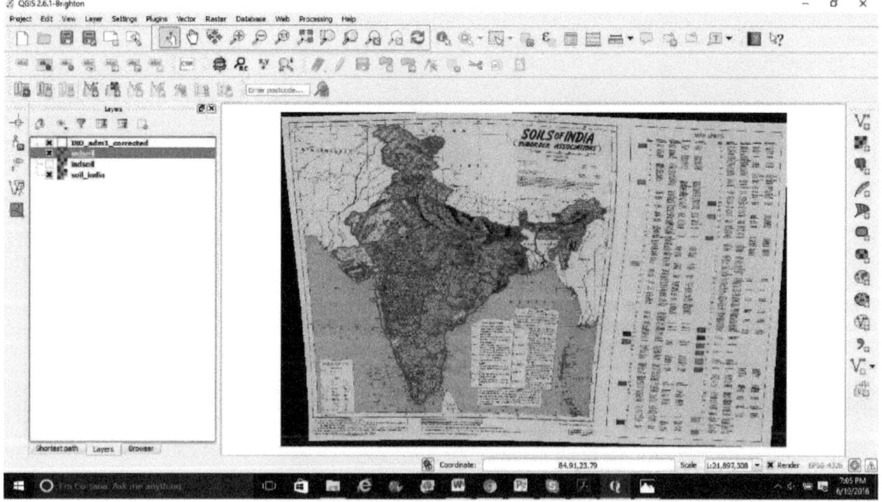

Fig. 11.5: Georeferenced raster map of soil suborder association of India and the state boundary of India overlayed on it.

Creation of vector layer

Vector layer in GIS is represented by points, lines and polygons. These vectors files may be the soil sampling locations as point data; roads, irrigation canal etc. as line data and soil mapping unit boundary as polygon data. Vector files may be available from other secondary sources, otherwise it needs to be created as new object. The spatial information source for creating a new vector object may be the scanned and georeferenced copy of soil maps or field collected data on spatial coordinates. In the following section, we will discuss how vector file may be created from these two sources in QGIS. Nowadays, with the advancement of digital information technology, every bit of information on earth surface is accessible to user through 'Google Earth' interface. Vector file in key markup language (kml) format may be created by digitizing or marking the desired spatial information on google earth images itself, which later on can be converted to desired format for GIS environment. The format for a vector file in GIS is shapefile with an extension of *.shp. It is to be noted here that every shapefile in QGIS is associated with few other accessory files also along with *.shp and these are *.shx, *.prj, *.dbf, *.qpg and *.cpg. Therefore, whenever a vector file is copied from a computer directory to a separate directory, care should be taken so that all accessory files are also copied otherwise it cannot be opened in QGIS. However, if the shapefile is copied in GIS environment itself, by default all accessory files are copied.

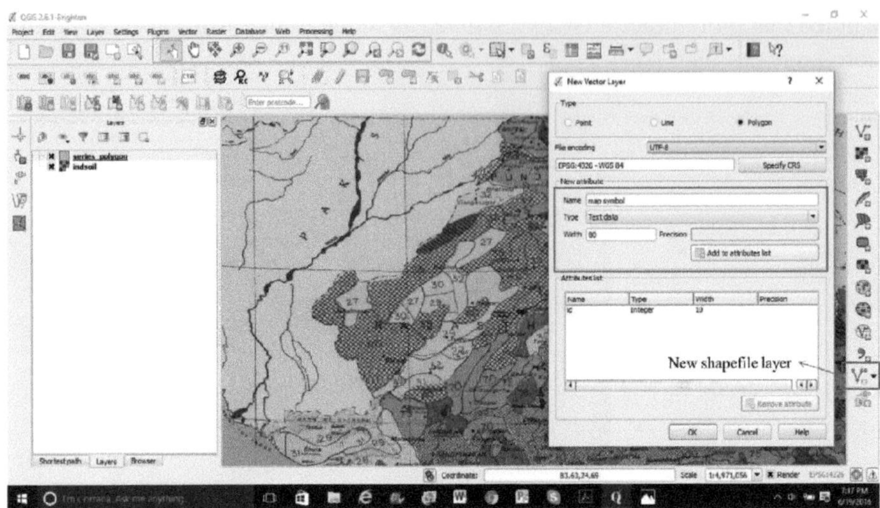

Fig. 11.6: Creation of a new polygon layer in QGIS

Here, we discuss how a vector polygon layer can be created from a scanned and digitized soil map. In previous section, we have presented the georeferencing steps of the soil suborder association map of India. Now, we will try to create vector polygons of soil mapping units as suborder association from the scanned map. For example, in arid western India, we found different mapping units *e.g.* 27, 32 *etc.* (Fig. 11.6). To start with, a vector layer is to be created first by clicking on 'New shapefile icon' on right vertical panel in QGIS window as depicted in Fig. 11.6. It will ask for type, coordinate reference system, and attribute list. Here we will click on polygon as we are going to create soil map polygons. The CRS may be specified, which has been selected as EPSG:4326–WGS 84 by default because a raster map with same CRS has already been opened. It is to be noted EPSG:4326 – WGS 84 CRS indicates Geographical coordinate system (lat/long) with WGS 84 as the datum and most commonly used CRS in GIS. By default, the 'id' attribute will be created as integer data type with a width of 10. Additionally, new attributes may be added for further characterization and description of the vector object. For example, an attribute 'map symbol' is added here with text data type and width as 80. It will help to identify the polygon and its characteristics features later on. Similar to this, point and line files may be created by selecting either point or line in the new vector layer sub-window. After assigning the type, format and attributes of the vector layer, it will ask for the name of the vector layer in a directory.

Once the file is created, we can start to create the polygon by digitizing the polygon boundary. Digitization tool can be accessed by clicking the 'Toggle editing' symbol. After clicking the 'Toggle editing' symbol, digitization tools *e.g.* 'Add feature', 'Move feature', 'Node tool' etc. become active and also see the changes in status of the file name at left vertical panel. Clicking second time on the 'Toggle

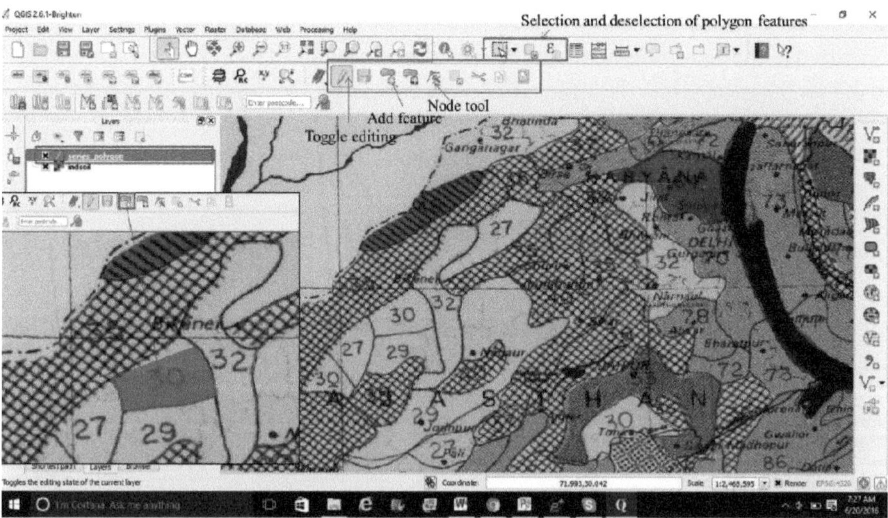

Fig. 11.7: Digitization tools in QGIS

editing' symbol will lead to deactivation of editing stage. New polygon features can be added by selecting the 'Add feature' tool. Left clicks will create nodes of the polygon whereas right click ends the process. Once the polygon is created, attributes of that polygon needs to be assigned. For example, map symbol attribute for two digitized polygons in Fig. 11.7 was entered as 27 and 32. After assigning the attributes, the edit needs to be saved before deselecting the 'Toggle editing' symbol. In the editing stage of the shapefile, wrong polygons can be deleted or modified. The polygons can be deleted by selecting the features as appeared in the top menu bar as shown in Fig. 11.7. The nodes of the polygon can be modified by using 'Node tool'. It is necessary to place the digitized lines exactly on the existing polygon boundary of background raster map. The polygon can also be moved from one place to another by clicking on the 'Move feature'. In the figures, it is shown that two polygons e.g. mapping unit number 27 and 32 were created, whereas mapping unit number 30 was in the process of digitization. It is to be noted that more number of nodes you create by zooming more accurate polygon map you can generate.

Once the polygons are created, any overlap of polygons or gaps between polygons (also known as slivers) needs to be checked and corrected. Number of digitized polygons can be checked from the attribute table from the file name as appeared in the left vertical panel. Attribute table can be accessed by right clicking on the file name and also shown in Fig. 11.8. These errors in digitization process can be corrected by setting the snapping options and the parameters may be fixed before starting the digitization process. Otherwise, it can be accessed from settings menu as depicted in Fig. 11.8. The tolerance limit in terms of map unit or pixel number can be set in snapping options which work on vertex and segments as shown in mode column. Also please check whether the file is selected on which snapping options will work. If the overlap or sliver area lies within the tolerance limit as specified in the snapping options, the errors are automatically corrected. Otherwise, if the overlap or sliver is large, it can be corrected by moving the nodes or by creating new nodes as discussed earlier. Once all errors are corrected, the file should be saved before deselecting the 'Toggle editing' symbol.

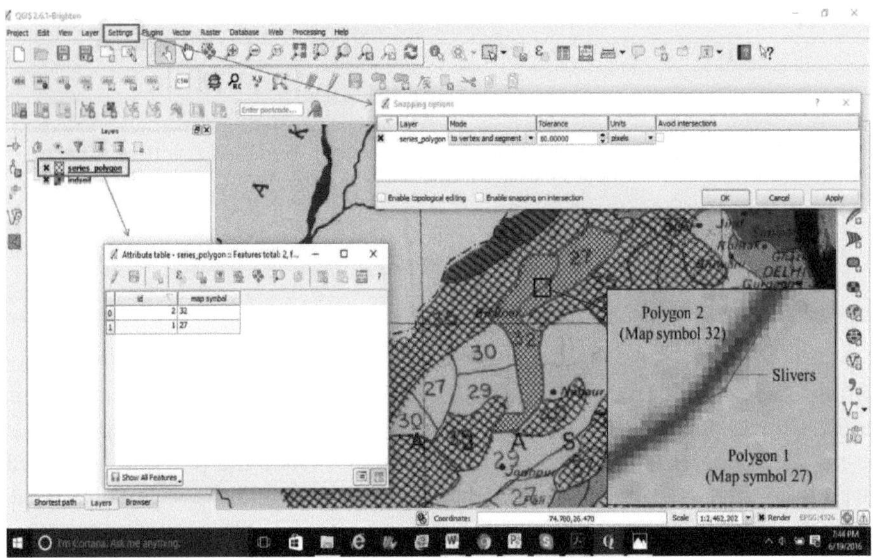

Fig. 11.8: Corrections of digitized polygons in QGIS

Creation of a point vector data

Point vector file or shapefile can be created from the field data collected through field survey. For example, the data given in Appendix-1 corresponds to legacy soil data of arid western India covering Rajasthan and Gujarat state. The geographical coordinates of each soil profile location representing a soil series is given along with the series name as primary key or identifier. At the first step, a comma separated value (CSV) formatted file containing minimum three columns (id, long and lat) should be prepared, which can be created by 'save as' tool in excel spreadsheet and selecting the file type as *.csv type. In the second step, the *.csv file can be opened in QGIS by clicking on 'Add tab delimited layer' as depicted in right vertical pane. After selecting the *.csv file, x-field and y-field of the data needs to be specified, which is generally the longitude and latitude of the sampling locations, respectively. The next step is defining the coordinate reference system of the data. In the example dataset it is geographical coordinates (lat/long), which is referred as 'WGS84: EPSG 4326' in QGIS (Fig. 11.9). Once the CRS is defined, a temporary shapefile is created and the locations are displayed in QGIS canvas. This temporary shapefile needs to be saved as *.shp file.

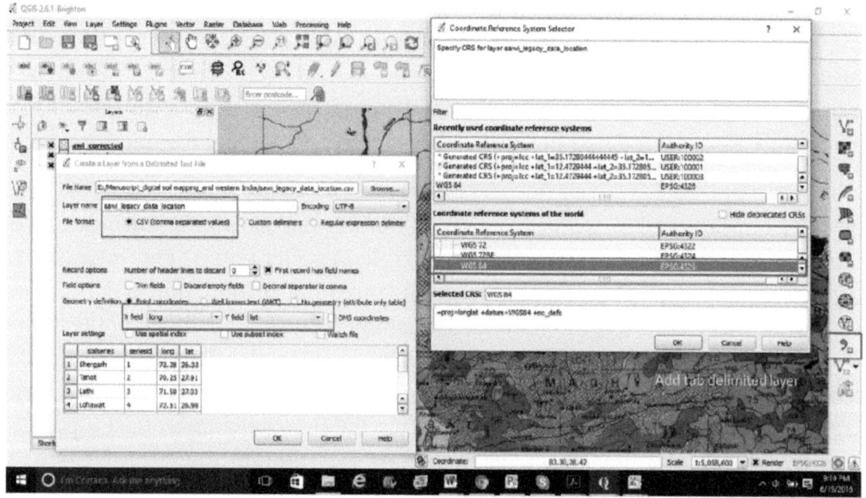

Fig. 11.9: Creating a point shapefile in QGIS from a spatial data

Conversion of coordinate reference system

Often the geographical coordinates of sampling locations are collected in Geo (lat/long) format. However, for geostatistical analysis, it is always better to use metre coordinate system for better interpretation of spatial variation of soil properties, especially the range parameter of semivariogram. Therefore, the degree coordinates system may be converted to a coordinate system in which meter is used as unit for defining distance e.g. Universal Transverse Mercator (UTM), Lambert Conformal Conic (LCC) projection etc. In QGIS, the coordinates system of a shapefile can be changed to a different system by saving as the file with a

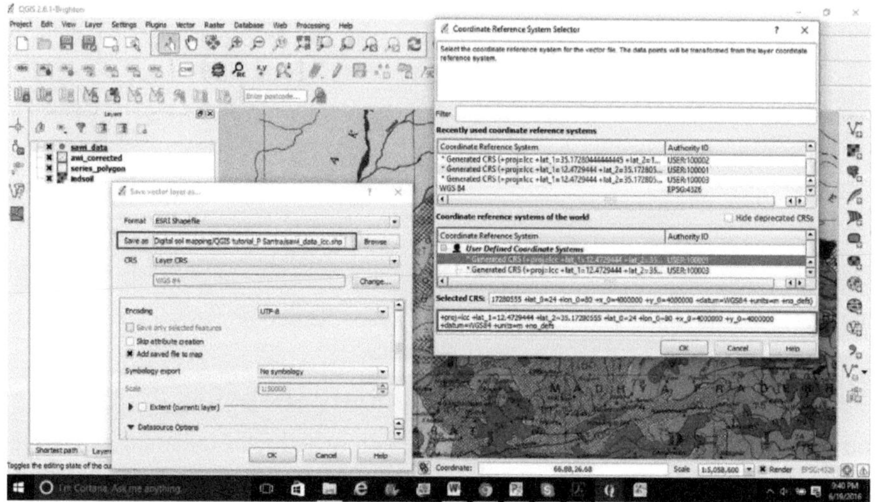

Fig. 11.10: Conversion of coordinate reference system of shapefile in QGIS

new shapefile and then changing the CRS as shown in Fig. 11.10. The coordinates can also be straightway changed; however, the original coordinate system will be lost in this case. In the example given in Fig. 11.10, CRS is changed from WGS 84 to LCC with a specific user defined CRS. Otherwise, EPSG definition for a number of CRSs is available in QGIS, from which a suitable CRS may be selected as a new CRS of the shapefile.

The next important step is the creation of easting and northing attributes of the shapefile in which coordinates will be available in meter unit. If the shapefile was originally created with geographic latitude and longitude information as coordinates, the spatial attributes available with the shapefile will be in decimal degree coordinates. The new attributes are created by clicking the field calculator tab in attribute table dialog box, which can be accessed by right clicking on the target shapefile and example is given in Fig. 11.11. In the field calculator, the output field name is defined and then from the function list, suitable function needs to be chosen. For example, if 'easting' is defined as the new field name or attribute name, $x may be selected from geometry tools to convert decimal degree in meter unit and automatically will be save as an additional attribute in the list of attribute table. Similarly, 'northing' attribute can be created by choosing $y in geometry tools. Other than these two distance attributes, other features of a shapefile can also be created e.g. area, perimeter etc.

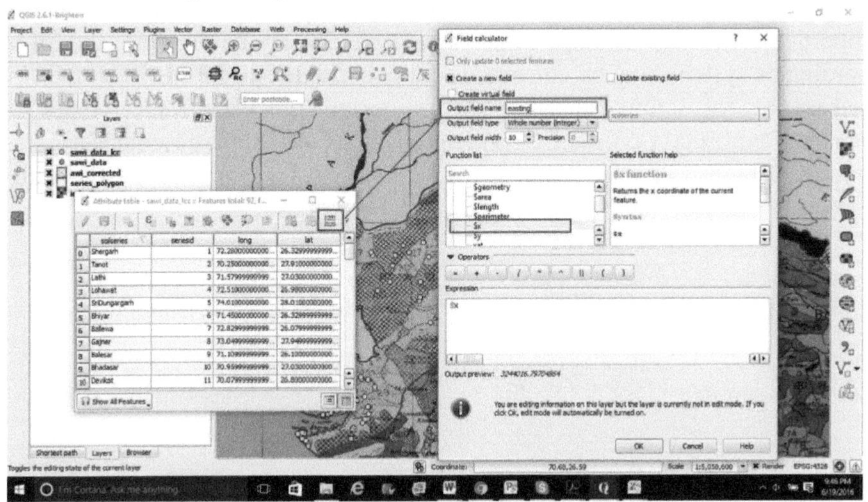

Fig. 11.11: Creation of additional attributes of shapefile defining meter coordinates of point shapefile

Creating mask raster layer from boundary shapefile

The raster grid of the study area is an important input in digital soil mapping through kriging. It is also called as mask grid where kriging predictions are to be made. Raster grid or mask layer of the study area can be prepared in QGIS from its boundary shape file. In Fig. 11.12, an example of preparing raster grid from boundary

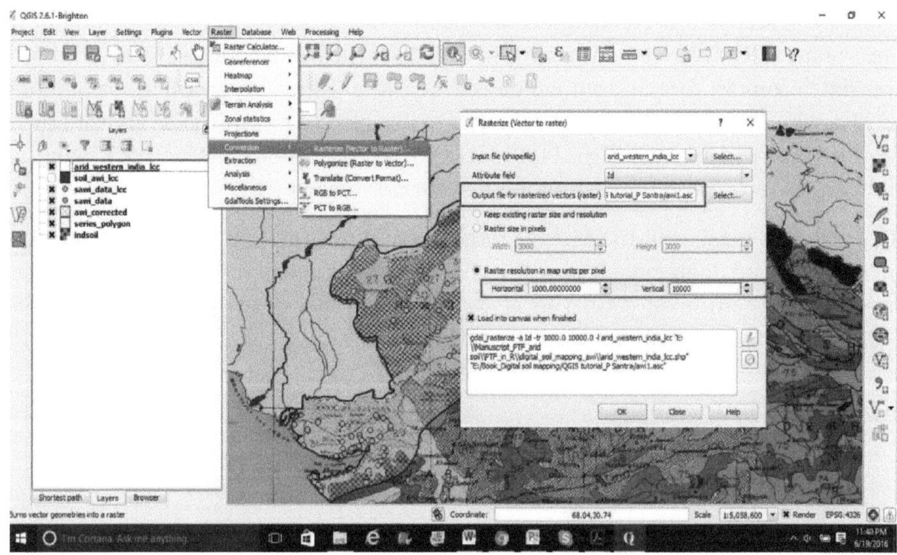

Fig. 11.12: Conversion of boundary shapefile in a raster format using QGIS

shapefile of arid western India is shown. In the first step, the shapefile of the study area are converted to a raster file using conversion tool available in raster menu of QGIS. By selecting the option Rasterize (vector to raster) of conversion tool, the input file, output file and cell resolution of the raster file are defined. In the example given below, the cell resolution is chosen as 1000 m × 1000 m. It indicates that each pixel of the prediction raster grid or mask layer is 100 ha or 1 km^2.

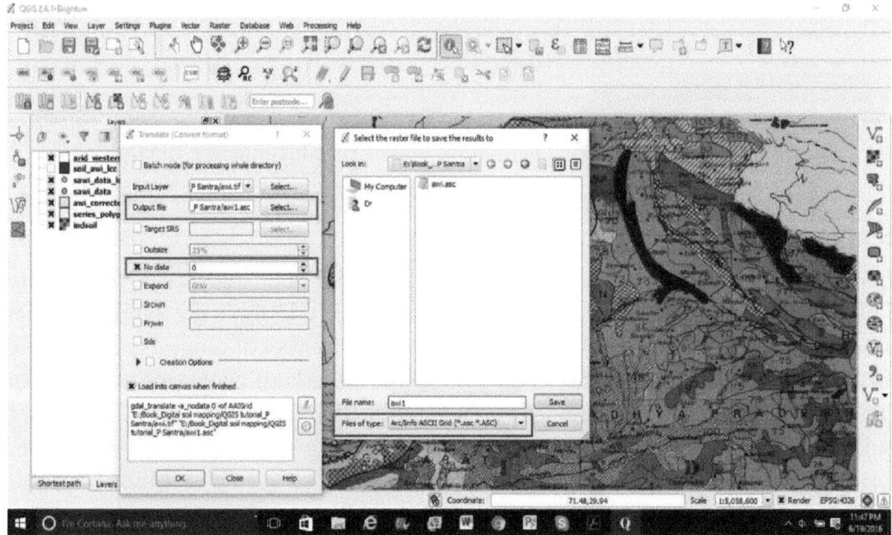

Fig. 11.13: Translation of raster grid file to an ascii grid file using QGIS

Once the raster grid is created, it is to be converted to ascii file to be handled in R environment. Conversion of raster grid to an asci file can be done in QGIS through translate (convert format) tool which is accessible from conversion tool available in raster menu. Example of converting a raster grid (e.g. awi.tif) to an ascii grid (e.g. awi.asc) is given in Fig. 11.13. It is to be noted here that no data value is defined as zero while translating the raster grid to ascii grid.

To create the mask grid file in *.txt format, ascii grid file (e.g. awi.asc) is opened in notepad and then saved as a text file (e.g. awi_mask.txt) (Fig. 11.14). The header information in *.txt file is written in a specified format as given in Fig. 11.14 so as to be handled in R environment.

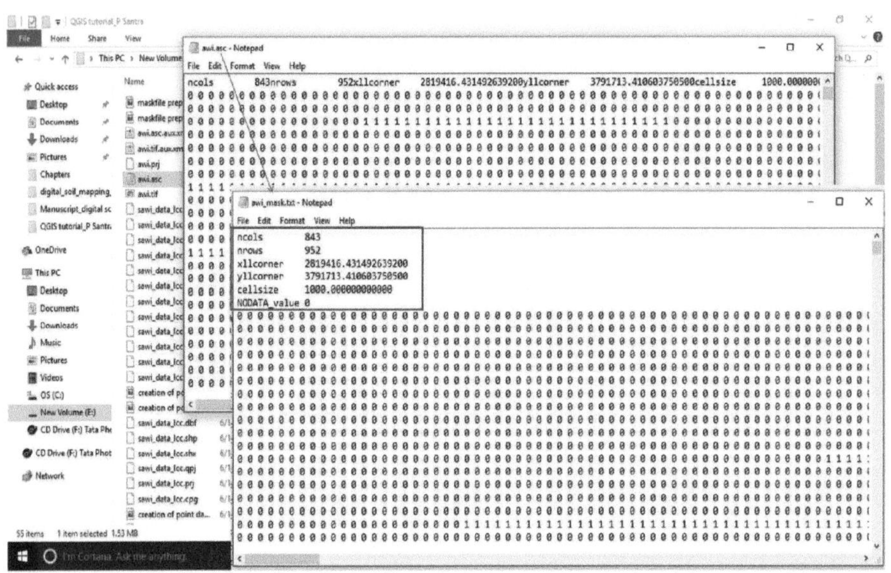

Fig. 11.14: Preparation of a prediction mask raster layer in text file format

Summary

QGIS is an open source GIS environment for spatial data analysis. It can handle a variety of spatial data format and most of the basic GIS tools are available in its base installation. Apart from the base installation, a number of plugins are available for specific and advanced spatial data analysis. Details of each tool available in GIS environment can be found in GIS training manual, user manual and user guides. In this chapter, few basic GIS analysis e.g. digitization of scanned map in GIS format, creation of shapefile, coordinate reference system definition, vector to raster conversion etc. have been discussed with example study. However, a large number of tools are available in QGIS, which can be fully utilized for different GIS analysis.

References

QGIS training manual (https://docs.qgis.org/2.2/en/docs/training_manual/)
QGIS user manual (http://docs.qgis.org/2.6/en/docs/user_manual/)
QGIS user guide (http://docs.qgis.org/2.8/pdf/en/QGIS-2.8-UserGuide-en.pdf).

12

Processing Covariate Data for Their Use and Interpretation in GIS Environment

Pratap Chandra Moharana

ICAR-Central Arid Zone Research Institute, Jodhpur, Rajasthan, India

Introduction

GIS (Geographical Information System) based mapping takes cogence of three major geometrical dimensions, point, line and polygons and in GIS these three features correlate their relationship through a concept called Topology. Among the various applications where covariate processing requires specific attention, the landuse/land cover and Digital Elevation Models/surfaces or terrains are the major area of interest. Use and processing of data for these covariates carries importance as accuracy of both these aspects would influence the derivatives extracted out of them further. For example, if landuse is improperly defined, the site for which planning is proposed may not sustain, similarly, point data/elevation data in case of DEMs would decide the exact topography of any landforms for their further utilization. For their use in GIS, while DEMs depend upon point data, the landuse/land cover will be taken care by polygons. For the purpose of the study, we are going to discuss about these two covariates in the following paragraphs.

Generally, Covariate (covariable) (koh-vair-iãt) in statistics is defined as a continuous variable that is not part of the main experimental manipulation but has an effect on the dependent variable. In other words, as per ANCOVA (Analysis of covariance) definition: a covariate is a control variable.

Processing covariates like landuse/land cover units

Generally, land cover refers to the surface cover on the ground like vegetation, urban infrastructure, water, bare soil while land use refers to activity or the purpose the land serves, for example, agriculture, recreation etc. An ideal classification of land use and land cover units seemed difficult because of its broad pattern. However, Andersons (1976) came out with a classification system which is being frequently used by many. This classification system consisted of Level -1 and

Level - II units which can be easily interpreted from remote sensing images. In India, NRSC (National Remote Sensing Centre) at Hyderabad has come out with a LU/LC classification system amenable for 1:50,000 scale map at country level. This system addresses 9 classes at Level-1, 35 classes at Level-2 and 79 classes at Level-3. Landuse / land cover units are polygon based features in GIS environment. Such units can be delineated through digitization from the standard FCCs or using remote sensing techniques. In the former case, it depends entirely upon the interpreter's sole wisdom and field experience. The drawback of this technique is that the image cannot be expanded (zoom) or reduced in case of any doubt. However, the accuracy is good enough. Accuracy of mapping depends upon the scale × 0.25, for example if the scale is 1: 50k, the accuracy should be 50000 × 0.25 mm = 12.5m. Minimum Mapping Unit would also depend upon users choice but calculated on the basis of scale (for 50,000 map it is 2.25 ha). With the advent of high resolution satellite images (LISS-III 23.5 m, LISS-IV, 5.8 m of IRS series) and revisit period ranging from 5 days to 24 days and swath ranging from 70 km to 800 km, digital interpretation and assessment has become easier and accurate (NRSC 2010). Then there is a system of calculating the accuracy.

Remote sensing approach

In remote sensing techniques, following two basic classification systems are used for landuse / land cover assessment and mapping.

Supervised classification

- Supervised classification uses the spectral signatures obtained from training samples to classify an image.
- The user has a background or prior knowledge of local level landuse / land cover types to be mapped.

Unsupervised classification

- Unsupervised classification finds spectral classes (or clusters) in a multiband image without the analyst's intervention.
- Computer can generate its own samples and the technique uses iso-data classifier

While processing data for supervised or unsupervised classification system, a basic knowledge of classification system requires to be remembered. For example according to many well adopted classification system there are three levels of landuse/land cover units (Level-I, -II and -III). All these levels correspond to scale and mapping units. For example, for a first order (Level-I, classification system such as croplands, even a very coarse resolution satellite data like AWiFs can be used while for Level III, like paddy field we need a high resolution image like LISS-4 (MX) image. The problem that user generally faces is the aspect of tonal mix ups, for example, for sandy area of western Rajasthan, there will be similar tone for features like waterbody, sand ridges or hill ridges. Therefore, while considering unsupervised classification, the analyst would first prepare a list of features indicating similar tone, then follow a systematic way to reclassification of each unit or group of units for a systematic separation of each unit. It is therefore, advisable to go for a bigger number of classes (~40) at the beginning of unsupervised classification system.

Vegetation index

Over the last 20 years, coarse resolution satellite sensors are being used routinely to monitor vegetation and detect the impact of moisture stress on vegetation, which can be summarized under, (1) AVHRR on NOAA's polar orbiting satellites, (2) MODIS (since 2000) provides improved radiometric, geometric, and spatial characteristics, (3) VIIRS (aboard NPP and NPOESS) – next generation operational instrument for monitoring.

1. Advanced Very High Resolution Radiometer (AVHRR) on board the National Oceanic and Atmospheric Administration's (NOAA'S) polar-orbiting satellites are the instrument of choice for collecting coarse-resolution imagery worldwide due to its twice-daily coverage and synoptic view.
2. MODIS data: In December 1999, NASA launched the Terra spacecraft, the flagship in the agency's Earth Observing System (EOS) program. Aboard Terra flies a sensor called the Moderate-resolution Imaging Spectroradiometer, or MODIS, that greatly improves scientists' ability to measure plant growth on a global scale. Briefly, MODIS provides much higher spatial resolution (up to 250-meter resolution), while also matching AVHRR's almost-daily global cover and exceeding its spectral resolution.
3. Enhanced Vegetation Index (EVI): The MODIS science team also worked on new data product–called the Enhanced Vegetation Index (EVI) that will improve upon the quality of the NDVI product. The EVI will take full advantage of MODIS' new, state-of-the-art measurement capabilities. While the EVI is calculated similarly to NDVI, it corrects for some distortions in the reflected light caused by the particles in the air as well as the ground cover below the vegetation.

The Normalized Difference Vegetation Index (NDVI) derived from AVHRR data, has been extensively used for vegetation monitoring, crop yield assessment, and drought detection. Researchers use this index to extract vegetation abundance from remotely sensed data (Tucker 1979). The NDVI is calculated as (NIR - Red)/(NIR + Red), where NIR is the reflectance radiated in the near-infrared waveband and Red is the reflectance radiated in the visible red waveband of the satellite radiometer. The result is a single band dataset with values remaining within -1 and +1. Higher NDVI (towards +1) indicates a greater level of photosynthetic activity. It has been demonstrated that multi-temporal NDVI derived from AVHRR data is useful for monitoring vegetation dynamics on a regional and continental scale. NDVI products are also generated from the data of following satellite sensor systems; MODIS NDVI of 250 m and 1000 m, SPOT VGT NDVI of 1000 m. NOAA AVHRR NDVI of 1000 m, IRS WiFS NDVI of 188 m and AWiFS NDVI of 56 m are widely used for drought monitoring (NRSC 2010). It can be used to prepare a number of maps like weekly to seasonal vegetation maps as well as monitoring the vegetation cover / density of any region.

Spatial distribution of NDVI over Jaisalmer District in western Rajasthan during drought and normal years: a case study

This study was carried out using NASA MODIS CMG Monthly NDVI (Global) data, processed by NASA Goddard from the Terra sensor projected on a 0.05 degree (5.5 km resolution). IRS - AWiFs data were used as secondary data for checking vegetated area. The data has been downloaded for the years, 2000 to 2012 and for monsoon period (July, August and September) for observation. For generation of yearly average scenes, we used raster calculator (July+August+ September/3) for each year. The maximum and minimum range of NDVI values fell within a range of 0 to 0.65 which were further classified into three categories (0.65/3) as 0-0.2 (Low), 0.2-0.4 (Moderate) and 0.4-0.7 (High) using ArcGIS spatial analysis tool. Area under high category was only experienced during 2003 (4.8% area) and in 2001 and 2006 with <1% area under high categories. In the moderate situation, during not so good rainfall years, the areas remained at par with 10-12% while in the low categories, the percent area varied from 89 to 100 (2002), except in 2003 (36%) indicating very bad situation of vegetation (Moharana et al., 2013).

Processing Digital Elevation Models (DEMs/DSMs/DTMs)

In general, there are few fundamental elevation representations

- DEM (Digital Elevation Model) stands for topographical representation of a surface or relief or the elevation of the ground (bald earth). Here, surfaces involve a third 'z' dimension (height/elevation/magnitude, quantity) in addition to (x,y). Continuous data as a surface (ground elevation, barometric pressure, rainfall, crop yield, noise levels, population density)
- DSM (Digital Surface Models): The elevation defined by the earth and the things upon it, including buildings, the canopy of trees, bridges, and so on (vegetation surface, land use / land cover etc.) (Fig. 12.1)
- DTM: It is a hybrid model (DSM draped over DEM). The digital terrain model refer to DEM data stored and modelled directly from points.
- A fourth representation is a hydrographically enforced DEM.

Elevation models can be either created or can be downloaded from authentic websites, for examples from BHUVAN site of NRSC (bhuvan.nrsc.gov.in). Using ARC-INFO's TIN modules one can create DEMs. Such models can also be used to simulate and extract hydrological parameters like flow direction, flow accumulation, stream network with stream orders. Similarly, many such studies have been made for morphological assessment of catchment sites. Moharana and Kar (2002) used TIN modules and prepared a DEM under ARC/INFO GIS environment for watershed delineation and reconstruction of flow paths of streams in the sandy areas in Jodhpur district of Rajasthan.

However before starting any project on DEMs / DSMs we need to ask some basic questions

- What sensor do you propose to use for acquisition of source data to be used for DEM production? Cartosat-1
- What data source is being used? DEMs produced from the Digital Surface Model (DSM) or DEMs from the Digital Terrain Model (DTM) or a combination or other process?
- What is the model you intend to use? Stereo pairs (aerial photographs)
- What is the vertical accuracy of the model?
- What the type of surface being interpolated? GRID / TIN etc.

Now a days, a wide but useful elevation models are available with standard software's (Table 12.1). In the following table, the author has listed some of them for ready reference of the readers.

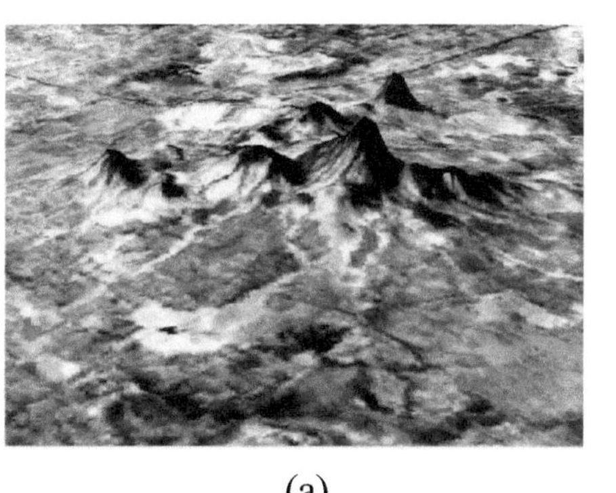

(a)

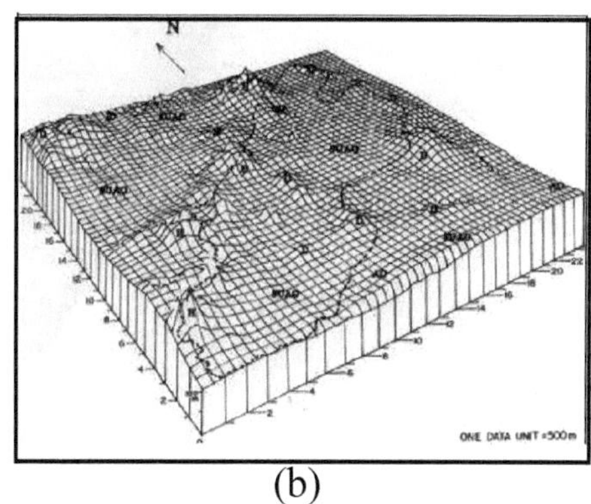

(b)

Fig. 12.1: (a) & (b) Raster and Vector representation of surfaces

(a) A Digital Surface Model of Agolai terrain of Jodhpur district using IRS-LISS-III image draped over SRTM DEM, *Source*: Author

(b) A DEM of Agolai area, generated in Arc/Info GIS using elevation data available in Survey of India toposheets (Moharana and Kar 2002) *Source*: Survey of India toposheets (Moharana and Kar 2002)

Table 12.1: Types and source of Elevation data models

Types of Elevation data available Data	Description
GTOPO	GTOPO is a global elevation dataset with resolution of 30 arc seconds (approximately 1 km), available for download at http://www1.gsi.go.jp/geowww/globalmap-gsi/gtopo30/gtopo30.html
SRTM	The Shuttle Radar Topography Mission (SRTM) is elevation data on a near-global scale, acquired from the Space Shuttle, to generate the mostcomplete high-resolution digital topographic database of Earth. http://srtm.usgs.gov/index.php
NED 10, NED 30	The National Elevation Dataset (NED) was created by the USGS for the USA. NED data is available nationally at resolutions of 1 arc-second (about 30 meters, "NED 30") and 1/3 arc-second (about 10 meters, "NED 10") : Source : http://ned.usgs.gov/
Lidar	The lidar data may come from a variety of sources and can be used to provide both a DTM and DSM. LIDAR, which stands for Light Detection and Ranging, is a remote sensing method that uses light in the form of a pulsed laser to measure ranges (variable distances) to the Earth. These light pulses, combined with other data recorded by the airborne system generate precise, three-dimensional information about the shape of the Earth and its surface characteristics.

Which Data Format ?

- ESRI : TIFF 32-bit float format, using LZW compression. This format is the easiest to use and maintain, while providing the best performance.
- ESRI GRID : Older format that is no longer recommended for storing elevation data. The data manager should consider converting to TIFF to improve performance.
- FLT : floating-point binary simple format, similar to 32-bit floating-point TIFF files , recommended only for small extents.
- ASCII DEM : This is a plain ASCII data file that may be a regular raster structure or irregularly gridded data. In the latter case, the file explicitly lists x,y,z values. It is inefficient for storage, reading, and writing, but it is a universal storage format. It is highly recommended that this data be converted to TIFF to improve performance.
- IMG from ERDAS: Elevation data may be stored in the IMG format, which is supported by ArcGIS.
- BAG (Bathymetry Attributed Grid): This format is used for bathymetric data, and is partially supported in ArcGIS 10. The software properly reads the raster elevation data, but does not completely support all format components.

- DTED (Digital Terrain Elevation Data): This is a format specification with specific aspects about resolution and accuracy of elevation data, defined by the NGA (National Geospatial Intelligence Agency). The DTED format data will generally perform adequately; therefore, conversion is not required.
- ESRI's terrain dataset: A terrain dataset having multi-resolution, TIN-based surface built from measurements stored as features in a geodatabase. They are typically made from lidar, sonar, and photogrammetric sources. They must be converted to a raster dataset, the TIFF is recommended.
- HRE (High Resolution Elevation: This is a relatively new format for storing high resolution elevation data. It is intended for a wide variety of National Geospatial-Intelligence Agency (NGA) and National System for Geospatial Intelligence (NSG) partners and members, and customers external to the NSG, to access and exploit standardized data products.
- LAS format LIDAR data:This format supports three-dimensional point cloud data and is designed by ASPRS (American Society of Photogrammetry and Remote Sensing). It can be supported directly by a mosaic dataset, or by creating a LAS (The LAS file format is a public file format for the interchange of 3-dimensional point cloud data between data users). Although developed primarily for exchange of lidar point cloud data, this format supports the exchange of any 3-dimensional x,y, tuplet) dataset (source:http://www.asprs.org/Committee-General/LASer-LAS-File-Format-Exchange ctivities.html).

Accuracy of data

While mapping from remotely sensed data, specially using manual interpretation, no two interpretation may look same. Therefore, there are admissible errors which would influence accuracy of the data. There are few means of expressing classification accuracy; one of them is the preparation of a classification error matrix also called a confusion matrix. In this table/matrix, a comparison is made on a category-by- category basis, the relationship between known data (field truth) and corresponding results of an automated classification. The overall accuracy is computed by dividing the total number of correctly classified to total number of reference pixels.

Another most common system is the find out the kappa value, It is a GIS based system where, a classification Table 12.2 is created from raw data in the spreadsheet, for two observers and inter-raster agreement statistic (Kappa) is calculated to evaluate the agreement between two classifications on ordinal or nominal scales. Alternatively, it's a measure of agreement that compares the observed agreement to agreement expected by chance if the observer ratings were independent. It also expresses the proportionate reduction in error generated by a classification process with the error of a completely random classification (Cohen 1960). Basically, accuracy is measured by the probability distribution that a value has from the true value. These can be circular error and linear error.

- Horizontal spatial accuracy: the circular error of a dataset's horizontal coordinates at a specified percentage level of confidence.
- Vertical spatial accuracy: the linear error of a dataset's vertical coordinate at a specified percentage of confidence, such as an elevation measurement.
- An accuracy of 90% confidence level means that 90% of positional accuracies will be equal to or smaller than the reported accuracy value.

Storing, converting and processing surface data

3-D surfaces are normally stored in one of two forms within ArcGIS as a GRID, which is ArcInfo's general raster format or as a TIN which is a vector format for surfaces. However, when one downloads data from the Internet, surface data may be in following formats, such as

- DEM format, as originally developed by USGS.
- SDTS (Spatial Data Transfer Standard) format, which is an FGDC (Federal Geographic Data Committee) standard.
- E00 which is ESRI's text formatted for distributing coverage and GRIDS
- Points and breaklines.

All these data needs to be converted to GRID or TIN, generally required for display or analysis within the ArcGIS system (Arc Toolbox). Other data formats are,

- Contour lines: stored as vector lines in coverage, shape file, or geodatabase, but these can only be used for map display but not analysis, so this is not a recommended format for surface storage.

Basic methods representing the surface or digital elevation model (DEM)

- GRID : uniformly spaced, 3-dimensional cartographic representation (x, y, z) in a grid or raster format.
- Lattice: each point represents a value on the surface only at the centre of the grid cell. Surface grid considers each sample as a square/rectangular cell with a constant surface value.
- Triangulated Irregular Network (TIN): Formulation of non-overlapping triangles from irregularly spaced x, y, and z points (vector-based) and topological relationship between the triangles and their adjacent neighbours.
- Contour lines: lines of equal elevation drawn at a given interval (e.g. every 10, 20, 100 m).
- Viewing & Processing Surface Data in Arc-GIS software requires additional procurement of some extensions like 3-D Analyst or Spatial Analyst extensions.

Table 12.2. A comparison of most commonly used Elevation formats

Comparing the surface representations TIN	GRID	Contouring
Advantages Can capture significant slope features (ridges) Efficient since require few triangles in flat areas.Easy for certain analyses: slope, aspect, volume	***Advantages*** Simple conceptual model Easy to relate to other raster data. Irregularly spaced set of points can be converted to regular spacing by interpolation.	***Advantages*** Presents a quick mental picture of surface (Close lines indicate steep slope)
Disadvantages Analysis involving comparison with other layers difficult.	***Disadvantages*** Does not conform to variability of the terrain.Linear features not well represented.	***Disadvantages*** Must convert to raster or TIN for analysis. Contour generation from point data requires sophisticated interpolation routines *(Surfer* from Golden Software, Inc., or ArcView Spatial Analyst extension)

Role of interpolation in the data analysis

Interpolation

Interpolation is the process of estimating unknown values that fall between known values. The user working on TIN or GID based models uses value points (attributes, say elevation /rainfall etc). Sometimes, values may not be available for all places. For example, for western parts of Rajasthan, rainfall data may not be available for many terrain conditions. In that case, the analyst would look for using interpolation. The primary assumption of spatial interpolation is that points near each other are more alike than those farther away; therefore, any location's values should be estimated based on the values of points nearby. The principle underlying spatial interpolation is the first law of geography. Formulated by Waldo Tobler, this law states that everything is related to everything else, but near things are more related than distant things. The formal property measures the degree to which near and distant things is related is spatial autocorrelation. Maps as representations of real world: Creating maps involves three major steps, (i) sampling (ii) measurement of variables and (iii) estimating values at un-sampled locations for which we need some form of interpolation.

Point Interpolation

- Lines to points (contours to elevation lines)

Areal Interpolation

- Given a set of data mapped on one set of source zones, determine the values of the data for a different set of target zones.

Principles of Interpolation

As described above, for interpolation, we look for a number of sample points, which may be adequate or may fall short of adequate data. Sometimes, even a lone sample point may represent the surface efficiently. Colin Childs (https://www.esri.com/news/arcuser/0704/files/interpolating.pdf), advocated that the spatial correlation is decided upon certain understanding like whether there is similarity of objects within an area, (ii) what is the level of interdependence between the variables and (iii) what can be the nature/strength of dependency?

In the following paragraphs we discuss utility of two commonly used interpolation techniques for processing point data.

Inverse Distance Weighting

Inverse Distance Weighting (IDW) is a type of deterministic method for multivariate interpolation with a known scattered set of points. It is based on extent of similarity of cell. The assigned values to unknown points are calculated with a weighted average of the values available at the known points. As the name suggests, it resorts to the inverse of the distance to each known point when assigning weight. IDW is a good interpolator for a phenomenon whose distribution is strongly correlated with distance. IDW does less well with phenomena whose distribution depends on more complex sets of variables. One potential advantage of IDW is that it gives you explicit control over the influence of distance; an advantage you don't have with Spline or Kriging. You can create a smoother surface by decreasing the power, increasing the number of sample points.

Kriging

Kriging is another popular method of interpolation technique but based on Geostastical interpolations. Here, the interpolated values are modeled by a Gaussian process governed by prior covariance, as opposed to a piecewise-polynomial spline chosen to optimize smoothness of the fitted values. This technique is more or less used for more advanced prediction of surfaces. Under suitable assumptions on the priors, Kriging gives the best linear unbiased prediction of the intermediate values. The method is widely used in the domain of spatial analysis and computer experiments. The technique is also known as Wiener–Kolmogorov prediction, after Norbert Wiener and Andrey Kolmogorov. Kriging is a weighted average technique, except that the weighting formula in Kriging uses much more sophisticated math measures distances between all possible pairs of sample points (that's right, all of them) and uses this information to model the spatial autocorrelation.

Kriging methods

Ordinary and Universal Kriging-Universal Kriging assumes that there is an overriding trend in the data. For example, in case of land salinity, there can be a possible reason, whether prevailing land use or use / over use of ground water etc. Such information may be existing or may be assumed. In case of Ordinary kriging, variables may not follow a trend of Mean value. An exercise was attempted to compare the distribution of micro-nutrient (Fe) in Jhunjhunun district of western Rajasthan using both techniques. Thematic maps (Fig. 12.2) and statistical outputs (Table 12.3 & 12.4) are presented below.

The surface created with Kriging exceeded the known value range, and does not pass through any of the sample points and with IDW, the surface did not exceed the value range. The extracted values based on Kriging (GIStable) were found to be more than IDW values.

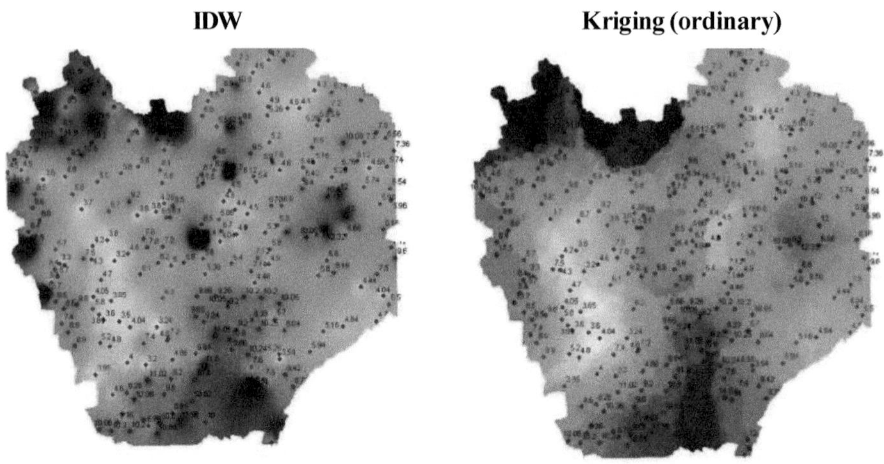

Figure 12.2: Points interpolated to surfaces showing distribution of Fe in Jhunjhunun district (*Source*: P.C. Moharana *et al.*, 2012)

Table 12.3: Values of Feas interpolated using IDW

X	Y	FE_IDW
75.364027777778	28.067027777778	4.601691
75.392166666667	28.027	6.098075
75.432027777778	28.048611111111	6.201128
75.461666666667	28.042777777778	5.024974
75.442083333333	27.979666666667	6.199246
75.490833333333	28.048055555556	6.803331
75.44325	28.080083333333	7.192841
75.4075	28.079194444444	7.799782
75.4	28.096833333333	7.78281
75.428944444444	28.147666666667	3.815891
75.451444444444	28.136861111111	6.517059
75.461166666667	28.137333333333	8.045119
75.467833333333	28.159833333333	8.473618
75.443777777778	28.204916666667	5.803406
75.439666666667	28.229666666667	6.101176
75.389611111111	28.235972222222	5.803
75.351222222222	28.22425	5.800511
75.296027777778	28.208055555556	5.605419
75.369138888889	28.169361111111	9.18266
75.387055555556	28.141833333333	3.826337
75.333333333333	28.054	3.211774
75.301166666667	28.016666666667	4.699948
75.279666666667	28.057611111111	4.331921
75.263888888889	28.0585	7.422687
75.283333333333	28.083333333333	4.210518
75.316666666667	28.094583333333	3.800967

Records (of 252)

Table 12.4: Values of Iron (Fe) as interpolated using kriging

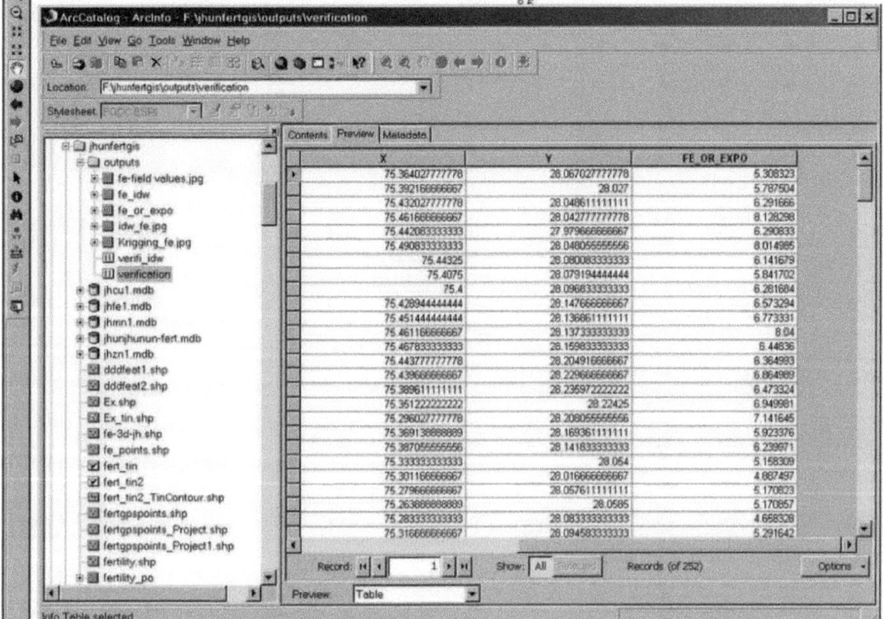

X	Y	FE_OR_EXPO
75.364027777778	28.067027777778	5.308323
75.392166666667	28.027	5.787504
75.432027777778	28.048611111111	6.291666
75.461666666667	28.042777777778	8.128298
75.442083333333	27.979666666667	6.290833
75.490833333333	28.048055555556	8.014985
75.44325	28.080083333333	6.141679
75.4075	28.079194444444	5.841702
75.4	28.096833333333	6.281684
75.428944444444	28.147666666667	6.573294
75.451444444444	28.136861111111	6.773331
75.461166666667	28.137333333333	8.04
75.467833333333	28.159833333333	6.44636
75.443777777778	28.204916666667	6.364993
75.439666666667	28.229666666667	6.864989
75.389611111111	28.235972222222	6.473324
75.351222222222	28.22425	6.949981
75.296027777778	28.208055555556	7.141645
75.369138888889	28.169361111111	5.923376
75.387055555556	28.141833333333	6.239971
75.333333333333	28.054	5.158309
75.301166666667	28.016666666667	4.887497
75.279666666667	28.057611111111	5.170823
75.263888888889	28.0585	5.170857
75.283333333333	28.083333333333	4.658328
75.316666666667	28.094583333333	5.291642

References

Anderson, J. R., Hardy, E.E., Roach, J. T. and Witmer, R.E. 1976. A Land Use and Land Cover Classification System for Use With Remote Sensor Data, Geological Survey Professional Paper 964.

Cohen, J. 1960. A coefficient of agreement for nominal scales. Educational and Psychological Measurement 20: 37-46

Moharana, P.C. and Kar, A. 2002. Watershed simulation in a sandy terrain of the Thar Desert using GIS. Journal of Arid Environments 51: 489-500.

Moharana, P.C., Soni, S. and Bhatt, R.K. 2013. NDVI Based Assessment of Desertification in Arid Part of Rajasthan in Reference to Regional Climate Variability. Indian Cartographer 33: 319-325.

Moharana, P.C, Kumar, M. and Kar, A. 2012. Methodology for soil fertility mapping using GIS. In 17th Annual Conference and National symposium on "Application of Clay sciences in Agriculture, Environment and Industry" held at NBSS & LUP, Kolkata organized by Indian Society of Clay Mineral Research.

NRSC. 2010. Remote sensing applications (Eds. P.S.Roy, R.S.Dwivedi, D.Vijayan). National Remote Sensing Centre, Hyderabad. 377p.

Tucker, C.J. 1979. Red photographic infrared linear combinations for monitoring vegetation. Remote Sensing Environment 8: 127-150.

13

Accuracy and Uncertainty Analysis of Spatial Prediction

Priyabrata Santra

ICAR-Central Arid Zone Research Institute, Jodhpur, Rajasthan, India

Introduction

Accuracy and uncertainty analysis is an important step in preparation of surface map through geostatistical techniques or digital soil mapping approach. It is because that we rely on a map based on how much accurate it is and the degree of uncertainty associated with the estimates while the map was prepared. Any stakeholder for a soil map especially policy makers is interested in accuracy and uncertainty of the map before taking suitable land management decisions involving soil map as an input. Again the level of accuracy and uncertainty depends on the properties for which the map has been developed. If the property of interest is sensitive to environment or human consumption, then the accuracy level should be very high while the uncertainty level should be very low e.g. heavy metal concentration in soil, quality parameters of water etc.

Evaluation of accuracy

Accuracy of a soil map is generally evaluated through validation approach by which independently observed data are compared with their estimated values through digital soil mapping approach.The independently observed data are often called as the validation data whereas the dataset used in preparing digital soil map through modeling approach is called the training data. In case, independent observed data are not available, the dataset is divided into several subsets and then cross-validated to judge the accuracy. There are broadly two types of cross-validation approach followed in digital soil mapping methodology. In the first category, a dataset is first randomly partitioned into *k* subsets and then one subset is kept aside as validation data while (*k*-1) subsets are used to prepare soil map through digital soil mapping approach, which was further tested on the validation set. This validation step is repeated k times (or k-fold) so that each of k subsets get a chance to be a validation data once. At the last, the outputs from several steps are averaged to calculate the accuracy level of a map. The first approach is called as the k-fold cross-validation approach and a value of k as 10 is most commonly used. When the value of k is taken as number of samples in the dataset the cross-validation is called the leave-out-one cross validation approach. In both the approaches, when the developed model based on training data is applied on a validation data, a pair of observed and estimated values is created for each sample and the difference between them is calculated which is also called as error. Smaller the value of error larger is the accuracy of the map.

Different statistics are used to quantify the error or to judge the accuracy. Following three indices are commonly used: mean error (ME), root mean squared residual (RMSR) and mean squared deviation ration (MSDR).

$$\text{ME} = \frac{1}{n}\sum_{i=1}^{n}\left[Z(x_i) - \hat{Z}(x_i)\right] \tag{1}$$

$$\text{RMSR} = \sqrt{\frac{1}{n}\sum_{i=1}^{n}\left[Z(x_i) - \hat{Z}(x_i)\right]^2} \tag{2}$$

$$\text{MSDR} = \frac{1}{n}\sum_{i=1}^{n}\frac{\left[Z(x_i) - \hat{Z}(x_i)\right]^2}{\hat{\sigma}^2(x_i)} \tag{3}$$

Where $Z(x_i)$ is the observed values of the variable at the location x_i, $\hat{Z}(x_i)$ is the predicted values with variance σ^2 at the location x_i, and n is the number of sampling location. The ME and RMSR estimates the accuracy of prediction (e.g., larger ME and RMSR values indicate less accuracy of prediction). The MSDR measures the goodness of fit of the theoretical estimate of error (Bishop and Lark 2008). If the correct semivariogram model is used, the MSDR values should be close to 1 (Lark 2000).

Apart from the above mentioned three indices, the G measure gives an indication of how effective a prediction might be, relative to that which could have been derived from using the sample mean alone

$$G = \left(1 - \frac{\sum_{i=1}^{N}\left[z(x_i) - \hat{z}(x_i)\right]^2}{\sum_{i=1}^{N}\left[z(x_i) - \bar{z}\right]^2}\right) \times 100 \tag{4}$$

where $\overline{z}$ is the sample mean. If G is equal to 100, it indicates perfect prediction while negative values indicate that the predictions are less reliable than using sample mean as the predictions.

Recently, Lin's concordance correlation coefficient (LCCC) has been used to evaluate the agreement of observed values with predicted values. Calculation of LCCC is based on the orthogonal distance from the 1:1 line of observed and predicted values passing through origin. The LCCC is calculated as follows:

$$LCCC = \frac{2S_{obs.pred}}{(\overline{z}_{obs} - \overline{z}_{pred}) + S_{obs}^2 + S_{pred}^2} \tag{5}$$

$$\overline{z}_{obs} = \frac{1}{n}\sum_{i=1}^{n} z(x_i) \tag{6}$$

$$\overline{z}_{pred} = \frac{1}{n}\sum_{i=1}^{n} \hat{z}(x_i) \tag{7}$$

$$S_{obs}^2 = \frac{1}{n}\sum_{i=1}^{n}\left[z(x_i) - \overline{z}_{obs}\right]^2 \tag{8}$$

$$S_{pred}^2 = \frac{1}{n}\sum_{i=1}^{n}\left[\hat{z}(x_i) - \overline{z}_{pred}\right]^2 \tag{9}$$

$$S_{obs.pred} = \frac{1}{n}\sum_{i=1}^{n}\left[z(x_i) - \overline{z}_{obs}\right]\left[\hat{z}(x_i) - \overline{z}_{pred}\right] \tag{10}$$

Value of LCCC may vary from -1 to +1. A LCCC value equals to 1 indicate perfect positive agreement between observed and predicted values whereas -1 indicate perfect negative agreement. A LCCC value equals to zero indicates that there is no agreement between observed and predicted values.

In the following example, observed and predicted values of SOC, pH and EC content in a horticultural orchard from ICAR-Central Arid Zone Research Institute is shown (Fig. 13.1). These predictions were made using k-fold cross-validation approach using k = 10. ME was found positive for SOC whereas they were negative for pH and EC. It indicates that predicted SOC values are slightly lower than observed values specifically higher SOC contents (Fig. 13.1a). In case of pH, prediction was found slightly higher than observed values in low pH range (7-8), resulting into positive ME. Similarly, in case of EC, predicted values are slightly higher than the observed values for log [EC (mS cm^{-1})] >-1.

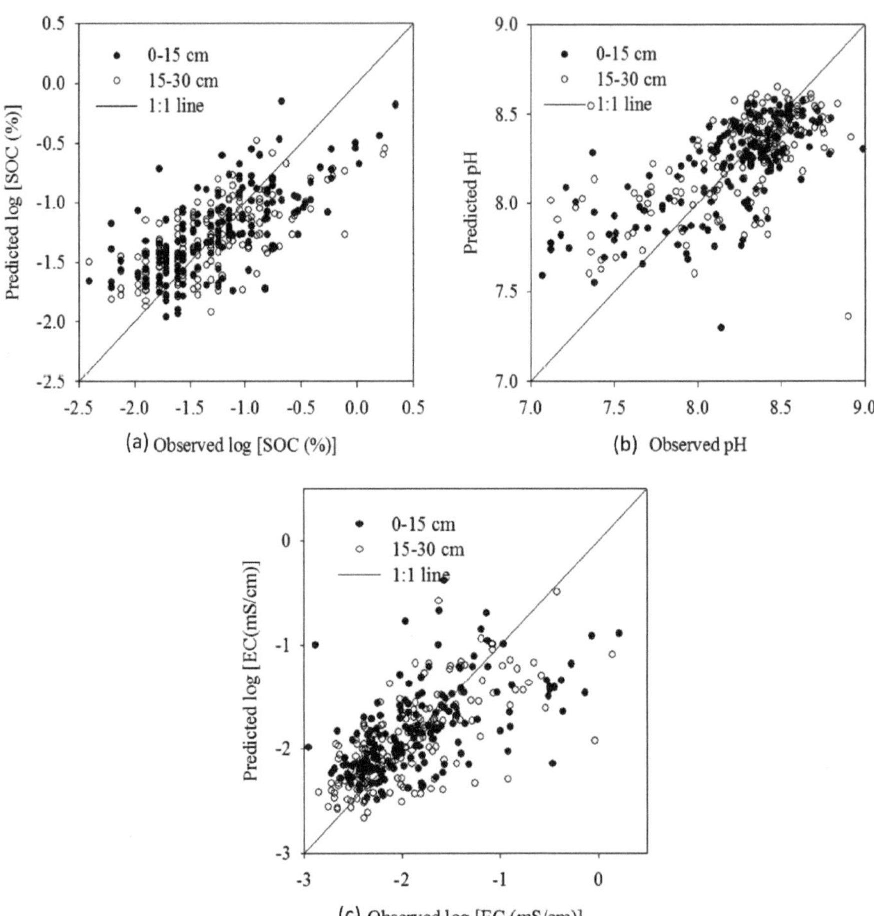

Fig. 13.1: Observed and predicted values of SOC content, pH and electrical conductivity; (a) log [SOC (%)], (b) pH and (c) log [EC (mS cm^{-1})]

RMSR of log [SOC(%)] prediction was 0.37 and 0.35, respectively for 0-15 cm and 15-30 cm soil layer. In case of pH, RMSR values were 0.27 and 0.30, respectively for 0-15 cm and 15-30 cm soil layer, whereas for log[EC (mS cm^{-1})], these were 0.49 and 0.44, respectively. The MSDR values of log[SOC (%)] prediction were 1.04 and 0.95 for surface and subsurface soil layer, respectively, which were very close to 1 and indicated that fitted semivariogram model parameters were very good to represent the spatial variation of SOC content in field. In case of pH, MSDR values were 1.09 and 1.23 for surface and subsurface soil layer, respectively. However, the deviation of MSDR from 1 was slightly higher in case of log[EC (mS cm^{-1})] prediction and were found 1.34 and 1.29, respectively. It indicates that the variation in predicted values of log[EC] were slightly higher than the variation observed in the measured values.

Evaluation of uncertainty of prediction

Apart from accuracy, the uncertainty of digital soil map is also very important and helps in judging the reliability of the developed map. The uncertainty can be estimated based on the objective of the study. If the target is to assess the uncertainty at a particular location of a region, it can be done by local uncertainty assessment. Otherwise, if the target is to quantify uncertainty simultaneously at many locations within the study area, the spatial uncertainty is to be assessed.

The probabilistic approach of local uncertainty assessment can be achieved by plotting conditional cumulative density function (ccdf).

$$F(u;z/(n)) = \Pr ob\{Z(u) \leq z/(n)\} \tag{11}$$

Where the notation "/(n)" refers to the conditioning of local information, for example, the number of neighborhood points. The ccdf fully models the uncertainty at u since it gives a probability that the unknown is no greater than any given threshold z. Uncertainty assessment can be done through ccdf using two approaches: parametric approach (assuming multi-Gausssian distribution model of random function Z(u)) and non-parametric approach (indicator or disjunctive kriging). In the parametric approach, the ccdf at any location u is Gaussian and fully characterized by its mean and variance, which correspond to the simple kriging estimate and variance at u. In the non-parameteric approach, each observation in a dataset is transformed into a set of K indicator values corresponding to K threshold values. Then, K ccdf values at u are estimated by a kriging of these indicator data, and the complete function is obtained by interpolation or extrapolation of the estimated probabilities. Further details of these approaches for uncertainty estimation is available in Goovaerts (1999) and Goovaerts (2001)

The usual approach for modeling uncertainty consists of computing a kriging estimate $z^*(u)$ and the associated error variance, which are then combined to derive a Gaussian-type confidence Interval. In kriging approach, how the estimates [$\hat{z}(u)$] are made with minimization of error variance is discussed below.

$$\hat{z}(u) = \sum_{i=1}^{n} \lambda_i z(x_i) \tag{12}$$

in which the λ_i are weights and $z(x_i)$ are neighborhood data. The weights are chosen as such so that they minimize the prediction error variance, provided they sum to 1 and thereby ensure that prediction is unbiased. The prediction error variance is calculated as

$$\sigma^2(u) = 2\sum_{i=1}^{n} \lambda_i \gamma(x_i - u) - \sum_{i=1}^{n}\sum_{j=1}^{n} \lambda_i \lambda_j \gamma(x_i - x_j) \tag{13}$$

Where, $z(x_{i/j})$, i/j = 1,2,…, N are observed locations, λ is the kriging weight and γ is the semivariogram value.

Therefore, kriging estimates are always associated with kriging variance and most times we neglect it. An estimate may be highly accurate but may be less uncertain which means that the estimated value is very close to observed value but the confidence interval of the estimate is very large. For example, sand content estimate for a particular grid point of a digital soil map is 70% with the variance of estimate as 15%. Thus, the 95% confidence interval of this estimate will be $70 \pm 1.96 \times \sqrt{15} \cong 70 \pm 7.6$. It indicates that out of 100 times, 95 times the estimate will be in the range 62.4-77.6 and 5 times it may be outside the above range. Therefore, if the range is quite high or the interval is large, the estimate is highly uncertain. Therefore, one should be careful before presenting the digital soil map products and it is always advisable to produce estimate map along with standard deviation map or variance map.

In the following example, as depicted in Fig. 13.2, predicted map of sand content in arid western India through kriging is shown at left hand side (Fig. 13.2a) whereas the standard deviation of the prediction is shown at right hand side (Fig. 13.2b). It is observed from the map of standard deviation that it is lower surrounding observed or measured locations than the locations which are away from measured locations. It indicates that uncertainty varies spatially and if the sampling locations fully cover the whole study area, the overall uncertainty component will be smaller. In the given map, the sampling density was higher at lower part of the map, and hence the standard deviation of prediction is lower at those areas. However, it is always not possible to collect very densely distributed soil samples because of several constraints and therefore a compromise has to be made between sampling design and the predefined level of uncertainty.

Once the standard deviation of prediction is known, the Gaussian type 95% confidence interval of estimate at may be prepared using the following equation.

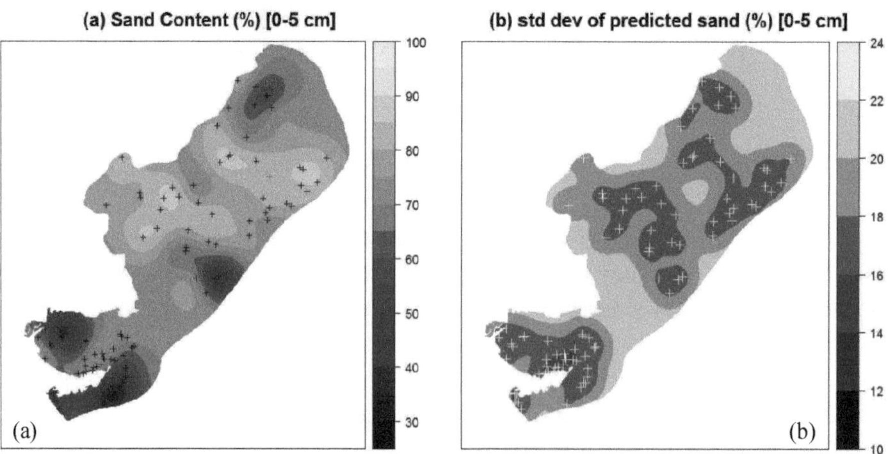

Fig. 13.2: Ordinary kriging estimate of sand content and the associated standard deviation of prediction

$$\hat{z}(u)_{\text{lower limit}} = \hat{z}(u) - 1.96 \times \sigma(u) \tag{15}$$

$$\hat{z}(u)_{\text{upper limit}} = \hat{z}(u) + 1.96 \times \sigma(u) \tag{16}$$

In the following example, lower limit and upper limit map of the 95% confidence interval of the kriged map of sand content is presented (Fig. 13.3).

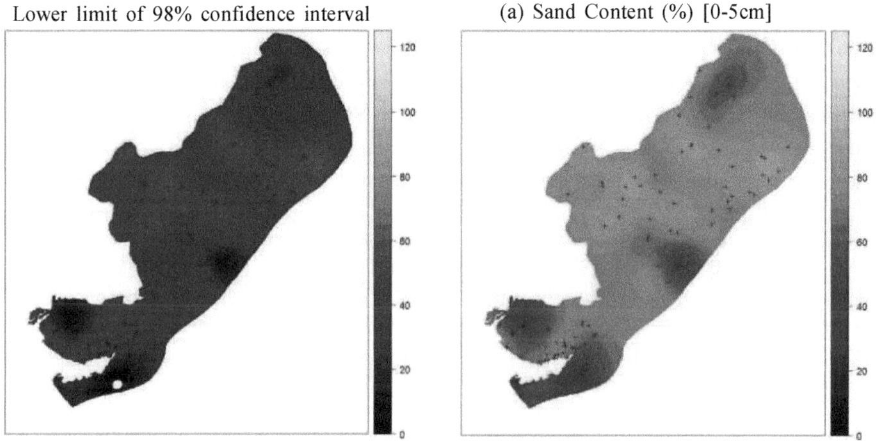

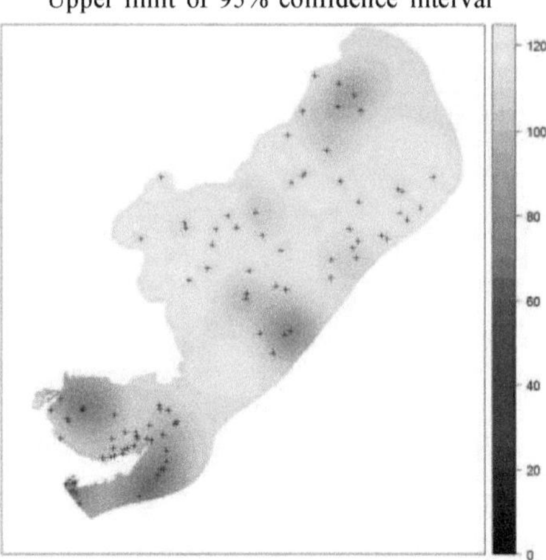

Fig. 13.3: 95% confidence interval maps of sand content

Wider the difference between lower and upper limit values for a particular location in the map, larger uncertainty is associated for that particular location. The difference between lower and upper limit was narrower surrounding sampling points or in the region where sampling locations are dense than those areas where sampling locations are sparsely distributed or not available at all. It indicates that the quality of the map in terms of uncertainty may be improved by increasing the number of sampling points within the study region. Otherwise, it can be said that sampling density need to be increased to lower the uncertainty of sand content map.

Multiple realization of a variable can also be created to evaluate the uncertainty with a condition of neighbourhood data through sequential simulation. Two major classes of sequential simulation algorithms can be used, depending on whether the series of conditional cdfs are determined using the multi-Gaussian approach (sGs=sequential Gaussian simulation) or the indicator approach (sis=sequential indicator simulation). In the following example, 6 realization of sand content has been created through conditional Gaussian simulation approach from which the uncertainty can be assessed (Fig. 13.4). In theory, when the number of realization is very large, the local distribution of simulated values almost reaches to the kriging estimate with its variance and therefore the kriging-based and simulation-based approaches yield similar uncertainty models. Hence, if only a location-specific assessment of uncertainty is required *e.g.* derivation of probability maps or the prediction interval maps is sufficient and the use of stochastic simulation entails an unnecessary waste of CPU time and disk storage.

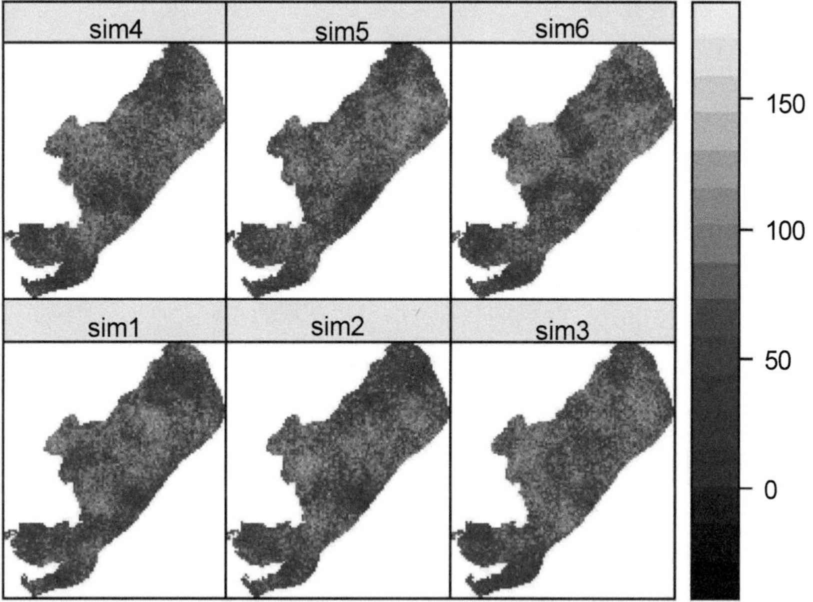

Fig. 13.4: Multiple realization of sand content using conditional Gaussian simulation

Scale issues on uncertainty assessment

Another important aspect of DSM is the scale, which is generally defined by the triplet of spacing, extent and support. The scale of a digital soil product is quite different from the mapping scale. Soil maps prepared through soil survey efforts are generally available at a particular scale *e.g.* 1 million scale (1:10,00,000) or ten thousands scale (1:10,000) etc. It indicates that 1 mm in a printed map represents 1 km in the real field for a 1 million scale map. However, such representation of scale for a digital soil product is irrelevant because it is occasionally or never printed as hard copy. Apart from it, the digital products are defined with a certain spatial resolution. If the spatial resolution is fine, pixel numbers will be large which means it can be zoomed to a very high level otherwise if it is coarse resolution; pixel numbers in the image will be less and cannot be zoomed to a high level. The resolution of the spatial product generally indicates the spacing in scale triplets. Triplets of scale in a digital soil map can be easily understood from the Fig. 13.5, which is nicely described in Blöschl and Sivapalan (1995). The spacing defines the separation distance between two grid points of a map in X-axis and Y-axis *e.g.* resolution of a digital soil map is 5 m×5 m. The extent defines the difference between minimum and maximum length of the map in pixel numbers along both X-axis and Y-axis direction e.g. the size of a digital image is 1000×1500 indicating 1000 pixels in X-axis direction and 1500 pixels in Y-axis direction. The support defines the surface area for which the estimates or values are given e.g. in most digital soil product support is point which is generally the centre point of a pixel. However, in case of block kriging products, the support is block size.

If the support of a map is quite large than the support of measurement points, the uncertainty component will be very small. Therefore, it is very important to know at which support it is intended to prepare a map and the applications of it. Often we are interested in mapping homogeneous regions within a study area rather than each possible location and thus support becomes a block. Secondly, the sampling density, which is defined by spacing and extent play a major role on uncertainty. If the sampling density is quite high the uncertainty becomes low as also discussed above. However, based on the pre-defined uncertainty level and limitations of sampling efforts, an optimal sampling design may be formulated to derive a map with desired level of uncertainty.

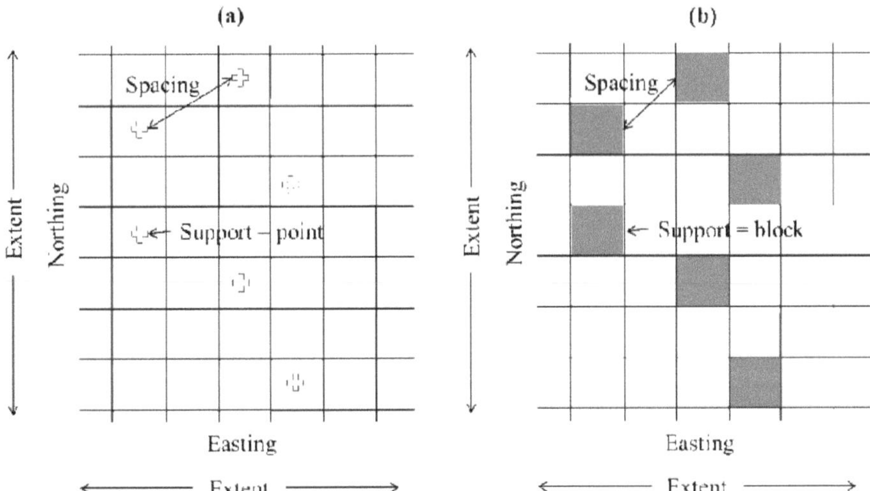

Fig. 13.5: Definition of the scale triplet (spacing, extent and support); (a) sampling design with point support (b) similar sampling design but with block support

Conclusion

Digital soil maps are often used as input in different transfer functions e.g. pedotransfer function or in different process based models e.g. water flow model, hydrological model, crop growth model etc. Therefore, the uncertainty associated with the maps will propagate through these transfer function or models leading to uncertain response. If the uncertainty component is known, Monte-Carlo simulations or latin hypercube sampling can be made to create several set of inputs, which are then fed to transfer function to know the uncertainty in response. In case of spatial uncertainty, multiple realizations are created through sequential Gaussian simulation or sequential indicator simulation, which are further fed in to the model and a set of responses are created to assess the uncertainty. Therefore, the assessment of accuracy and uncertainty of a map is very important for its final application in decision making process.

References

Bishop, T.F.A. and Lark, R.M. 2008. Reply to "Standardized vs. Customary ordinary CoKriging…" by A. Papritz. Geoderma 146(1-2): 397–399.

Bloschl, G. and Sivapalan, M. 1995. Scale issues in hydrological modelling: A review. Hydrological Processes 9: 251-290.

Goovaerts, P. 2001. Geostatistical modelling of uncertainty in soil science.Geoderma 103:3-26.

Goovaerts, P. 1999. Geostatistics in soil science: state-of-the-art and perspectives. Geoderma 89: 1-45.

Lark, R.M. 2000. Estimating semivariograms of soil properties by the method of moments and maximum likelihood.European Journal of Soil Science 51: 717–728.

14

Spectral Signatures of Soils and Its Controlling Factors with Focus on Spectral Behaviour of Various Soil Types of India

A.K. Bera[1], *Sushil B. Rehpade*[1], *Sagar S. Salunkhe*[1] *and S. Rama Subramoniam*[2]

[1]*RRSC (West), NRSC/ISRO, Jodhpur, Rajasthan, India*
[2]*RRSC (South), NRSC/ISRO, Bangalore, Karnataka, India*

Introduction

Soil is one of the vital resource, and is the basis of the existence of mankind. Management of this resource including its conservation and utilization is of crucial importance. It is a base on which all life depends. In the recent past, with burgeoning populations and the national goals of seeking self-sufficiency in food and fibre production, this resource base is slowly being stripped. While natural systems often adapt to stress in a remarkable fashion, some relationships - once destroyed - can never be restored. Soil resources can be assessed and monitored through its spectral signatures gathered through remote sensing.

The term 'remote sensing' is commonly restricted to methods that employ electromagnetic energy (such as lights, heat, micro wave) as means of detecting and measuring target characteristics. Sensors acquire data as various earth surface features reflect or e mit electromagnetic energy (EM energy). Remote sensing is the technology that is now the principal method (tool) by which the earth's surface and atmosphere (as targets or objects of surveillance) are being observed, measured, and interpreted from vantage points. Remote sensing of the earth traditionally has used reflected energy in the visible and infrared, and emitted energy in the thermal infrared and microwave regions to gather radiation that can be analyzed numerically or used to generate images whose tonal variations represent different intensities of photons associated with a range of wavelengths that are received at the sensor.

Electromagnetic Spectrum

EM radiation can be produced at a range of wavelengths, and can be categorized according to its position into discrete region, which is generally referred as 'Electro-magnetic Spectrum'. Electro-magnetic Spectrum is a continuum sequence of EM energy arranged according to wavelength or frequency (Fig. 14.1).

The earth's atmosphere absorbs energy in the Gamma ray, X-ray and most of the ultra-violet region; therefore these regions are not used for remote sensing. Remote sensing deals with energy in visible, infrared, thermal and microwave regions. These regions are further sub-divided into bands such as blue, green, red (in visible region), near infrared, mid infrared, thermal, microwave etc.

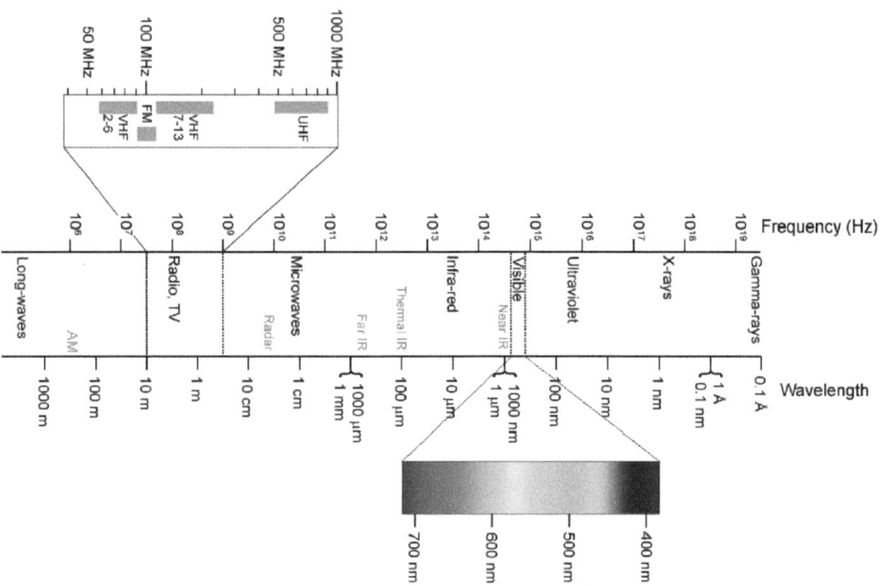

Fig. 14.1: Electromagnetic spectrum

Interaction of EM Radiation with Earth's Surface

When EM radiation strikes a surface, it may be reflected, scattered, absorbed and/or transmitted (Fig. 14.2). These processes are not mutually exclusive: EM radiations may be partially reflected and partially absorbed. Which processes actually occur depends on the wavelength of radiation, angle of incidence, surface roughness, condition and composition of surface material. Interaction with matter can change intensity, direction, wavelength, polarization, and phase of incident EM radiation. The science of remote sensing detects and records these changes. Applying the principle of conservation of energy, the energy balance equation for radiation at a given wavelength (λ) can be expressed as:

$$EI = ER + EA + ET$$
$$(\text{Incident}) = \text{Reflected} + \text{Absorbed} + \text{transmitted}$$

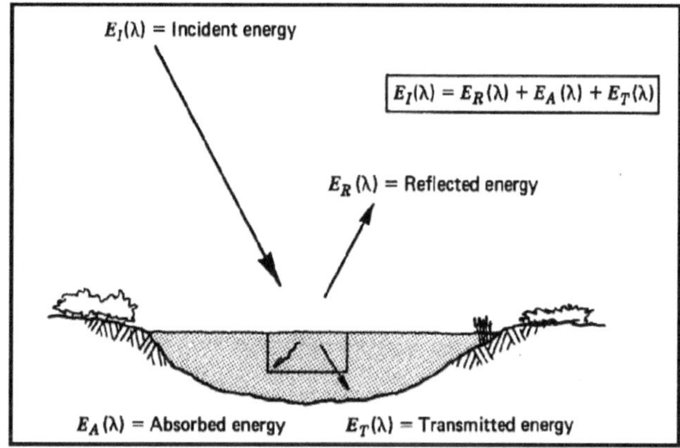

Fig. 14.2: Interaction mechanism of EM radiation with earth's surface

Spectral signature

The reflectance characteristics of earth surface features can be quantified by measuring the portion of incident energy that is reflected. It is measured as a function of wavelength and is termed as spectral reflectance, ρ_λ. It is mathematically defined as:

$$\rho_\lambda = \frac{\text{energy of wavelength } \lambda \text{ reflected from object}}{\text{energy of wavelength } \lambda \text{ incident on object}} \times 100$$

Remote sensing systems often operate at wavelengths, which cannot be detected visually, and in order to understand the signature obtained by sensors require knowledge of reflectance and absorption properties of different features. A graph of spectral reflectance of an object as a function of wavelength is termed as spectral reflectance curve and gives an insight into the spectral characteristics of an object.

Sensors are available to measure all types of electromagnetic energy as it interacts with objects. The ability of sensors to measure these interactions allows us to use remote sensing to measure features and changes on the Earth and in our atmosphere. Differences among spectral signatures are used to help in classifying remotely sensed images into classes of landscape features since the spectral signatures of like features have similar shapes.

The term spectral signature refers to the relationship between the wavelength (or frequency) of electromagnetic radiation and the reflectance of the surface. The signature is affected by several things including the material composition and structure.

A spectral signature is some measurable quantity (*e.g.*, reflectivity, emissivity), which varies as a function of wavelength and can be used to identify a material. To obtain a signature, the quantity must be measured at a sufficient number of wavelengths (and at fine enough spectral resolution) such that the material can be discriminated from other materials. For example, an RGB image (converted to reflectance), provides reflectivity information at three wavelengths (red, green, & blue); however, that typically wouldn't be considered a spectral signature because it doesn't provide adequate information to discriminate various materials (*e.g.*, pixels containing grass, artificial turf, or green tennis court can look nearly identical in RGB imagery). Spectral signatures are typically obtained either from hyperspectral images or with handheld spectrometers.

A feature on the other hand is simply an object in landscape. For example, a feature may be a field of uniform crop, a road, or building, or any other part of the landscape. We often try to identify features by using their spectral signatures, assuming uniformity, which is not always the case. Sometimes, rather than classifying pixels based on their spectral signatures alone, we also try to account for spatial relations such as the proximity of similar pixels. This is common for example with object-based image segmentation, which attempts to identify features using a combination of spectral and spatial characteristics.

The term "feature" can have multiple meanings. While it can refer to a spatial characteristic (or object), in the spectral domain it usually means something quite different. A spectral feature could be the original spectral measurement data (e.g., reflectivity) but it is often something derived from the spectral measurements, typically by creating linear or nonlinear transformations of the original data values. Spectral features are often created to reduce the dimensionality of the spectral data prior to further processing. Examples of linear features are those obtained from Principal Components Analysis (PCA) or Linear Discriminant Analysis (LDA). An example of a nonlinear feature is the Normalized Difference Vegetation Index (NDVI), which is a scaled difference between the red and near-infrared bands in an image pixel. There are numerous types of spectral features that can be extracted from spectral data and the best one depends on the details of what you are trying to accomplish.

Measurement of soil spectra

Current laboratory instruments such as spectroradiometer are used for soil spectral reflectance measurements. Soil samples are more readily prepared for scanning with a spectroradiometer than rocks or plant samples. After air-drying, a simple sieving is generally sufficient for a first run. A more rigorous approach requires grinding of the soil sample down to a standard grain size.

In the field, spectral data can also be easily recorded over soils with hand-held devices, whereas measurement of canopies often require more complex attachments such as cherry picker booms or truck mounted cranes. By ratioing the radiances observed over soils by the one measured over a reference target (compressed halon plate for instance) reflectance factors can be computed.

The recent advent of new type of portable instrument has brought field spectra measurements to a new era. After the generation of broad band instruments such as portable radiometers, array detector technology and high density computer storage of data has allowed the development of portable spectrometers. They currently perform very satisfactorily in the visible to near-infrared range. Newer mid-infrared detectors will allow for coverage of the full optical domain in the near future. Technically, laboratory spectroradiometers measure the diffuse reflectance (they use an integrating sphere) while field instruments measure bidirectional reflectance factors which vary with the geometry of the sun-target system. In the case of soils, this latter effect is obviously more pronounced for soils with rough surfaces. Although these geometrical conditions affect the apparent brightness of the soils, the spectral shape remains almost constant as reported by various researchers. Thus, field recorded reflectance factor curves retain the spectral features and can be compared to laboratory spectra of pure minerals, for instance.

Spectral signature of soils and controlling factors

Soil reflectance curve shows considerably less "peak-and-valley" variation in reflectance (Fig. 14.3). Majority of the incident radiation is either reflected or absorbed and little transmitted. As the wavelength increases, reflection steadily increases.

Some of the factors affecting soil reflectance are moisture content, soil texture (proportion of sand, silt and clay), surface roughness, presence of iron oxide and organic matter etc. These factors are complex, variable and interrelated (Govardhan 1991, Sinha 1986, Sinha 1987).

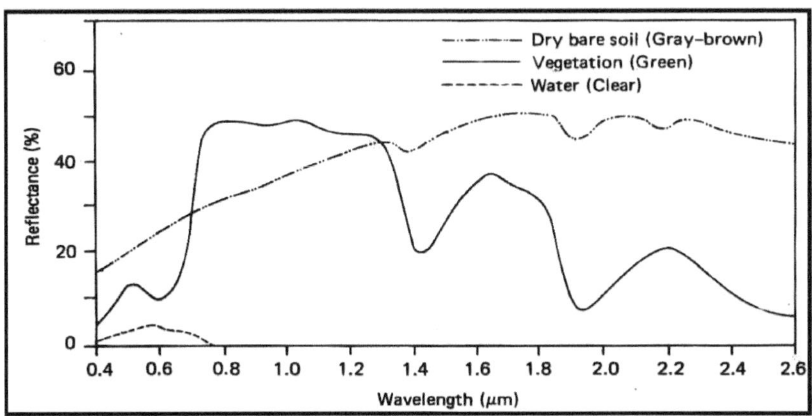

Fig. 14.3: Spectral reflectance curve of three important earth surface features

With the increase in the moisture content, reflection decreases. As with vegetation, the effect is greater in water absorption bands at 1.4, 1.9 and 2.7 μm. Clay soils also have hydroxyl absorption bands at 1.4 and 2.2 μm (Fig. 14.4).

Soil moisture is strongly related to soil texture: coarse, sandy soil has low moisture content and high reflectance; poorly drained fine textured soil will have low reflectance (Fig. 14.5). In the absence of water, however, the soil will exhibit the reverse tendency. Coarse texture soil will have less reflectance than fine textured soil.

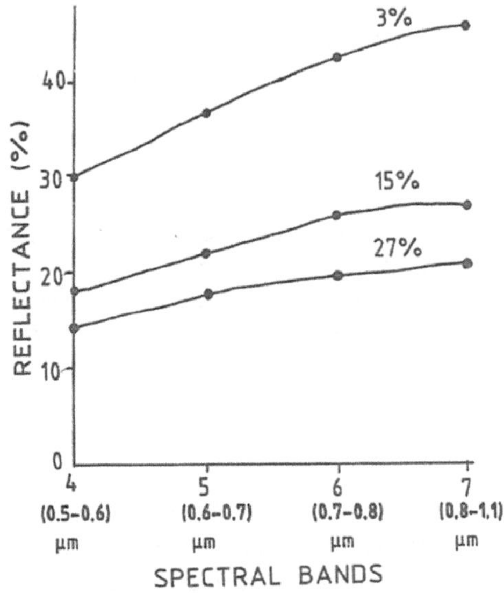

Fig. 14.4: Effect of moisture on spectral reflectance of coastal sandy soils (Sinha 1986)

There is a steady decrease of reflectance from 55 to 25% as the organic matter content increases from none to 5%, thereafter the decrease is gradual (Fig. 14.6). Organic matter present in soil is dark in colour, its presence decrease the reflectance.

The presence of iron oxide in soil will also significantly decrease reflectance (Fig. 14.7). Iron oxide gives soil rusty red colour and reflect red (0.6-0.7 μm) and absorbs green (0.5-0.6 μm).

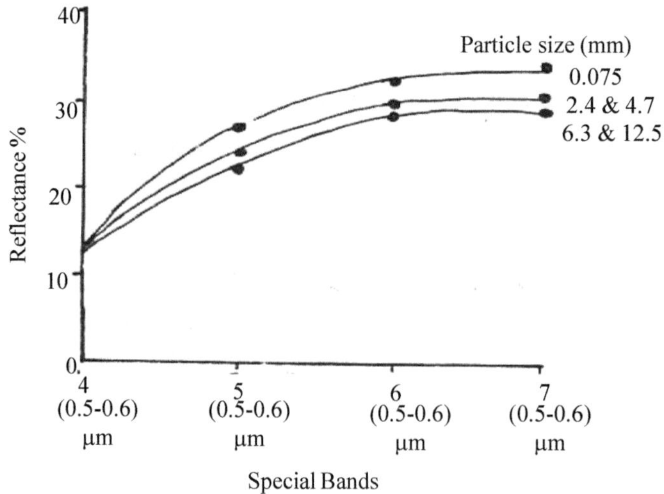

Fig. 14.5: Effect of particle size on spectral reflectance of laterite soils (Sinha 1986)

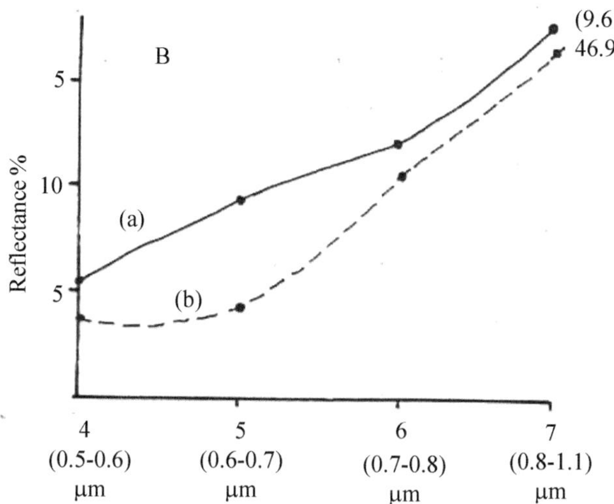

Fig. 14.6: Spectral reflectance curves for (a) partially decomposed compost (b) fully decomposed compost (Sinha 1987)

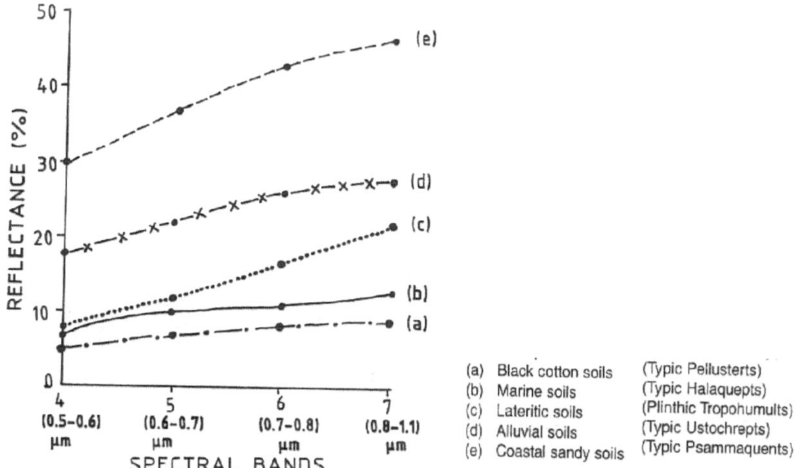

Fig. 14.7: *In-situ* Spectral reflectance curves in five soil types (Sinha 1986)

Satellite Data Utilization for Soil Resources Studies in India

Before the launch of Landsat-1 in 1972, aerial photographs were being used as a remote sensing tool for soil mapping, and, exhibited their potential in analyzing physiography, land use and erosion status. Subsequently, 1972 onwards satellite data in both digital and analog have been utilized for preparing small scale soil resource maps showing soil sub-groups and their association. The high resolution Landsat TM and Indian Remote Sensing Satellite (IRS) LISS II data which became available during mid-eighties, enabled soil scientists to map soils at 1:50,000 scale, which is used for district level planning. At this scale soils could be delineated at association of soil series/family level. The SPOT and IRS PAN data offered stereo capability, which has improved the soil mapping efforts. Indian Remote Sensing satellites (IRS-1A, 1B,1C and 1D) provide state-of-the-art database for natural resources inventories. Many studies have been conducted to explore the potential of LISS-I and LISS-II data for soil resource mapping both at 1:250,000 and 1:50,000 scale. Several studies have been initiated on potential use of IRS-1C, LISS III and PAN data for mapping soils and it is expected that information on scale 1:25,000 to 1:12,500 scale could be generated through combination of these data (Kudrat et al., 2000; Manchanda et al., 2002). Studies are also being carried out to explore potential of Synthetic Aperture Radar (SAR) data for soil moisture (Mohan *et al.,* 1990) estimation.

Multi-temporal satellite imageries from satellites like Landsat, SPOT and IRS have been operationally used in India to derive information on degraded lands, and monitor them periodically in time and space domain (Sharma and Bharghava 1987, NRSA 1981, Venkataratnam 1984, Karale et al., 1988, Rao et al., 1991, Venkataratnam and Ravisankar 1992).

Conclusions

Soil resources of our country have witnessed significant alterations over the past few decades because of increase in population, changes in land use and climate variability. Floods, soil erosion, and deterioration of soil health due to intensive agriculture are the main factors that have modified soil surface characteristics. Conventional soil sampling and laboratory analyses cannot efficiently provide this information on dynamic change because they are slow, expensive, and also could not map all temporal and spatial variability. Remote sensing tool has shown a high potential in soil characteristics retrieving in the last few decades. A better knowledge of soil spectral properties will be needed, which means collecting more spectral measurements over soils and standardizing the techniques to establish a reference database.

References

Govardhan, V. 1991. Spectral reflectance characterisation of different soils of Deccan plateau. Journal of Indian Society of Remote Sensing 19(3): 175-1859.

Karale, R.L., Saini, K.M. and Narula, K.K. 1988. Mapping and monitoring ravines using remotely sensed data. Journal Soil Water Conservation India 31(1,2): 76-82.

Kudrat, M., Sinha, A.K. and Manchanda, M.L. 2000. Multi-level soil mapping using IRS1C, WiFS, LISS III and Pan Data. Indian Space Research Organisation, Bangalore. India.

Manchanda, M.L., Kudrat, M. and Tiwari, A.K. 2002. Soil survey and mapping using remote sensing. Tropical Ecology 43(1): 61-74.

Mohan, S., Mehta, N.S. and Patel, P. 1990. Radar Remote Sensing for Land Resources - A Review. Scientific Report No.ISRO-SAC-SR-36-91. Space Applications Centre, Ahmedabad.

NRSA. 1981. Satellite remote sensing survey for soil and land use in part of Uttar Pradesh, Project Report, National Remote Sensing Agency, Hyderabad.

Rao, B.R.M., Dwivedi, R.S.,Venkataratnam, L., Ravi Sankar, T., Thammappa, S.S., Bhargawa, G.P. and Singh, A.N. 1991. Mapping the maginitude of sodicity in part of Indo-Gangetic Plains of Uttar Pradesh, using Landsat-TM data. International Journal Remote Sensing 12(3): 419-425.

Sharma R.C. and Bhargava, G.P. 1987. Operational use of SPOT-1 image in mapping and management of alkali soils. Proc. National Symposium on Remote Sensing in Land Transformation and Management organised by Indian Society of Remote Sensing held at Hyderabad during December 21-23, 1987.

Sinha, A.K. 1986. Spectral reflectance characteristics of soils and its correlation with soil properties and surface conditions. Journal of Indian Society of Remote Sensing 14(1): 1-9.

Sinha, A.K. 1987. Variation in soil spectral reflectance related to soil moisture, organic matter and particle size. Journal of Indian Society of Remote Sensing 15(2): 7-12.

Venkataratnam. L. and Ravisankar, T. 1992. Digital analysis of Landsat TM data for assessing degraded lands In: 'Remote sensing applications and Geographical Information Systems-Recent Trends', Published by Tata McGraw Hill Publishing Company Ltd., New Delhi, pp. 87-190.

Venkataratnam, L. 1984. Monitoring and managing soil and land resources in India using remotely sensed data. Proc. 3rd Asian Agricultural Symposium, Chiang Mai, Thailand.

15

Soil Resources Assessment: Scope of Hyperspectral Imaging

A.K. Bera[1], Sagar S. Salunkhe[1], Sushil B. Rehpade[1] and S. Rama Subramoniam[2]

[1]*RRSC (West), NRSC/ISRO, Jodhpur, Rajasthan, India*
[2]*RRSC (South), NRSC/ISRO, Bangalore, Karnataka, India*

Introduction

Over the past few decades, soil resources have witnessed many changes driven by demographic pressure and climate change. It is necessary to have up to date soil information to monitor the state of soils health with time. Soil resource inventory provides information about the potentialities and limitation of soils for its proper utilization. Soil survey offers an accurate and scientific inventory of different soils, their properties and extent. An in-depth knowledge of the type of soils and their spatial distribution is a prerequisite for developing rational land use planning. Remote sensing is the science of obtaining information about the earth's surface without actually being in contact with it. On the other hand, proximal sensing refers to remotely sensed measurements that are taken at the field or laboratory level. Both these processes involve an interaction between incident radiation and the targets of interest. Fundamentally, remote sensing data products are a representation of the functional response of objects to energy sources. These are typically spectra, which are acquired as a discrete series of reflectance measurements taken at different wavelength intervals, or bands. In the shorter wavelengths of the electromagnetic spectrum (visible part), features can be detected by virtue of reflected solar energy, while in the longer wavelengths (microwave, thermal parts), sensing of emitted energy works. Conventional soil sampling, and subsequent laboratory analysis are generally time consuming and costly. In this context, remote sensing plays a crucial role for studying soil resources. The term remote sensing is generally used for airborne and space borne acquisitions, whereas proximal sensing refers to ground-based laboratory and field measurements.

Advances in the quantitative disciplines like remote sensing and proximal sensing have laid the foundations for a spatial exploration of soil-system dynamics within a landscape context (Pennock and Veldkamp 2006).

Remote sensing of soils covers various spatial, spectral and temporal scales using sensors that utilize different parts of the EM spectrum. Resolution of sensors and the platform control the product's accuracy. Air or space borne sensors has a low signal-to-noise ratio as compared to ground-based sensors. This is due to the larger atmospheric path length, decreased spatial resolution, geometric distortions, and spectral ambiguity caused by recording multiple signals from adjacent features. Furthermore, differences between sensors in available wavelength bands and in the mechanics of imaging influence the accuracy (Kasischke et al. 1997). A number of in-depth reviews have been done regarding application of remote sensing to soil mapping and related issues, which demonstrated a significant increase in the efficiency of conventional soil survey methods when remote sensing data were used (Anderson and Croft 2009, Ben-Dor et al. 2002, Dwivedi 2001, Joyce et al. 2009, Metternicht et al. 2010, Mulder et al. 2011). Remote and proximal sensing methodologies hold considerable potential to facilitate soil mapping at larger temporal and spatial scales as achievable with conventional soil mapping methods. Existing remote and proximal sensing methods support three main components in Digital Soil Mapping (DSM) namely, (a) remote sensing data support the segmentation of the landscape into homogeneous soil-landscape units whose soil composition can be determined by sampling, (b) remote and proximal sensing methods allow for inference of soil properties using physically-based and empirical methods and (c) remote sensing data supports spatial interpolation of sparsely sampled soil property data as a primary or secondary data source (Mulder 2013). In general, remote and proximal sensed data are an important and essential source for DSM as they provide valuable data for soil mapping in a time and cost efficient manner.

Multispectral Remote Sensing

Multispectral remote sensing refers to acquiring image data at several discrete, discontinuous regions across the visible and infrared electromagnetic spectrum. Multispectral sensors record data in less number of bands (resulting in a coarser spectral resolution) as compared to hyperspectral sensors. Multispectral data has been used generally to derive information on land use/land cover, land degradation and terrain characteristics. Retrieval of soil parameters through remote sensing has progressed over the years with the availability of advanced multispectral sensors. Remote sensing imagery has been used to discriminate between crop residues and soil, distinguish iron oxides, iron hydroxides and iron sulphates, and differentiate between clay and sulphate mineral species (Abrams and Hook 1995, Hubbard and Crowley 2005).

Hyperspectral imaging

Hyperspectral imaging, also known as, imaging spectroscopy, is defined as a passive remote sensing technology that acquires simultaneous images in many spectrally contiguous, registered bands such that for each pixel a reflectance spectrum can be derived (Goetz et al. 1985, Schaepman et al. 2007). Soils are complex dynamic systems, which are formed and developed as a result of the combined effects of climate, biotic activities, and topography. Soil genesis modifies the chemical, physical, and mineralogical properties of soil surfaces. This process results in distinct spectral absorption features, which can be detected using high-resolution reflectance spectra (Leone and Sommer 2000). The relatively high spectral resolutions as well as contiguous placement of narrow bands covering wider region of electro-magnetic spectrum provide better chances for assessment of soil properties. Satellites namely, ASTER, Landsat, MERIS, MODIS, AVIRIS, HyMAP etc. have several advanced sensors for soil assessment (Table 15.1).

Important absorption features happen in NIR and SWIR range. These absorption characteristics can vary in their spectral depth, width, and location, and therefore serve as diagnostic indicators, which enable us to characterize soil properties. In particular, the amount of organic matter and iron content, particle size distribution, clay mineralogy, water content, soil contamination, CEC and calcium carbonate content, can be determined with imaging spectroscopy (Ben-Dor et al. 2009).

Spectral signature of soils is influenced by its chemical compositions due to absorption. In near infrared (NIR) and middle infrared (MIR) regions, absorption feature of soil components in solid phase originate mainly from the vibrations of bounded nuclei. Vibration features and corresponding absorption wavelengths of various constituents are given in Table 15.2. In addition to vibrations, molecular rotation and transition may also occur in the pores where gas and water molecules exist, which also cause higher absorption in MIR region. Soil water shows absorption peaks at about 1450 nm, 1880 nm and 2660 nm (Hoffer 1978).

Ferric or ferrous iron causes absorptions in the VNIR spectra (at about 860 nm) but organic matter results in an overall reduction of the reflectance. Clay minerals namely, montmorillonite, kaolinite, illite, smectite and carbonates have distinctive narrowband absorption features in the SWIR region between 2000 and 2500 nm.

Diverse combination of mineral, organic substances, soil moisture and various salts control the spectral properties of soils which contain the valuable information related to the vital soil parameters namely, humic substances, soil texture, thermal properties and primary and secondary minerals. Table 15.3 shows the potentiality of narrow band imaging spectroscopy in characterizing soil mineralogy (Manchanda et al. 2002).

Table 15.1: Advanced sensors available from various space borne and airborne platforms

RS Technology	Sensor Type	No. of Spectral Bands	Spectral Class	Spectral Range (μm)	Spatial Resolution (m)	Revisit Time (days)	Spatial Coverage
Optical - Spaceborne							
Landsat	MS	11	VNIR-TIR	0.43-12.51	15-60	16	Global
MODIS	MS	36	VNIR-TIR	0.40-14.40	250-1000	1	Global
MERIS	MS	15	VNIR	0.39-1.40	300	3	Global
ASTER	MS	15	VNIR-TIR	0.52-11.65	15-90	16	Global
Hyperion	IS	242	VNIR-SWIR	0.40-2.50	30	Irregular	Regional
ALOS/PRISM	MS	1	VIS	0.52-0.77	2.5	46	Local
Sentinel-2	MS	13	VNIR-SWIR	0.44-2.19	10-60	5-10	Global
Optical - Airborne							
AVIRIS	IS	224	VNIR-SWIR	0.38-2.50	4-20	Irregular	Local
HyMAP	IS	128	VNIR-SWIR	0.45-2.48	2-10	Irregular	Local
APEX	IS	300	VNIR-SWIR	0.40-2.50	2-5	Irregular	Local
DAIS-7915	IS	79	VNIR-TIR	0.45-12.00	3-10	Irregular	Local
LiDAR - Spaceborne							
ICEsat GLAS	Active	2	VNIR	0.53,1.06	70	91	Global
LiDAR - Airborne							
ALTM Germini	Active	1	NIR	1.06	2-3.5	Irregular	Local
ALTM Orion	Active	1	NIR	1.06	<1.5	Irregular	Local
RADAR - Spaceborne							
SRTM	Active	2	C,X	2.5-8.0	30	None	Global
RADARSAT2	Active	1	C	2.5-4.0	3-100	24	Global
ASAR Envisat	Active	1	C	2.5-4.0	30-1000	35	Global

Contd.

PALSAR	Active	1	L	15-30	100	46	Global
SMOS	Active	1	L	15-30	50000	3	Global
SMAP	Active/Passive	1	L	15-30		2-3	Global
Sentinel-1	Active	1	C	2.5-4.0	1000	3-6	Global
RADAR - Airborne							
E-SAR	Active	4	X,C,L,P	2.5-85	2-4	Irregular	Local
GeoSAR	Active	2	X,P	2.5-85	3	Irregular	Local
MIRAMAP	Active	3	X,C,L	2.5-30	5-50	Irregular	Local

Note: Optical sensor types distinguish between multispectral (MS) sensors and imaging spectroscopy (IS) sensors. Spectral classes distinguish between visible near-infrared (VNIR), shortwave infrared (SWIR), thermal infrared (TIR) and different radar bands (X, C, L, P). Spatial coverage distinguishes local = $<10^4$, regional = $>10^4$ - $<10^7$, and global = $>10^7$ scales.

Table 15.2: Vibration features and corresponding absorption wavelengths of various constituents

Constituents	Reactions	Absorption wavelength (nm)
H_2O	• Symmetric stretch	3106
	• Asymmetric stretch	2903
	• H-O-H bend	6080
	• Stretching fundamental	2770
	• Al-or Mg-OH bend	2200 or 2300
Oxides	• Fundamental stretching	5000
Hematite	• Fe-O fundamental stretching	20000
Carbonates		7000, 11000-12000,13000-15000
Phosphates		9250, 10300,18000, 28500
Sulphates		9000, 10200, 16000, 22200
Gypsum	• Overtones and combination of OH stretching in molecular water	1750, 2300
	• Fundamental bending mode of constitutional water	6000
	• Si-O bending	around 5000
Silicates	• Si-O stretching	1000
	• Si-O-Si, A1-O-Si stretches	12000-15000
	• (Si, Al)-O-(Al, Si) stretch	15000-20000
	• Deformation and bending modes of O-(Al, Si)-O, (Si, Al)-O-(Si, A1), O-(A1,Si)-O	20000-40000
	• Al, Si-O metal valence stretching	20000-40000

(Source: Bear 1968)

Table 15.3: Spectral/Dielectric/lonic conductivity features of soil

Objective/Phenomena	Relevant Spectral Region (nm)	Optimum Spectral Resolution (nm)
VIS (445 to 700 nm)		
• Soil Color	445	10-20
• Organic matter	490	10-20
• Broadband absorption due to		
Fe-bearing minerals	565	10-20
Carbonates	620	10-20
Sulphate	670	10-20
• Erosional features	683	10-20
NIR (745 to 1035 nm)		
• Vegetative cover	711	10-20
• Humus content	720	10-20
• Weak absorption due to		
Fe-bearing minerals	880	10-20
SiO_2	960	10-20
Erosional features	1035	10-20
SWIR-I (1500 to 1800 nm)		
• Soil moisture	1500	350
• Weak absorption bands of carbonates	1740	10-20

Contd.

SWIR-II (2000 to 2400 nm)		
• Characteristic absorption bands of $-CO_3$, -OH and	2320	10
• SO_4 bearing minerals *e.g.* Carbonates & layer silicates	2310	10
• -OH and ALOH bearing minerals *e.g.* kaolinite	2200	10
• Muscovite/Alunite	2300	10
• MgOH bearing minerals *e.g.* Talc & Brucite	2400	10
• Soil moisture	2160/2040	20

(Source*:* Manchanda et al. 2002)

Remote Sensing Data Utilization for Soil Studies

Present research using remote sensing data products usually produce qualitative outputs. Various methods that are available for spectroscopy and geostatistics have made limited use for digital soil mapping (Ben-Dor et al. 2009, Dewitte et al. 2012). Recent studies demonstrated that at the moment, remote sensing data does not provide the high spectral resolution that is needed to quantitatively map soil mineralogy (van der Meer et al. 2012). Use of geostatistical approach along with a small representative sample considerably increases the feasibility to quantitatively map mineralogy (Mulder et al. 2013). The overall accuracy of model reduced with increasing scale of the study area due to incompatibility between the remote sensing data and the available sample data. According to the perception of soil scientists, the major gap area is the lack of readily available remote sensing based soil products.

It is now necessary to fully develop methodologies for large scale soil mapping. Estimating properties of the topsoil using various proximal sensing methods were found to be satisfactorily accurate as compared to chemical soil analysis. The residual inaccuracies in estimated soil properties of the topsoil, such as soil organic matter, have often been contributed to other constituents in soil samples (Bartholomeus et al. 2008). This indicates that the inaccuracies are the result of overlapping absorption features, which need to be accounted for detailed analysis (Mulder 2013).

Other aspects of digital soil mapping that deserve further attention in the development of soil products are: (a) the interpolation of gaps in the spatial coverage due to cloud cover, vegetation or other obscuring areas; (b) the combination of proximal sensing and remote sensing data with geostatistics to address the lack of soil property data at regional to global scales; and (c) the improved understanding and multi-temporal monitoring of processes related to soil property changes to model future variations (Wulf Hendrik et al. 2015). Despite all initiatives, it is to be likely that the existing soil data have insufficient coverage, and thematic variability for regional and global models.

Future digital soil map products

The global consortium (Global Soil Map) has been formed in 2008 that aims to make a new digital soil map of the world using state-of-the-art and emerging technologies for soil mapping and predicting soil properties at fine resolution. This new global soil map aims to predict mainly functional soil properties that define soil depth, water storage, texture, fertility and carbon at fine spatial resolution (~100 m) for most of the ice-free land surface of the globe over the next five years. These maps will be supplemented by interpretation and functionality options to support improved decisions for a range of global issues such as food production, climate change, and environmental degradation (Arrouays et al. 2014). Global Soil Map will be freely available, web-accessible, and widely distributed.

Conclusions

The retrieval of soil attributes with remote sensing technology has made significant progress, particularly since the launch of advanced multispectral sensors and imaging spectrometers such as ASTER and Hyperion, which have made it possible to detect subtle differences between spectral signatures. Various indices, proxies, quantities and patterns have been derived from remote sensing in order to map soil and terrain attributes. However, remote sensing technology still needs to catch up with proximal sensing in terms of number and feasibility of derived soil attributes. Due to the heterogeneity of landscapes and the spatial resolution of the imagery, it is often difficult to find pure pixels representing soil or bare rock. Advanced unmixing tool methods are needed to extract sub-pixel soil and rock composition. In general, the use of spectral imagery for the spatial prediction of soil properties is based on the spatial relation between existing soil data and observed patterns in the imagery, and not on physically-based retrievals, such as soil moisture (Dobos et al. 2000, Stoorvogel et al. 2009). Depending on spatial and spectral resolution, spatial coverage and the availability of legacy data, remote and proximal sensing data are either used as primary or secondary data source for the spatial prediction of soil properties.

Future research will aim for the integrated use of remote sensing methods for spatial segmentation, as well as measurements and spatial prediction of soil properties to achieve complete area coverage. Currently, most imaging spectrometers are airborne sensors (e.g. AVIRIS, HyMAP, APEX, AISA, HySPEX), in contrast to few spaceborne prototypes (e.g. Hyperion, HICO). Analysing hyperspectral data is challenging because: (a) the file size of multidimensional imaging spectroscopy data increases linearly with the number of spectral bands, (b) atmospheric absorption affects particularly hyperspectral data, due to the selective absorption of atmospheric gases and water vapour across the spectral range, which requires sophisticated pre-processing, (c) an overall lower signal-to-noise ratio as compared to multispectral data is another issue related to narrow spectral bandwidths and atmospheric attenuations, (d) a significant band-to-band correlation results in dimensionality issues and consequently reduces the total amount of available bands, and (e) imaging spectroscopy data needs to be corrected for BRDF (Bidirectional Reflectance Distribution Function) effects, which vary as a function of illumination and viewing geometry and depend on the wavelength as well as structural and optical properties of the surface. The soil science community is aware of these challenges and current efforts are on data harmonization (Global Soil Partnership 2011, Panagos et al. 2011, Sulaeman et al. 2012) while research efforts are initiated for temporal modelling of soil properties (Banwart, 2011).

References

Arrouays. D., Grundy., M.G., Hartemink, A.E., Hempel, JW., Heuvelink, G.B.M., Hongjj, S.Y., Lagacherie, P., Lelyk, G., McBratney, A.B., McKenzie, N.J., Mendonca-Santos, M,D.L., Minasny, B., Montanarella, L., Odeh, I.O.A., Sanchez, P.A., Thompson, J.A. and Zhang, G.L. 2014. Global Soil Map. Toward a Fine-Resolution Global Grid of Soil Properties, edited, pp. 93-134.

Anderson, K. and Croft, H. 2009. Remote sensing of soil surface properties. Progress in Physical Geography 33(4): 16.

Abrams, M. and Hook, S.J. 1995. Simulated ASTER data for geologic studies. IEEE Transactions on Geoscience and Remote Sensing 33(3): 692-699.

Banwart, S. 2011. Save our soils. Nature 474 (7350): 151-152.

Bartholomeus, H.M., Schaepman, Kooistra, M.E.L., Stevens, A., Hoogmoed, W.B. and Spaargaren, O.S.P. 2008. Spectral reflectance based indices for soil organic carbon quanitification. Geoderma 145: 9.

Ben-Dor, E., Patkin, K., Banin, A. and Karnieli, A. 2002. Mapping of several soil properties using DAIS-7915 hyperspectral scanner data—a case study over clayey soils in Israel. International Journal of Remote Sensing 23(6): 19.

Ben-Dor, E., Chabrillat, S., Demattê, J.A.M., Taylor, G.R., Hill, J., Whiting, M.L. and Sommer, S. 2009. Using Imaging Spectroscopy to study soil properties. Remote Sensing of Environment 113 (Supplement 1): S38-S55.

Bear, F.I. 1968. A Text Book of Soil Chemistry. John Wiley Publication, New York.

Dewitte, O., Jones, A., Elbelrhiti, H., Horion, S. and Montanarella, L. 2012. Satellite remote sensing for soil mapping in Africa: An overview. Progress in physical geography 36(4): 514-538.

Dobos, E., Micheli, E., Baumgardner, M.F., Biehl, L. and Helt. T. 2000. Use of combined digital elevation model and satellite radiometric data for regional soil mapping. Geoderma 97(3-4): 367-391.

Dwivedi, R.S. 2001. Soil resources mapping: A remote sensing perspective. Remote sensing reviews, 20(2): 89-122.

Global Soil Partnership. 2011. Global Soil Partnership, edited, Food and Agriculture Organization of the United Nations.

Goetz, A.F.H., Vane, G., Solomon, J.E., Rock, B.N. 1985. Imaging Spectrometry for Earth Remote Sensing. Science 228(4704):1147-1153.

Hoffer, R.M. 1978. Biological and physical considerations in application computer aided analysis techniques to remote sensing. pp. 237-286. In: (P.H. Swain and S.M. Davis, Eds.) Remote Sensing: Quantitative Approach. McGraw-Hill International Book Co.

Hubbard, B.E. and Crowley, J.K. 2005. Mineral mapping on the Chilean-Bolivian Altiplano using co-orbital ALI, ASTER and Hyperion imagery: Data dimensionality issues and solutions. Remote Sensingof Environment 99 (1-2): 173-186.

Joyce, K.E., Belliss, S.E., Samsonov, S.V., McNeill, S.J. and Glassey, P.J. 2009. A review of the status of satellite remote sensing and image processing techniques for mapping natural hazards and disasters. Progress in physical geography 33(2): 183-207.

Kasischke, E.S., Melack, J.M. and Dobson, M.C. 1997. The use of imaging radars for ecological applications - A review. Remote Sensing of Environment 59(2): 141-156.

Leone, A.P. and Sommer, S. 2000. Multivariate analysis of laboratory spectra for the assessment of soil development and soil degradation in the southern Apennines (Italy). Remote Sensing of Environment 72(3): 346-359.

Manchanda, M.L., Kudrat, M. and Tiwari, A.K. 2002. Soil survey and mapping using remote sensing. Tropical Ecology 43(1): 61-74.

Metternicht, G., Zinck, J.A., Blanco, P.D. and Del Valle, H.F. 2010. Remote sensing of land degradation: Experiences from Latin America and the Caribbean. Journal of Environmental Quality 39(1): 42-61.

Mulder, V.L., de Bruin, S.,Schaepman, M.E. and Mayr, T.R. 2011. The use of remote sensing in soil and terrain mapping - A review. Geoderma 162(1-2): 1-19.

Mulder, V.L. 2013. Spectroscopy-supported digital soil mapping, 188 pp, Wageningen University, Wageningen.

Mulder, V.L., Plotze, M., de Bruin, S., Schaepman, M.E., Mavris, C., Kokaly, R. and Egli, M. 2013. Quantifying mineral abundances of complex mixtures by coupling spectral deconvolution of SWIR spectra (2100-2400 nm) and regression tree analysis. Geoderma 207–208(1): 279-290.

Panagos, P., van Liedekerke, M. and Montanarella, L. 2011. Multi-scale European Soil Information System (MEUSIS): A multi-scale method to derive soil indicators. Computational Geosciences 15(3): 463-475.

Pennock, D.J. and Veldkamp, A. 2006. Advances in landscape-scale soil research. Geoderma 133(1-2): 1-5.

Schaepman, M.E., Wamelink, G.W.W., Van Dobben, H.F., Gloor, M., Schaepman-Strub, G., Kooistra, L., Clevers, J.G.P.W., Schmidt, A. and Berendse, F. 2007. River floodplain vegetation scenario development using imaging spectroscopy derived products as input variables in a dynamic vegetation model. Photogrammetric Engineering and Remote Sensing 73(10): 1179-1188.

Stoorvogel, J.J., Kempen, B., Heuvelink, G. B. M. and de Bruin, S. 2009. Implementation and evaluation of existing knowledge for digital soil mapping in Senegal. Geoderma 149(1-2): 161-170.

Sulaeman, Y., Minasny, B., McBratney, A.B., Sarwani, M. and Sutandi, A. 2012. Harmonizing legacy soil data for digital soil mapping in Indonesia. Geoderma 192: 77-85.

van der Meer, F.D., van der Werff, H.M., van Ruitenbeek, F.J., Hecker, C.A., Bakker, W.H., Noomen, M.F.,van der Meijde, M., Carranza, E.J.M., de Smeth, J.B. and Woldai, T. 2012. Multi- and hyperspectral geologic remote sensing: A review. International Journal of Applied Earth Observation and Geoinformation 14(1): 112-128.

Wulf Hendrik, Mulder,Titia Schaepman, Michael, E., Keller Armin. and Jörg Philip. 2015. Remote Sensing of Soils. Remote Sensing Laboratories, Dept. of Geography, University of Zurich, Winterthurerstrasse 190 (Doc. Ref:00.0338.PZ / L435-0501, Version: 5.2.

16

Soil Fertility Management to Combat Desertification in Hot Arid Rajasthan

Mahesh Kumar, Priyabrata Santra and C.B. Pandey

ICAR-Central Arid Zone Research Institute, Jodhpur, Rajasthan

Introduction

In India, hot arid ecosystem covering an area of 31.7 million ha lies at north western part of the country and 62% of this area falls in western Rajasthan covering 12 districts i.e. Jodhpur, Jaisalmer, Jhunjhunu, Sikar, Hanumangarh, Sri Ganganagar, Bikaner, Barmer, Pali and Sirohi. The major landforms in arid western Rajasthan are dune and inter-dune plains having sandy to sandy-loam textured soils. Temperature remains high for most part of the year reaching up to 42-45°C during prolonged summer months (April to July), whereas annual rainfall is very low and highly sporadic (100-400 mm) mainly occurs during August and September. Annual potential evapotranspiration ranges from 1500-2000 mm. Loss of fertile top soil from denuded soil surface through wind erosion is frequent during summer months. The region is subjected to frequent occurrence of droughts, once in every three years, which further intensifies the process of desertification (Samra et al. 2006). Shorter growing period, sandy soils, very low available water capacity, inadequate nitrogen and phosphorus availability are the major constraints. Soils are prone to severe wind erosion, terrain deformation and nutrient depletion. The region is mostly mono cropped and double cropping is being practiced extensively in canal and tubes well command area. Extensive fallowing in the past even in very good rainfall years because of uneven undulating topography and severe windstorm are the cause of concern. Ground water is deep and is usually brackish, declining at alarming rate due to excessive withdrawal for irrigation and other purposes. The region, however, sustains on perennial vegetation of *Prosopis cineraria*, *Tecomella undulata,* arid shrubs and grasses, which are rich resources of fodder for animal because cropping is affected by vagaries of monsoon. During last two to three decades, a 2 to 3-fold increase in irrigated areas with cultivation of both *kharif* and *rabi* crops has been recorded and most of these increments are resulted through conversions from rainfed areas, grazing lands and sand dunes into irrigated areas.

Soils

The soils of the hot arid region of Rajasthan have been classified in Entisols and Aridisols soil orders. The Entisols cover maximum 51.84% area followed by Aridisols covering 41.05% area. The suborders psamments, cambids, calcids, gypsids, salids and fluvents covers 51.72%, 30.83%, 9.34%, 0.43%, 0.45% and 0.12% area of western Rajasthan, respectively (Narain and Joshi 2001). These soils are low in organic carbon and deficient in nitrogen. These soils are vulnerable to wind erosion, have poor moisture retention capacity (50-80 mm water m^{-1} soil depth), high infiltration, low nutrient retention and shallow depth due to underlying hard pan.

Fertility status of arid soils

Physical properties

Soils under Entisols are generally found in those parts of arid regions where high aeolian activity is observed. The average soil depth of this type of soil is 118 cm. Average bulk density of these soils is 1.58 Mg m^{-3} (Kumar et al. 2009). Surface horizon is rich in sand content in comparison to subsurface horizons. Soil texture is sandy with average sand content of 89%. In the extreme arid part at Jaisalmer, Bikaner and Barmer sand content is >90% for most of the places. Soil color is mainly yellowish brown. The hue of the soil is 10YR, value ranges from 5-6, and chroma ranges from 3-4. Surface as well as subsurface soils are single fine grain in structure. Soils are mainly loose, non-sticky and non plastic in consistency. Most of the soils are well drained and falls under very rapid (>25.4 cm hr^{-1}) class of O'Neals permeability class. Water retention at 1/3 bar (field capacity) is on an average is 7% (0.07 cm^3 cm^{-3} or 70 mm of water per meter of soil depth). Similarly, the water retention at 15 bar (permanent wilting point) is also very low (2.62%). Aridisols are dominant in those parts of arid regions where aeolian activity is comparatively less. The average soil depth of this type of soil is 100 cm with well demarcated horizons. Concretions of calcite below the soil profile are a common feature of these soils. Average bulk density of these soils is 1.43 Mg m^{-3}. Average sand content of Aridisols is lesser than Entisols and is around 70%. Soil color is mainly Yellowish brown to dark brown or pale brown. The hue of the soil is 10YR, value ranges from 5-6, and chroma ranges from 4-5. Surface as well as subsurface soils are single fine grain in structure. At few places, fine weak subangular blocky structure is found at subsurface soil layers. Soils are mainly loose, non-sticky and non plastic in consistency. At few places, presence of high clay and organic carbon at subsurface layers make the soil massive, friable, and slightly sticky. Most of the soils are well drained and falls under either rapid (12.5-25.4 cm hr^{-1}) or very rapid (>25.4 cm hr^{-1}) class of O'Neals permeability class. Water retention at field capacity (FC) and permanent wilting point (PWP) in Aridisols are comparatively higher than Entisols and are 13% and 5%, respectively (Kumar et al. 2009).

Soil organic carbon and nitrogen

Arid zone soils are low in organic matter because of low vegetation cover, high temperature and coarse texture. Low amount of organic matter in these soils has been attributed to high temperature, low rainfall, scanty and scrub vegetation cover and sandy texture of soils (Kumar et al. 2009). About 93.0% area was found deficient (<0.50%) in organic carbon and only 7.0% area is medium (0.50-0.75%) in organic carbon content (unpublished data). Organic carbon content in soils below 300mm rainfall zone ranges between 0.05-0.2% in coarse textured soils 0.2-0.3% in medium textured and 0.3-0.4% in fine textured soils (Mahesh Kumar et al. 2009, Kumar et al. 2009). Dhir (1977) and Mahesh Kumar et al. (2011) has shown that even in low range of organic carbon (>0.5%) increase in clay content is associated with rise in organic carbon. With increase in clay content there is increase in organic carbon content. Joshi (1990a) has reported that bulk of organic carbon (67-89%) is present in non humic form followed by fulvic and humic form and mean value of HA-carbon/FA carbon increased increase with increase in silt+clay. Nitrogen in soils is present as organic (95-98%) and inorganic but in arid region soils, the inorganic form constitutes abot 5 to 18% of the total N. Agarwal and Lahiri (1981) reported 46 to 67% N as inorganic or mineralized from in unstabilized sand dunes of western Rajasthan. Total N content of arid soils varied widely, as coarse textured Chirai, Pal, etc. contained 0.028 to 0.050%, medium textured Pipar and Soila (0.042 to 0.056%) and fine textured Asop contained 0.042 to 0.056% N. Agarwal et al. (1975) reported relatively highest concentration amino acid bound N amongst the various form in arid soils of Rajasthan.

Phosphorus

Phosphorus present in soils as organic and inorganic forms but organic form constitutes hardly 10-20% of the total phosphorus. Total phosphorus content in soils range between 300-1500 $\mu g\ g^{-1}$ (Choudhari et al. 1979) and about 80% of the inorganic P remain bound with Ca. Al-P was generally higher than Fe-P. Total and inorganic phosphorus were irregularly distributed in the soil profile of arid soils but organic phosphorus decreased with depth (Talati et al. 1975). 15-20% (97-110 kg ha^{-1}) of the total P is present in organic form as phytin, lecithin, phospholipids and other unidentified compounds (Tarafdar et al. 1989). In arid western Rajasthan about 45 and 44% area was found low and medium in available phosphorus content and remaining 11% area of arid western Rajasthan is high available phosphorus. The available phosphorus content varies widely in different soils and is about 2.4 to 3.9% of total P and the mean content in different soil series is less than 10 μg g^{-1} (Mathur et al. 2006). Even though soils are often medium to low in available P, response of P-fertilization in arid soils is generally observed only in the years of good rainfall (Aggarwal et al. 1979). The available phosphorus in different soils series ranged from 7 to 80 kg P_2O_5 ha^{-1} with a pooled mean value of 18.6 kg P_2O_5 ha^{-1} in Churu district of arid Rajasthan (Mahesh Kumar et al. 2011). About 67.5 % samples contained low (<11.2 kg P ha^{-1}), 27.5 % medium (11.2-25 kg P ha^{-1}) and only 5.5% samples showed high content of available phosphorus (>25 kg P ha^{-1}). P deficiency in soils (2.7-18.4 kg ha^{-1}) of north eastern parts of arid Rajasthan was also reported by Mahesh Kumar and Sharma (2011 and 2011) and Singh et al. (2009). Gupta et al. (2000) also reported that the available phosphorus content varies widely in arid soils and the mean content in different soil series is less than 20 kg P_2O_5 ha^{-1}, but the alluvial soils contain slightly higher amount. The higher amount of phosphorus in these alluvial soils or soils under irrigated land use system inherited from the parent material from which the soils have been formed and also may be due to carry over effect of continuous application of phosphate fertilizers for the past 2-3 decades (Singh et al. 2007, CAZRI 2004).

Available potassium

The arid soils are well distributed with available potassium (70-890 kg ha^{-1}). About 10, 68 and 22% area of arid western Rajasthan was found low medium and high in available potassium. In the irrigated north-west plain soils of Rajasthan Mathur et al. (2006) reported wide variation of potassium content within the soil group i.e. 277 to 488 kg K_2O ha^{-1} in Torrifluvent, 215 to 389 kg K_2O ha^{-1} in Torripsamment and 200 to 374 kg K_2O ha^{-1} in Calciorthids soil group. Mean values of soil available K were 212, 161 and 156 kg ha^{-1} respectively under grazing lands, agriculture (Rainfed and irrigated agriculture) and sand dunes in north eastern parts of arid Rajasthan (Mahesh Kumar et al. 2011). The total K content in arid region soils ranged between 980-1890 mg $100g^{-1}$ with an average value of 1489 mg $100g^{-1}$ soil (Kumar et al. 2009). Major proportion of total potassium in arid soil is present as mineral form followed by interlayer, non-exchangeable and water-soluble form. Reserve K constituted 8 to 15% of the total K. Arid soils of Rajasthan contain (30 to 1270 mg kg^{-1}) HNO_3-soluble and (20 to 1120 mg kg^{-1}) fixed potassium. Some dune and interdunal soils show negative K fixation (Aggarwal et al. 1979, Joshi et al. 1982). The K fixation capacity was related with the clay content, K-saturation and weathered K-bearing minerals (Talati et al. 1974, Dutta and Joshi 1993).

Micronutrients

Micronutrients have an important role in balanced plant nutrition and stabilization of crop production. Their availability is influenced mainly by the soil properties, particularly pH, organic matter, $CaCO_3$, soluble salts, cation exchange capacity and soil texture. Micronutrient deficiencies in soils get enhanced due to mining by the plants (Rattan and Sharma 2004, Shukla 2011). Micronutrients in soils are derived almost entirely from the parent material. Each of the four micro-elements (Zn, Mn, Fe and Cu) occurs in variety of minerals in both igneous and metamorphic rocks, with Fe being by far the most abundant and Cu is least. Soils rich in ferromagnesian minerals contained higher amount of total iron (Joshi and Dhir 1981) and the soils associated with augite and hornblende is associated with higher amount of total Zn. In arid region of western Rajasthan, the variety of soils reported (Dhir 1977) have been found to be associated with their physiographic occurrence. The soils of the alluvial plain of the mid west of Rajasthan contained comparatively greater amount of HCl soluble and DTPA extractable Mn and Fe than sandy arid plains (Typic Torripsamment), Ghaggar plain (Calciorthid) and plains of interior drainage (Torripsamment). Dune soils occurring in better rainfall zones (400-450 mm) contained higher amount of HCl soluble and free iron (Joshi and Dhir 1983a). Free and reducible Mn was lower in dunes with low rainfall areas as compared to the dunes with high rainfall (Joshi and Dhir 1983a). Free iron and exchangeable Fe were higher in interdune soils. Among the four micronutrients, Zn, Mn and Fe appeared marginal in their deficiencies in arid soils of Bikaner (Soni et al. 2006).

In the arid soils of Rajasthan about 40 per cent soils samples contained 2-5 ppm and 54% samples had 5 to 10 ppm DTPA-Fe (Joshi and Dhir 1983b). Iron in arid soils is comparable with hilly, red loam, brown and grayish brown soils of the Rajasthan (Choudhary et al. 1979). Based on the critical limit for Fe as 4.5 mg kg^{-1} soils of Jhunjhunu district are well supplied with available iron and only 15% soil sample were found deficient in available Fe (Mahesh Kumar et al. 2011). The coarse textured soils associated with Dunes, Molasar, Modasar and Chirai series were found deficient at places in Churu district (Mahesh Kumar et al. 2011b). DTPA extractable Fe constitutes about 3.71 to 4.17% of total Fe. Joshi and Dhir (1981) however reported that in extreme arid parts total HCl soluble, exchangeable Fe ranged between 1.2 to 3.4%, 0.5 to 2.08%, 33 to 21.5 ppm and 3.4 to 8.8 ppm, respectively.

Available Mn content in soils of arid Rajasthan ranged between 1.52–57.5 mg kg^{-1}. Taking 2.5 mg kg^{-1} as threshold value for available manganese most of the soils are adequate in north eastern parts covering Churu and Jhunjhunu in its content. The medium to moderately fine textured soils of Masitawali and Naurangpura series generally had higher content than the other soils (Mahesh Kumar et al. 2011). However, lower content were reported for Jaisalmer and Barmer soils. DTPA extractable Mn in arid soils of Rajasthan varied from 1.1 to25 $\mu g\ g^{-1}$ (Sharma et al. 1985). Total manganese content in arid soils ranged between 250 to 875 $\mu g\ g^{-1}$. Reducible and active forms varied between 3 to 123 and 4 to 128 $\mu g\ g^{-1}$ respectively (Johari et al. 1978).

Arid soils of Rajasthan are in general fairly medium with available zinc while deficiency could be encountered only in more 53% area of arid western Rajasthan and the deficiency was mainly present in soil samples of Barmer and Jailsalmer, Bikaner, Churu, Jodhpur and Nagaur districts. The DTPA extractable zinc content in arid region soils ranged from 0.08 to 4.7 mg kg^{-1} (Mahesh Kumar et al. 2011 and CAZRI 2011). Further the dune field soils showed higher content compared Chirai, Khajwana, Gajsinghpur and Pipar soils. Wide variations in the contents of exchangeable (0.24-1.28 µg g^{-1}) and DTPA soluble (0.27-2.36 µg g^{-1}) have been reported (Sharma et al. 1983). The exchangeable and DTPA soluble from were not associated with clay, silt and organic carbon. Exchangeable and DTPA soluble forms of zinc showed irregular distribution with depth (Joshi and Dhir 1981). The Zn adsorption by sandy soils followed the Langmuir adsorption equation. The quantity of Zn adsorbed by sandy soils was much less than the medium fine textured soils (Joshi et al. 1983a). Significant relationships between the quantity, intensity and supply parameters indicated that sandy soils could maintain higher level of available Zn. However, the calcareous soils had low adsorption maxima and high bonding energy constants than the non calcareous soils. Strong affinity of Zn with $CaCO_3$ indicated low availability of Zn in these soils (Joshi, 1996).

Most of the arid soils of Rajasthan are adequate in copper content (Dhir et al. 1983). DTPA extractable Cu status of soils ranged from (0.06–7.68 mg kg^{-1}) with a mean value (0.74– 0.78 mg kg^{-1}) in arid Rajasthan. Considering the critical limits (0.20 mg kg^{-1}) for copper the soil of Churu and Jhunjhunu district have sufficient amount of available copper, about 96% of samples have more than 0.20 mg kg^{-1} in which 20% of the samples have more than 1.5 mg kg^{-1} of Cu. The higher content of Cu was recorded in medium to moderately fine textured soils and lowest in Dunes (Mahesh Kumar et al. 2011). The exchangeable (0.23-1.56 µg g^{-1}) DTPA extractable forms of copper showed wide variations in different arid soils (0.28-1.25 µg g^{-1}) have been reported (Sharma et al. 1983).

Soil fertility management in arid regions of Rajasthan

Integrated nutrient management

Majority of the arid region soils are deficient in organic carbon and nitrogen, low to medium in phosphorus and medium to high potassium. Hence a balanced fertilization is necessary to increase the productivity. The combined application of nitrogen and phosphorus @ 40 kg ha^{-1} increased the yield of pearl millet by 35 to 150% (Mann and Singh 1977) during good rainfall years. The application of potassium @ 15 kg ha^{-1} along with 60 kg N and 30 kg P_2O_5 further enhanced the yield of pearl millet. The fertilizer requirement of local varieties is less as compared to hybrid of pearl millet. Combined application of chemical fertilizers and organic manures was suggested for reducing the cost of production. Application of sheep manure in arid region increased the yield of pearl millet substantially as compared to urea alone (Aggrawal and Venkateswarlu 1989). Substitution of 50 % fertilizer dose by FYM was equally good to the application of inorganic fertilizer alone (Gupta et al. 1983, Rao and Singh 1993). Aggrawal and Kumar (1996) concluded the beneficial effect of FYM and its synergistic impact on the efficacy of inorganic fertilizers by improving soil moisture status and micro flora. Crop residues incorporation in soil has an important role to play in improving the productivity of soil, beside conservation effects. A higher yield of pearl millet with cluster bean residues and FYM application was reported due to increased nitrogen use efficiency. Incorporation of crop residue into soil has been reported to increase the organic matter content of soil (Aggrawal et al. 1996). The carbon, phosphorus and potassium stock have been increased after 5 continuous year of manuring.

Biofertilizers such as *Azospirillum, Rhizobium*, blue green algae and VAM fungi further improve the fertilizer use efficiency. Biomass production and grain yield of pearl millet were enhanced on seed inoculation with efficient strains of *A. brasilense*. Inoculation also reduced the doses of nitrogen by 13 to 20 kg ha^{-1} in pearl millet and increased the production of forage by 10 to 15% in arid region (Rao and Venkateswarlu 1987). Seed inoculation with efficient strains of Rhizobium, biomass production and grain yields of arid legumes significantly increased over non inoculated control (Rao and Venkateswarlu 1983). The phosphate solubilizing microorganisms enhanced the availability of phosphorus from phosphorus rich organic compound. Phosphate solubilizing fungi belonging to *Pencillium* and *Aspergillus* group enhanced the dry matter production and grain yield of clusterbean and green gram (Tarfdar et al. 1989). Similarly, VAM fungi inoculation in the soils at the time of sowing increased dry matter production and grain yields of arid legumes significantly besides improving nodulation and nitrogenase activity. Concentration of N, P, Zn and Cu in the shoot was found significantly higher in the inoculated plants (Tarafdar and Rao 1997). Thus combined application of organic, inorganic and biofertilizers increased the crop and forage production in the arid region.

Crop residue management

Incorporation of crop residues and natural vegetation in soil improve microbial activity during decomposition. Also adhesive action of decomposition products improves soil aggregation, hydraulic conductivity and moisture retention (Gupta and Gupta 1986). Leaving the crop residues in soil generally have a positive effect on grain yield (Aggarwal et al. 1996). Rao and Singh (1993) have reported crop residues to be as efficient source of nutrient as other organics like cattle manure and compost. However, Aggarwal et al. (1996) did not find any significant change in the yield of succeeding crop of pearl millet after addition of crop residues with wide C:N ratio, whereas it was significant after incorporation of residues with narrow C:N ratio.

Crop rotation inclusion of legumes in cropping systems

Legume rotations are an important practice for restoring soil fertility on larger land holdings. The amount of N returned from legume rotations depends on whether the legume is harvested for seed, used for forage, or incorporated as a green manure.

Mann and Singh (1977) and Singh et al. (1985) also reported that in arid soils pearl millet-clusterbean rotation gave higher yield than of continuous cultivation of pearl millet due to improved soil fertility (Praveen-Kumar et al. 1996). Singh et al. (1985) in a long term study found an increase soil organic carbon by 12% and available soil P by 25% after legume cultivation. Singh and Singh (1977) on the basis of a long term study reported that cultivation of green gram in rotation with pearl millet supplied with 20 kg N ha^{-1} and gave similar yield as with application of 40 kg fertilizer N ha^{-1}. Singh et al. (1985) observed that rotation of pearl millet with green gram or clusterbean was found better than its rotation with moth bean. Praveen-Kumar et al. (1996) reported that higher yield of pearl millet when it was preceded by clusterbean cultivation than green gram. Beneficial effect of legumes to pearl millet also depends on the number of seasons of their cultivation prior to pearl millet. The intercropping of pearl millet and legumes in arid soils has also shown promising results (Singh and Joshi 1980). This could be seen from a long-term trial conducted at CAZRI Jodhpur with Pearl millet-moth bean based cropping sequence in a rotation of four years. The sequence maintains initial SOC of 0.22 and 0.14% in surface and subsurface horizons. Another sequence with legume-legume-legume-pearl millet increased soil organic carbon. Other rotation consisting of fallow-legume-legume-pearl millet also gave the similar results (Praveen-Kumar and Aggarwal 1997).

Management of saline and high RSC water for irrigation- Saline water (EC 3.4-11.3 dS m^{-1} and SAR of 18 to 20), charged with chlorides and sulphates of sodium, calcium and magnesium, is commonly used for irrigation. Irrigation with saline water results in development of high salinity and sodicity that leads to poor physical condition of soils. However, the negative impact of saline water on infiltration and SAR could be mitigated, if irrigation is done after the gypsum treatment @ 50 and 100% of soil requirement. The improvements are reflected in terms increased yields in loamy sand soils of Balotra and Osia in Rajasthan (Joshi and Dhir 1992, Mahesh Kumar et al. 2016a and 2016b). The impact of carbonate and bicarbonate salinity present in irrigation could be effectively utilized by applying them in association with gypsum (Joshi and Dhir 1991). The higher plant density and tiller/plant could be attained in gypsum treated plot as compared to their non-treated counterpart. The quantity of gypsum required to neutralize RSC in excess of 5 meq l^{-1}, resulted an increase of 135-401 kg ha^{-1} grain yield of gram (Joshi and Dhir 1994). The quantity is more effective in lowering soil pH by 0.3 to 1.0 units and decrease of SAR by 6.4 to13.3 and also improvement in nutrient status could be attained (Joshi and Dhir 1994, Mahesh Kumar and Singh 2012).

Moisture conservation

Increased availability of good quality water for irrigation improves soil quality and productivity by increasing period of vegetative cover, vegetative input to the soils and microbial population. Storage of rainwater in tanks, creation of subsurface barrier for ground water recharge and khadin management are some of the techniques that can be helpful for increasing the availability of fresh water for irrigation. *In situ* moisture conservation including inter row water harvesting, field bunding, mulching, deep ploughing and other agronomic practices such as drought tolerant cultivars, optimum plant density, and proper sowing time, balance fertilization, use of sprinklers and drips for irrigation are some of other techniques that have beneficial effect on soil quality and productivity.

Agro forestry

Growing of crops with shrubs, herbs and grasses are the old age practice for providing fodder to the animals, timber to the farmers and for improving soil health. The practice simultaneously could enrich the soils by sequestering 121% higher organic carbon than the pearl millet-fallow system. Growing of trees with crops could sequester 6.29 kg m^{-2} higher atmospheric CO_2 than the cultivation of crops alone (Singh et al. 2007). This could be possible because of higher biomass production (Aggrawal et al. 1993) and higher soil moisture profile beneath the agro forestry system. Extensive research revealed that agro forestry including moth, cluster bean and local variety of pearl millet as a crop component with *Calligonum polygonoides* and *Lasiurus sindicus* as perennial trees and grass, respectively are more successful in 100 to 250 mm rainfall region of arid areas, while plantation of *Prosopis* and *Ziziphus species* in the field and *Capparis decidua* on the boundary with pearl millet, moth bean, sesame and clusterbean is beneficial in 250 to 350 mm rainfall areas. However, growing of *Prosopis cineraria* and *Tecomella undulata* with pearl millet, green gram, moth bean and cluster bean is advantageous in the area of 250 to 450 mm rainfall adjoining to the semi arid region. In the irrigated area of arid region plantation of *Prosopis cineraria* and *Acacia nilotica* with wheat, barley, mustard and gram in winter cotton, sorghum, pearl millet and sesame in summers are expected to improve soil health and total productivity.

Conclusions

Severe erosion, force fallowing, inadequate or imbalanced nutrient application, in sufficient moisture, lack of soil aggregation, mono-cropping, salinity, sodicity, gradual building of inorganic carbon, excessive tillage and use of marginal land for cropping are some of the factors that threat to the soil fertility vis-a vis productivity in the arid region. Crops respond favourably to application of nutrients in the dryland situations except in some areas received low rainfall. Organic sources of nutrients like FYM/compost in combination with chemical fertilizers should be applied as per the recommendation to maintain long term productivity and improved soil physical conditions. So in view of above a holistic approach dealing with agro-ecozoning, evaluation of land for ideal land utilization types, optimization of land use may improve the entire system in terms of better soil quality and enhanced productivity. Provided, these are supported with agro techniques which includes adequate area allocation for alternate land use, agroforestry, agro horticulture, integrated nutrient, water and pest management, conservation tillage, moisture conservation, water harvesting and special allowance for drought mitigation.

References

Aggarwal, R.K., Dhir, R.P., Bhola, S.N. and Kaul, P. 1975. Distribution of nitrogen fractions in Jodhpur soils. Annals of Arid Zone 14:183-190.

Aggarwal, R.K., Sharma, V.K. and Dhir, R.P. 1979. Studies on potassium fixation and its relationship with chemical characteristics of some desert soils of western Rajasthan. Annals of Arid Zone 18: 174-180.

Aggarwal, R.K. and Lahiri, A.N. 1981. Evaluation of fertility status of stabilized and unstabilized dunes of Indian desert. Agrochemica 25:54-60.

Aggarwal, R.K. and Venkateswarlu, J. 1989. Long term effect of manures and fertilizers on important cropping systems of arid region. Fertilizer News 34: 67-70.

Aggarwal, R.K., Praveen-Kumar and Raina, P. 1993. Nutrient availability from sandy soils underneath *Prosopis cineraria* compared to adjacent open site in an arid environment. Indian Forester 119: 321-325.

Aggrawal, R.K., Parveen-Kumar and Power, J.F. 1996. Use of crop residue and manure to conserve water and enhance nutrient availability and pearl millet yields in arid tropical region. Soil Tillage Research 41: 43-51.

Aggarwal, R.K. and Praveen-Kumar. 1996. Integrated use of farm yard manure and fertilizer N for sustained yield of pearl millet (*Pennisetum glaucum*) in arid region. Annals of Arid Zone 35: 29-35.

CAZRI 2011. Annual Report, Central Arid Zone Research Institute, Jodhpur, Rajasthan.

CAZRI 2004. Annual Report, Central Arid Zone Research Institute, Jodhpur, Rajasthan.

Choudhari, J.S., Pareek, B.L. and Jain, S.V. 1979. Distribution of different forms of iron in major soil groups of Rajasthan. Journal of Indian Society of soil Science 27: 338-340.

Choudhari, J.S., Jain, S.V. and Mathur, C.M. 1979. Phosphorus fractions in some soil orders of Rajasthan and their sequence of weathering stage. Annals of Arid Zone 18: 260-268.

Dhir, R.P. 1977. Western Rajasthan soils: their characteristics and properties. In: Desertification and its Control, ICAR, New Delhi. pp. 102-115.

Dhir, R.P., Sharma, B.K. and Joshi, D.C. 1983. Availability of iron, manganese, zinc and copper in some soils of western Rajasthan. Annals of Arid Zone 22: 343-349.

Dutta, B.K. and Joshi, D.C. 1993. Studies on potassium fixation and release in arid soils of Rajasthan. Transactions of Indian Society of Desert Technology 18: 191-199.

Gupta, J.P, Aggrawal, R.K., Gupta, G.N. and Kaul, P. 1983. Effect of continuous application of farm yard manure and urea on soil properties and production of pearl millet in western Rajasthan. Indian Journal of Agriculture Science, 53, 53-56.

Gupta, J.P. and Gupta, G.N. 1986. Effect of tillage and mulching on soil environment and cowpea seedlings growth under arid conditions. Soil Tillage Research 7: 233-240.

Gupta, J.P., Joshi, D.C. and Singh, G.B. 2000. Management of arid agro-ecosystem In: Natural Resources Management for Agricultural Production in India. (J.S.P. Yadav and G.B. Singh, Eds.), pp. 557-668.New Delhi, India.

Johari, S.N., Joshi, D.C. and Sharma, V.C. 1978. Study on the manganese status of some soils of western Rajasthan. Annals of Arid Zone 17: 133-135.

Joshi, D.C. 1996. Adsorption and desorption of Zn by calcareous Aridisols. Annals of Arid Zone 35: 319-324.

Joshi, D.C. 1990. Composition and nature of humus in typical aridisols of Rajasthan. Annals of Arid zone 29: 93-97.

Joshi, D.C. and Dhir, R.P. 1981. Distribution of different forms of copper and zinc in the soils of extremely arid part of western Rajasthan. Journal of Indian Society of Soil Science 29: 379-381.

Joshi, D.C. and Dhir, R.P. 1983a. Distribution of micronutrients forms along duny landscape. Annals of Arid Zone 22: 135-141.

Joshi, D.C. and Dhir, R.P. 1983b. Available forms of Maganese and iron in some arid soils and their relation with soil properties. Annals of Arid Zone 22: 7-14.

Joshi, D.C. and Dhir, R.P. 1991. Rehabilitation of degraded sodic soils in an arid environment by using residual Na-carbonate water for irrigation. Arid Soil Research and Rehabilitation 5: 175-185.

Joshi, D.C. and Dhir, R.P. 1992. Amelioration of degraded saline lands in the arid regions of Rajasthan. Current Agriculture 16: 33-44.

Joshi, D.C. and Dhir, R.P. 1994. Amelioration and management of soils irrigated with sodic water in the arid region of India. Soil Use and Management 10: 30-34.

Joshi, D.C., Gupta, B.S. and Dutta, B.K. 1982. Soil factors affecting forms of potassium and potassium fixation capacity. Annals of Arid Zone 21: 199-205.

Joshi, D.C., Dhir, R.P. and Gupta, B.S. 1983a. Influence of soil parameters on DTPA extractable micronutrients in arid soils. Plant and Soil 72: 31-38.

Kumar, P., Tarafdar, J.C., Painuli, D.K., Raina, P., Singh, M.P., Beniwal, R.K.,Soni, M.L., Mahesh Kumar, Santra, P. and Shamsudheen, M. 2009. Variability in arid soil characteristics. In A. Kar, B.K. Garg, M.P. Singh & S. Kathju (Eds.),Trends in Arid Zone Research in India. Central Arid Zone Research Institute, Jodhpur, India, pp 78-112.

Mahesh Kumar, Singh, S.K. and Sharma, B.K. 2009. Characterization, Classification and Evaluation of Soils of Churu District. Journal of the Indian Society of Soil Science 57: 253-261.

Mahesh Kumar and Sharma, B.K. 2011. Soil fertility appraisal under dominant land systems in North eastern parts of Arid Rajasthan. Annals of Arid Zone 50: 11-15.

Mahesh Kumar and Singh, R. 2012. On farm management of sodic soils in arid wester Rajasthan. In International conference on "Extension Education in the Perspectives of Advances in Natural Resource Management in Agriculture (NaRMA-IV)" during 19–21 December, 2012 at Swami Keshwanand Rajasthan Agriculture University, Bikaner. pp. 45.

Mahesh Kumar, Moharana, P.C. Raina, P and Amal Kar. 2014. Spatail distribution of DTPA-extractable micronutrients in arid soils of Jhunshunun district, Rajasthan, Annals of Arid Zone 53:9-15.

Mahesh Kumar, Singh, R. and Panwar, N.R. 2016. Interventions of high RSC-water degraded soils amelioration technology in Indian Thar desert. In 25th National conference on "Natural Resource Management in Arid and Semi-arid Ecosystem for Climate resilient Agriculture and Rural Development during 17–19 February, 2016 at Swami Keshwanand Rajasthan Agriculture University, Bikaner pp. 18

Mahesh Kumar, Amal Kar and Raj Singh 2016b. Interventions of high residual sodium carbonate cvator-degraded soils amelioration technology in indian thur Desert and Formers' response. national Academy science Letters 39: 245-249.

Mahesh Kumar, Singh, S.K., Raina, P. and. Sharma, B.K. 2011. Status of available major and micronutrients in arid soils of Churu district of western Rajasthan. Journal of the Indian Society of Soil Science 59: 188-192.

Mann, H.S. and Singh, R.P. 1977. Crop production in Indian Arid Zone. In: Desesrtification and its Control, ICAR, New Delhi. pp. 215-334.

Mathur, G.M., Ramdeo and Yadav, B.S. 2006. Status of Zn in irrigated north-western plain soils of Rajasthan. Journal of the Indian Society of Soil Science 54: 359-361.

Narain, P. and Joshi, D.C. 2001. Nature, distribution and management of desert soils in India. In Souvenir 2001, 66th Annual convention of the Indian society of soil Science (Eds. L.L. Somani and K.L. Totawat), Udaipur Chapter, Indian Society of soil Science, Department of Agricultural Chemistry and Soil Science, Maharana Pratap University of Agricultural and Technology, Rajasthan College of Agriculture, Udaipur. pp. 50-60.

Praveen-Kumar, Aggarwal, R.K. and Power, J.F. 1996. Cropping systems: Effect on soil quality indicators and yield of pearl millet in arid Rajasthan. American Journal of Alternative Agriculturc 14: 178-184.

Praveen-Kumar and Aggarwal, R.K. 1997. Nitrogen in arid soils: availability and integrated management for increased pearl millet production. In: *F*rom Research Station to Framer's field: Nutrient Management Research for Millet Based Cropping Systems for Western Rajasthan (Eds. B. Seeling, and N.L. Joshi), International Crop Research Institute for Semi-Arid Tropics, Hyderabad. pp. 25-42.

Rao, A.V and Venkateswarlu, B. 1983. Microbial ecology of the soils of the Indian desert. Agricultural Ecosystem and Management 10: 361-369.

Rao, A.V. and Venkateswarlu, B. 1987. Nitrogenase activity of pearl millet-Azospirillum association in relation to the availability of organic carbon in the root exudates. Proceedings of Indian Academy of Sciences 97: 33-37.

Rao, V.M.B. and Singh, S.P. 1993. Crop responses to organic sources of nutrients in dryland conditions. In: Advances in Dryland Agrilculture (Ed. L.L. Somani), Scientific Publishers, Jodhpur. pp. 287-304.

Rattan, R.K. and Sharma, P.D. 2004. Main micronutrients available and their method of use. In: Proceedings, IFA International Symposium on Micronutrients, pp. 1-10.

Sharma, B.K., Joshi, D.C., Gupta, B.S. and Raina, P. 1983. Influence of physiography soil association on the vertical distribution of different forms of copper and zinc in some soils of Rajasthan. Current Agriculture 7: 174-180.

Samra, J.S., Narain, P., Ratan, R.K. and Singh, S.K. 2006. Drought management in India. Indian Society of Soil Science, Bulletin No. 24: 1-82.

Sharma, B.K., Dhir, R.P. and Joshi, D.C. 1985. Available micronutrient status of some soils of arid zone. Journal of the Indian Society of Soil Science 33: 50-55.

Shukla, A.K. 2011. Micronutrient research in India: Current status and future strategies. Journal of the Indian Society of Soil Science 59: S88-S98.

Singh, P. and Joshi N.L. 1980. Intercropping of pearl millet in arid areas. *I*ndian Journal of Agricultural science 50: 338-341.

Singh, K.C. and Singh, R.P. 1977. Inter cropping of annual grain legumes with sunflowers. Indian Journal of Agricultural Science 47: 563-567.

Singh, S.D., Bhandari, R.C. and Aggarwal, R.K. 1985. Long term effects of phosphate fertilizers on soil fertility and yield of pearl millet grown in rotation with grain legume. Indian Journal of Agricultural Science 55: 274-278.

Singh, S.K., Mahesh Kumar and Sharma, B.K. 2009. Changes in soil properties in hot arid region of India. Journal of the Indian Society of Soil Science 57: 24-30.

Singh, S.K., Mahesh Kumar, Sharma, B.K. and Tarafdar, J.C. 2007. Depletion of organic carbon, phosphorus and potassium under pearl millet based cropping system in the arid region of India. Arid Land Research and Management 21: 119-131.

Soni, M.L., Beniwal, R.K., Mondal, B.C. and Yadava, N.D. 2006. Micronutrient status of arid soils under three predominant land use systems in arid zone. Current Agriculture 30: 53-56.

Talati, N.R., Mathur, G.S. and Attri, S.C. 1975. Distribution of various forms of phosphorus in north-west Rajasthan soils. Journal of Indian Society of soil Science 23: 202-206.

Talati, N.R., Mathur, S.K. and Attri, S.C. 1974. Behaviour of available and non exchangeable potassium in soils and its uptake in wheat crop. Journal of Indian Society of soil Science22: 139-144.

Tarafdar, J.C. and Rao, A.V. 1997. Response of arid legumes to VAM fungal vinoculation. Symbiosis 22: 265-274.

Tarafdar, J.C., Kiran Bala and Rao, A.V. 1989. Phosphatase activity and distribution of phosphorus in arid soil profiles under different land use patterns. Journal of Arid Environment 16: 29-34.

17

Desertification in the Arid Part of Rajasthan: Status Mapping Through Interpretation of Multi-date Satellite Images

Pratap Chandra Moharana

ICAR-Central Arid Zone Research Institute, Jodhpur, Rajasthan, India

Concept and Definition

Desertification is a major problem in the drylands of India, affecting the way of life for its inhabitants. The problem is more severe in the arid lands in the north-western part of the country. There have been a few numbers of definitions of the word "Desertification". However for simplicity of its origin and meaning we would refer to two of the frequently used ones; (1) proposed in 1991 by UNEP: "Land degradation in arid, semi-arid and dry sub-humid areas resulting mainly from adverse human impact and (2) prepared for the Earth Summit at Rio, held in June 1992, for inclusion in Agenda 21, as "Land degradation in arid, semi-arid and dry sub-humid areas resulting from climatic variations and human activities. (Anonymous 1992).

Comparing two of these initial definitions, it would show that first one indicated desertification as a state of natural resources like soil and vegetation, mainly resulting from human activity while the second one included climate as a major cause. However, it reduced the emphasis of human activities as being the primary responsible factor and tried to link the possible impact of climate change on desertification. But the simplicity of the 1992 definition of desertification helped researchers to bring in sharp focus on the special environmental and socio-economic problems of the drylands.

So over the years, the key words that explains the term "desertification" has been changing or are being amended for specific purposes. This would at least mean that the problem of desertification is still debated world-wide, and its impact is not limited to some nations/regions as was presumed and discussed in first of UNCCD platforms. Therefore, as a precaution and combating this process, few catchy words like non-sustainable management, productivity, over-exploitation of natural resources, dryland livelihood, droughts are found to surround the word desertification. Earlier, some, people equated desertification with advancing boundaries of the existing desert and in India, examples of expansion / contraction of geographical boundary of Thar desert was a public debate among various researchers and planners. In the same time, some of the definitions tried to broaden its geographical scope so that greater populations suffering from adverse effects of land degradation would benefit out of various remedial measures.

Need for Mapping

Mapping is one of the best ways of understanding both spatial and temporal dimensions of desertification. Since, first impression of land degradation is easily derived from the surface manifestations, mapping is the first of the technical steps in the assessment of desertification. With the revolution in the remote sensing platforms, sensors and resolutions of images produced by them, researchers have a wide number of options to choose scales for measurement of surface manifestations of desertification. Use of Geographical Information System is now mandatory, not only for creating a series of maps but for storing and displaying a digital database for ready reference.

But for all purpose, mapping requires few basic steps, (1) use of platforms which may vary from aerial photographs to satellite images, (2) a reference map like survey of India toposheets for Indian region, (3) interpreter's basic knowledge of the region.

Need for a Database at National Level

At our country level, a number of organizations like ICAR (Institutes: NBSS & LUP, CAZRI) and other Central Institutes like AIS&LUS, ISRO (NRSC and SAC) and National Wasteland Board have carried out mapping to estimate degraded areas. Based on methodology and the criteria, the figures indicating total degraded area vary. Very recently, a national attempt was made to harmonize the different datasets from different agencies on the status of degraded lands in India by a team of scientists from Indian Council of Agricultural Research (ICAR), namely CAZRI (Jodhpur), CSSRI (Karnal), CSWCRTI (Dehradun), NBSS&LUP (Nagpur), as well as from National Remote Sensing Centre (NRSC) of ISRO institutes. The data (Table 17.1) showed that 120.72 mha of the country is degraded. Wind erosion is estimated to affect 12.40 m ha area, the majority of such area falling within hot arid regions of western Rajasthan.

Table 17.1: Harmonized estimates of degraded and wastelands in India

Degradation category	Arable land (m ha)	Open forest (m ha)	Total (m ha)
Water erosion (>10 t/ha/y)	73.27	9.30	82.57
Wind erosion	12.40	-	12.40
Exclusively salt-affected	5.44	-	5.44
Salt-affected & water eroded	1.20	0.10	1.30
Waterlogging	0.88	-	0.88
Exclusively acidic soils	5.09		5.09
Acidic and water eroded soils	5.72	7.13	12.85
Mining/industrial wastes	0.19		0.19
Total affected area	104.19	16.53	120.72

Source: Anonymous 2010

Desertification Indicators

Researches in CAZRI indicate that desertification usually takes place when the rates of natural geomorphic processes in the drylands are accelerated by human action or through extreme natural events like droughts, high-intensity rains, tectonic disturbances, etc. The degraded areas, therefore bear the signatures of wind and water erosion, soil salinity, waterlogging, industrial effluent discharges, mine wastes, denuded pastures and forests, etc. Analysis of approaches to desertification studies in the world would indicate that there is a gradual shift from a dominantly biophysical-process-based enquiry to an amalgamated biophysical-cum-socio-economic analysis.Therefore, many of the recent approaches suggest identification and measurement of the biophysical and socio-economic factors (drivers). Reynolds et al. (2011) advocated an integrated assessment model for the human-environment dryland system, through integration of sub-models on environmental, social, economic and institutional variables, which may effectively link the scientific research with policy development.

Our experiences of working in arid western part of Rajasthan suggest that definitions have implications considering various scales of assessment which varies from farm plots to regional level and global. This concept of scale is appropriate since we are equating global climate change, occurrence of droughts and also impact of human activities on the occurrences of desertification. Generally, it is advocated that a better understanding of the incidences of land degradation, its causes and trends requires the interpretation of data collected over a period of several years, it sometimes becomes very difficult to draw broader conclusions from micro-level studies for understanding processes at a regional or global level.

Field level indicators in arid part of rajasthan

Wind erosion/deposition, water erosion, waterlogging, salinity/alkalinity and vegetation degradation are the major problems in western Rajasthan (Singh et al.1992). Besides, industrial effluents and mining are also gradually becoming important factors of desertification. For, field level measurements, we need to look at the landuse/landcover. According to Dregne (1991) the processes are to be monitored in rainfed croplands, irrigated croplands and rangelands. Unfortunately, there is hardly any consensus on the appropriate indicators and benchmarks that can be applied at global, regional, sub-regional, national and local levels. A selection of the case studies, and the indicators on which these are based, is provided in UNEP's Atlas of Desertification (Middleton and Thomas 1997).

CAZRI's studies in the arid lands of Rajasthan suggest that some terrain features (Ghose et al. 1977) can be used as state indicators of desertification, provided we understand the spatial pattern of process-form interactions, and vulnerability of the landscape (Kar et al. 2009). Kar (1993) developed a climatic wind erosion index (WEI) for Thar Desert. He correlated these index values with the spatial pattern of the degree of reactivation of the sandy terrain. The study showed that the areas having index values between 120 and 480 are very highly vulnerable to wind erosion where he found fields of barchans which are low dunes. Areas having >480 values corresponded to field of megabarchans in the south of Jaisalmer. In case of water erosion, researchers from CAZRI studied various dimensions from field and used DEMs (Digital Elevation Models) to assess flash floods in western Rajasthan. In July 1979, high intensity monsoon rainfall for 5 days at a stretch in the southern part of the desert affected nearly 108799 ha of agricultural land with either rill and gully formation or coarse sand deposition due to shifting channels. Crops were damaged over 87186 ha area. Similarly, two spells of high intensity monsoon rains in July and August, 1990, led to flooding and soil erosion in the same region (Sharma and Vangani 1992, Vangani 1997). In August, 2006 high-intensity rainfall for three days in Barmer-Jaisalmer tract not only resulted in year-long flooding of some interdune plains, but also led to erosion of topsoil and nutrients from the upper catchment and their deposition in the lower catchment (Fig.17.1). The stream network, revived after more than half a century, however, did not deviate from the simulated channel courses for the area (Kar et al. 2007, Kar 2011).

Fig. 17.1: Sediment deposits during Flash flood of 2006 in Barmer district

Sediments deposits in the low lands during Kawas flood event in 2006
Quality assessment of above sediment deposits.

- Concentration of zinc, iron and copper ranged from 0.01 to 0.12, 0.19-0.65 and 0.03 - 0.22 ppm respectively.
- Concentration of bicarbonate ranged from 48 to 122 ppm, while carbonate was in traces.
- Concentration of sodium, potassium and calcium ranged from 1.38 – 12.88, 0.05 – 6.63 and 13 – 40 ppm respectively.
- Electrical conductivity varies from 0.07 – 0.40 dS/m while pH ranged from 7.6 – 8.7.

Some of the critical field indicators for assessing desertification in the western part of Rajasthan state are presented in Table 17.2 and Fig.17.2. Since, wind erosion is the single major problem of desertification in western Rajasthan, some of its indicators are mentioned below.

Table 17.2: Field indicators for assessing wind erosion/ deposition in the Thar Desert

Terrain	Average rainfall (mm)	Major indicators for assessment	Severity
Flat sandy plains with dominantly loamy sand to sandy loam soil.	100-550	Fresh sand sheet up to 30 cm thick; few scattered new fence line hummocks and nebkhas up to 100 cm high.	Slight
Moderately sandy undulating plains and sand dunes with loamy sand soils; thickly sand sheeted plains.	Above 300	Presence of reactivated fresh sand of 50 to 150 cm thickness on stable dunes, sandy plains and fence line hummocks; many recently formed nebkhas.	Moderate
Moderately sandy undulating plains and sand dunes with sand to loamy sand soils.	Below 300	Reactivated and fresh sandy hummocks (nebkhas) and sand ridges of 90-300 cm height; sand sheets of 60-150 cm thickness between undulations; reactivated stable dunes with fresh sand deposits of 70 to 200 cm thickness; exposed plant roots to a depth of 40 to 100 cm in the sandy plains indicate erosion.	Moderate
Moderate to strongly undulating sandy. plains with closely spaced hummocks and high sand dunes with sand to loamy sand soils.	100-550	Closely spaced sandy hummocks and sand ridges of 1 to 4 m height with fresh sand cover; sand deposits of 100-300 cm thickness usually present between undulations; highly reactivated sand dunes are covered by fresh sand and superimposed by crescentic bedforms of 2 to 4 m height.	Severe
Barchan dunes and very thick sandy plains with loose sand throughout the profile.	100-550	Areas of drift sand, especially as fields of barchans of 2 to 5 m height, which encroach upon roads, settlements and agricultural fields; also areas with very closely spaced nebkhas of 2-5 m height.	Very severe

Wind deposition (Low) : Sandy plains, levelled and put under crops under conserved moisture

Wind erosion (Moderate): Stable sand dunes, barren crest zone with fine sand deposits.

Wind erosion (Severe) : Low , mostly barren and sparse or no vegetation cover (<10% area)

Wind erosion (sand sheet) : Typical wind deposition over a rocky/gravelly plain

Limestone mining, spoil deposits and land degradation

Action of wind, water and chemical weathering on a rocky landform

Rock mining : denudation, erosion and deposition over good croplands

Crops (Mustard) on dune slopes, reducing wind erosion/deposition

Gullied lands : moderate water erosion

Spheroidal rock weathering : action of water and wind

High rainfall events and water erosion

Flash flood and inundation in Malwa Barmer district in 2006

Fig. 17.2: Field photographs showing physical manifestations of desertification

CAZRI's role in desertification mapping and assessment

The first mapping of desertification by CAZRI was carried out in 1977 by the UN Conference on Desertification (UNCOD) held at Nairobi. The processes of desertification in arid western Rajasthan were mapped at 1: 2 million scale as per the guidelines and nomenclature supplied by the UNCOD. Estimates of land degradation in India, therefore, vary from 53 m ha to 187 m ha. It is because of the fact that criteria set up by various institutes / organizations for land degradation assessment (perceived degree of erosion/ deposition, etc. from terrain features; reliance on secondary data, etc. are different. In 1991 CAZRI prepared a 1:1 million scale map of desertification that was based on visual interpretation of 1:250,000-scale false colour composites of Landsat-TM images of western Rajasthan for 1989-1991, which was backed up by extensive field information. The mapping used a combination of legends instead of single process.

Methodology of a recent assessment of desertification through mapping

Central Arid Zone Research Institute was entrusted to be a part of a national level task by the Space Applications Centre (SAC), in association with several national institutes in 2003-04 to prepare a Desertification Status Map for the entire country at 1:0.5 m scale. A three-level, hierarchical classification system was prepared that included land use/land cover (level 1), processes of degradation (level 2), and severity of degradation (level 3). The criteria (SAC 2007a) for the assessment of different land degradation categories are presented in Tables 17.6 to 17.10. Since it is difficult to differentiate slight and moderate categories at this scale, only two categories of severity were identified and mapped: low (corresponding to slight and moderate degradation) and high (corresponding to severe and very severe degradation). Satellite data for *kharif*, *rabi* and summer seasons of 2003, 2004 and 2005 were used for visual interpretation of degraded areas. Ground truth was collected to verify the mapped features. Following Tables (Table 17.3, 17.4 and 17.5) shows the classification system used for the purpose.

Table 17.3: Level-1 mapping of land use/ land cover for nation-wide mapping of desertification

Land use category	Mapping code
Agriculture – Unirrigated	D
Agriculture – Irrigated	I
Forest	F
Grassland/Grazing land	G
Land with scrub	S
Barren / Rocky area	B/R #
Dune / Sandy area	E
Water body / Drainage	W
Glacial / Peri-glacial (in cold region)	C/L
Others (Urban, Man-made etc.)	T

Table17.4: Level-2 mapping of processes of degradation for nation-wide mapping of desertification

Degradation category	Mapping code
Water Erosion	w
Wind Erosion	e
Waterlogging	i
Salinization / alkalinisation	s/a
Vegetation degradation	v
Mass Movement (in cold areas)	g
Frost heaving (in cold areas)	h
Frost shattering (in cold areas)	f
Man-made (Mining/Quarrying, Brick Kiln, Industrial Effluents, City Waste, Urban Agg., etc.)	m

Table 17.5: Level-3 mapping of severity of degradation for nation-wide mapping of desertification

Slight	1
Moderate	2
Severe	3

Table 17.6: Criteria for nation-wide mapping of the severity of water erosion

Status	Desertification classes		
	Slight	Moderate	Severe
Non-arable land			
Type of erosion	Sheet erosion and/or single rills (depth = 0.5 m and width = 0.4-0.9 m)	Rill erosion, and/ or formation of gullies (depth= 0.6-3.0 m and width = 1.0-3.5 m)	Network of gullies / ravines (In Gullies: depth 3 m =10 m and width= 3.5-20.0 m; in Ravines: depth = >10 m, width = 20-40 m)
Density of channels, linear km per sq. km	<0.5	0.5-1.5	1.5-3.0
Removal of top soil horizon, %	<25	25-50	>50
Arable land			
Removal of top soil horizon	<25	25-50	>50
Loss of yield of main crop, %	<25	25-50	>50

Table 17.7: Criteria for nation-wide mapping of the severity of wind erosion

Status	Desertification classes		
	Slight	Moderate	Severe
Non-arable land			
Sand sheet in cm	< 30 cm per hummock up to 100 cm over plains	< 50-150 cm/stable dune and sandy hummock (east of 300 mm isohyet); < 90-300 cm/reactivated sand/plant roots up to 40-100 cm (west of 300 mm Isohyet)	< 1-4 m dunes/100-300 interdunal sand/ Barchans 2-4 cm (mostly west of 300 mm isohyet); 2-5 m active/ drifting dunes: very severe
Percent area covered with sand dunes	<30	30-70	>70
Percent area covered with sod forming plants	50-30	30-10	<10
Arable land			
Removal of top horizon, %	<25	25-50	>50
Blow-outs, percentage of area	<5	5-10	>10
Loss of yield of main crops, %	<25	25-50	>50

Table 17.8: Criteria for nation-wide mapping of the severity of salinity / alkalinity

Status	Desertification classes		
	Slight	Moderate	Severe
Soil salinity/ alkalinity	4-8 dS/m; <15	8-30 dS/m; 15-40	>30 dS/m; >40
Soil salinization, solid residue, %	0.20-0.40	0.40-0.60	>0.60
Salinity of ground water, g/litre	3-6	6-10	10-30
Salinity of irrigation water, g/litre	0.5-1.0	1.0-1.5	>1.5
Seasonal salt accumulation, t/ha	16-30	30-45	45-90
Loss of yield of main crop, %	<15	15-40	>40

Table 17.9: Criteria for nation-wide mapping of the severity of water logging

Status	Desertification classes		
	Slight	Moderate	Severe
Water logging	Seasonal (affecting one crop); 4-6 months	Affecting two crops; >6 months of submergence	Inland marshes

Table 17.10: Criteria for nation-wide mapping of the severity of vegetation degradation

Status criteria	Desertification classes		
	Slight	Moderate	Severe
Plant community	Climax or slightly changed	Long lasting secondary	Ephemeral secondary
Percentage of climax species	> 75	75-25	< 25
Decrease of total plant cover, %	< 25	25-75	> 75
Loss of forage, %	< 25	25-75	> 75
Loss of current increment of wood, %	< 25	25-75	> 75

Desertification in Rajasthan: Status mapping

The mapping revealed that leaving aside the islands, India had 105.48 mha (32.07% of country's area, excluding the islands) under different land degradation categories. Notable among these, water erosion covered 33.56 mha, followed by vegetation degradation (31.66 m ha) and wind erosion (17.56 m ha). The total area affected by desertification within arid region was 34.89 m ha, while that in the semi-arid region was 31.99 m ha (SAC 2007b). Summary of details of the mapped desertification units in arid western Rajasthan shows wind erosion/ deposition occurred in 159561 sq. km (75.83%), water erosion (3938 sq. km, 6.96%), salinity/alkalinity (4551 sq. km, 2.16%), waterlogging (183 sq. km, 0.08%), vegetation degradation (5854 sq. km, 2.78%), mining (202 sq.km, 0.09%) and rocky area (2129 sq. km, 1.01%).

Conclusion

The arid zone in western Rajasthan is among the most thickly populated drylands in the world. More than 3-fold increase in its human population and 2-fold increase in animal population during last few decades has resulted in excess pressure on the croplands. Apart from physical factors, human dimensions and factor has become a major driving force for desertification in western Rajasthan. There are now number of studies to indicate the state of desertification in western Rajasthan, but few of them are based on a village/ farm/ plot level information. It is highly desirable to assess the situation at this level, for which information can be derived using very high resolution satellite platforms so that local specific problems can be addressed and quantified. Using satellite images with very high resolution images, researchers have better advantages to find clues for few frequently asked questions like "where the problem is, what is the severity and what are the factors responsible for the situation". Once this is understood, using field data, situations for the future can be modelled. Then only, one can think of solution to "Zero degradation" slogans of UNCCD" for a land degradation neutral world.

References

Anonymous 1992. Managing fragile ecosystems: Combating desertification and drought. In,Chapter12, Agenda 21,United Nations Conference on Environment and Development, Rio de Janeiro. A/CONF.151/4 (Part II), United Nations, New York, pp. 46-66.

Anonymous. 2010. Degraded and Wastelands of India: Status and spatial distribution, Indian Council of Agricultural Research & National Academy of Agricultural Sciences, New Delhi, 133p.

Dregne, H.E. 1991. Global status of desertification.Annals of Arid Zone 30: 179-185.

Ghose, B., Singh, S. and Kar, A. 1977. Desertification around the Thar: A geomorphological interpretation. Annals of Arid Zone 16: 290-301.

Kar, A. 1993. Aeolian processes and bedforms in the Thar Desert. Journal of Arid Environments 25: 83-96.

Kar, A. 2011.Flood havoc in Thar Desert: Is nature to blame? In, Water Crisis in the Indian Subcontinent (Zahid Husain and Cajee L, Eds.), pp. 3-17. Bookwell, Delhi.

Kar, A., Moharana, P.C., Singh, S.K., Goyal, R.K. and Rao, A.S. 2007. Barmer flood, 2006: Causes and consequences. In, Flood of August 2006 in Arid Rajasthan: Causes, Magnitude and Strategies (A.S. Faroda and D.C. Joshi, Eds), pp. 26-39. State of Art Report, Scientific Contribution No.INCOH/SAR-29/2007; INCOH, Roorkee.

Kar, A., Moharana, P.C., Raina, P., Kumar, M., Soni, M.L., Santra, P., Ajai, Arya, A.S. and Dhinwa, P.S. 2009. Desertification and its control measures. In, Trends in Arid Zone Research in India (A. Kar, B.K. Garg, S. Kathju and M.P. Singh, Eds.), pp. 1-47. Central Arid Zone Research Institute, Jodhpur.

Middleton and Thomas. 1997. World atlas of Desertification (2nd edition).UNEP and Arnold, London, 182p.

Reynolds, J.F., Grainger, A., Stafford Smith, D.M., Bastin, G., Gracia-Barrios, L, Fernandez, R.J., Janssen, M.A., Jurgens, N., Scholes, R.J., Veldkamp, A., Verstraete, M.M., Von Matitz, G. and Zdruli, P. 2011.Scientific concepts for an integrated analysis of desertification. Land Degradation and Development 22: 166-183.

SAC.2007a. Desertification Monitoring and Assessment using Remote Sensing and GIS: A Pilot Project under TPN-1 UNCCD, Scientific Report, SAC/RESIPA/MESG/DMA/2007/01, 93 p.

SAC. 2007b. Desertification and Land Degradation Atlas of India. Space Applications Centre, ISRO, Ahmedabad.

Sharma, K.D. and Vangani, N.S. 1992. Surface water resources in the Thardesert and adjoining arid areas of India. In, Perspectives on the Thar and the Karakum (A. Kar, R.K. Abichandani, K. Anantharam and D.C. Joshi, Eds.), pp. 130-140. Department of Science and Technology, Govt. of India, New Delhi.

Singh, S., Kar, A., Joshi, D.C., Ram, B., Kumar, S., Vats, P.C., Singh, N., Raina, P., Kolarkar, A.S., Dhir, R.P. 1992. Desertification mapping in western Rajasthan. Annals of Arid Zone 31: 237-246.

UNEP.1977. UNEP and Fight Against Desertification. Nairobi, Kenya.

Vangani, N.S. 1997.Water erosion and flash flood hazards in Indian arid ecosystem. In, Desertification Control in the Arid Ecosystem of India for Sustainable Development (S. Singh and A. Kar, Eds.), pp. 131-138. Agro Botanical Publishers (India), Bikaner.

18

Role of Major Land Resource Units in the Natural Resources Assessment and Planning in Arid Rajasthan

Pratap Chandra Moharana and C.B. Pandey

ICAR-Central Arid Zone Research Institute, Jodhpur, Rajasthan, India

Land resources units or major land resources units are the suitable segmentations of land, based on available natural resources, their potential use and management. Irrespective of terms used by different organizations or institutes, these units have spatial dimension and represent a composite unit for interstate, regional, and national level planning. A typical meaning of these land units can be related to terms like land system. Christian and Stewart (1953) of the Commonwealth Scientific and Industrial Research Organization (CSIRO), Australia devised the 'Land system' as the composite mapping-unit which has been defined as "an area or group of areas, throughout which there is a recurring pattern of topography, soils and vegetation". This system is based on the premise that the land system is expressed on aerial photographs by a distinctive pattern and, as such, the land system can be directly mapped from aerial photographs (Christian and Stewart 1953, Mabbutt and McAlpine 1965). It is a fact that during those period, aerial photographs were more frequently used and studies related to land systems used technique of aerial photogrammetry. As per USDA (United States Developmental Agency), the typical segmentations are; (1) Land Resource Region (LRR), which is the highest level in the hierarchy (2) Major Land Resource Area (MLRA), the second highest level and generally representing broad landforms or a geologic region at a small scale and (3) the Land Resource Unit (LRU). The LRUs generally contains information on resource attributes. Typical examples of LRU can be a watershed, forest region or cropland. Therefore, it is considered to be the significant unit (www.nrcs.usda.gov/Internet/FSE.../nrcs142p2_052855.doc). There are terms like MLRAs or CRAs which have similar functional use and have been used as basic units for state land resource maps. The Common Resource Areas (CRAs) are created by subdividing MLRAs by resource concerns, soil groups, hydrologic units, resource use, and topography, other landscape features, and human considerations affecting use and treatment needs. The term "Component" of resource areas are used for individual units like land use, physiography, geology, climate, water, soils and biological resources. Biological resources are commonly defined by vegetation species, type, and phenology (http://www.nrcs.usda.gov/wps/portal/nrcs/detail/tx/home/?cid=nrcs142p2_053625). CSIRO, Australia came out with a new concept of composite mapping called the "Unique Mapping Area" or in short known as 'UMA'. In this system, the specialists individually draw the boundaries of soil, landform, vegetation, land use, water resources, etc. on the same aerial photograph, aided by the local knowledge of the terrain. The methodology was tested by CAZRI in the Indian arid zone (Abichandani and Sen 1977).

MLRU by Central Arid Zone Research Institute, Jodhpur : The institute came out with a model of composite mapping unit called MLRU (Major Land Resources Units) after much of the discussions with experts like Dr. H.Th.Verstappen, Dr. J.J. Nossin from ITC, Netherland, Dr. J.R. McAlpine and Professor, J.A. Mabbutt from Australia (Kar 2004). The methodology follows pattern and philosophy of Land systems and UMA. This is defined as an area or group of areas having recurring patterns of landform, soil, vegetation along with human activities and resource potential (Abichandani et al. 1975, Abichandani and Sen 1977). The basic purpose of MLRU was to evaluate the physical potential of land, potential hazards to agricultural and non-agricultural land uses as well as to find out suitability of land for developmental plans. For preparing a MLRU, the individual experts / subject matter specialists interpret remote sensing data (satellite images as well as other data products), make a traverse of the field for a correlation matrix and make an unbiased assessment of a particular land resource including land use. In the second step, all these thematic resource maps (landform, soil, water and vegetation and land use) are draped (kept one over another) to find out homogeneous units. Here, homogeneity would depend upon the pattern of recurrence. For example, for arriving at a MLRU of sandy undulating alluvial plains with moderately deep soil etc., one need to select those areas where the terrain should have a recurrent pattern having sand and alluvium over a contiguous area. At CAZRI, scientists of the Division of Natural resources and Environment, earlier known as BRS (Basic Resources Survey since beginning), and RSM (Resource Survey and Monitoring), advocated the idea of MLRU (as mentioned above) and used this concept for district level planning Bikaner (1974), and Jodhpur (1982) etc.

In the following Table 18.1, an explanation of MLRU mapped for Jaisalmer and Jalor district by CAZRI is presented for the better understanding of the concept.

Table 18.1: MLRU mapped for Jaisalmer and Jalor district

Jaisalmer district (10 MLRUs)	Jalor district (16 MLRUs) (only 10 are listed here)
Hills (360.0 km^2)	Hills (249 km^2)
Rocky/gravelly pediments (1353.2 km^2)	Rocky/gravelly pediments with coarse textured soils (110 km^2)
Gravelly pavements (352.8 km^2)	Moderately deep buried pediments with medium textured soils (398 km^2)
Flat collegial plains (10830.0 km^2)	Very deep nearly level older alluvial plains with coarse textured soils (682 km^2)
Saline flat collegial plains (65.8 km^2)	Very deep older alluvial plains with sparse undulations and coarse texture soils (1896 km^2)
Flat older alluvial plains (694.2 km^2)	Very deep older alluvial plains with medium to fine textured soils (2281 km^2)
Sandy undulating plains (3051.6 km^2)	Sand dunes (574 km^2)
Sand dunes (17192.4 km^2)	Non-saline younger alluvial plains (833 km^2)
Interdunal plains (3061.0 km^2)	Saline younger alluvial plains (346 km^2)
Saline depressions (125.4 km^2)	Saline depressions (148 km^2)

Comparision of MLRUs : In the above table, MLRUs for two districts have been presented. In total, there were 10 number of such units for Jaisalmer and 16 were suggested for Jalor district (Anonymous 1992 and Anonymous 1995). But for clarity of comparison, we have illustrated only 10 numbers for Jalor district.

It is observed that in case of Jaisalmer district, the units were absolutely landform based (Hills, Flat older alluvial plains etc.) while in Jalor district, the land units have been suffixed with soil association, soil depth and texture (very deep nearly level older alluvial plains with coarse textured soils). This indicates how, with time, there have been substantial development in the classification level. Thus, in the latter, there are attempts to use additional information like that of soil characteristics of the area too. Now, let us see what more information that were provided in the assessment of each MLRU.

For example, for MLRU "Saline depression" in Jaisalmer district, following details have been mentioned (Anonymous 1992).

1. Topography : Flat, saline hard surfaces when dry and pool of water after the rains, areas of internal drainage, having boundary of like rocky/soil cliff with an outlet at the other end to drain excess water.
2. Soil and land capability : Deep, fine textures saline soils, high pH.
3. Vegetation : On sand surface, *Tamarix dioica*, *Salvadora oleoides*, *Ziziphus numularia.*
4. Surface water : low infiltration, no potential due to salinity.
5. Groundwater : High saline groundwater at shallow depth.
6. Present land use : Saline waste, commercial exploitation of salt.
7. Villages assessment : Efforts to plant salt tolerant species along the margin for fuel wood purpose

Thus, considering the kind of information provided for any such MLRU, one can understand the potential / problems of the land unit and also get idea on how such land can be used effectively. However, studies show, similar units may have different potential or perspectives being located in different set up or climatic zones. This scope of such resource units will be different with scale (Table 18.1).

In the following diagram (Fig. 18.1), the typical model that is being used at CAZRI in Natural Resources Assessment is presented. This also shows how MLRU based information can be effectively used for land and water development action plans or for Land suitability analysis. The basic principle of such procedure or model followed up at CAZRI for natural resources assessment are:

(1) Land characterization includes systematic survey of natural resources (supported by information and analysis of climate data), their assessment based on field level observation, measurement and laboratory analysis. Use of remote sensing and GIS for data interpretation and delineation.

(2) Land quality assessment based on studies related to land degradation and land capability assessment.

(3) Delineation of MLRUs based on land characterization.

(4) Correlation with socio-economic characteristics of the unit area.

With the advent of GIS, there is a scope to integrate many attributes and model the MLRUs as may be required in any development plan. In the following table (Table 18.2), the authors describe a model of MLRU in relation to different scales of assessment.

Table 18.2: Scale and increasing scope for MLRUs (as proposed by Authors)

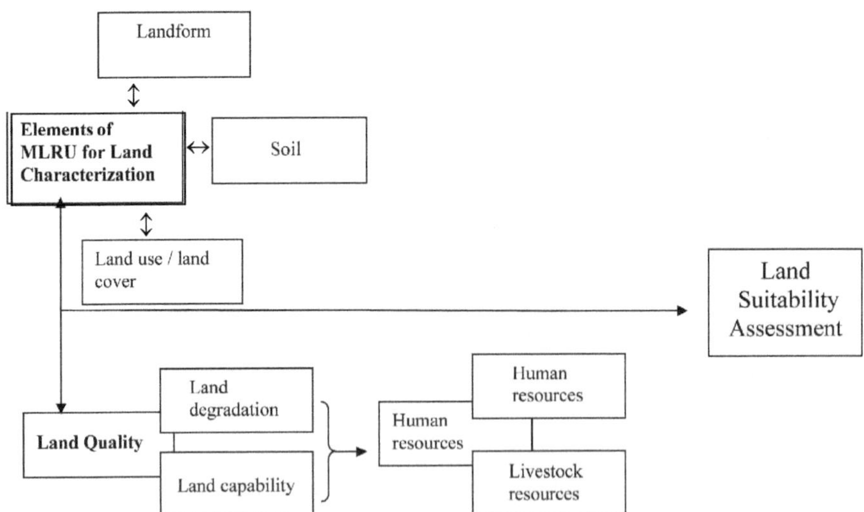

Fig. 18.1: Flow diagram for land resources assessment using MLRU
Source: Modified version (Kar 2004)

Scale	Typology of MLRU
Regional/National, 1:500,000	Physiographic unit/ agro climatic zones like Northern Alluvial plains /western Arid plain etc.
District (1:250 k* or 50 k)	Flat Alluvial plain with deep soil
Tehsil (1:50 k)	Moderately deep alluvial plain with very deep soil
Village/cadastral (1:10 k or higher)	Moderately deep alluvial soil, irrigated and wheat crop in >50% area with additional information on soil nutrients

*k=000

Mapping elements and minimum mapping units

With advancement in the technology, availability of data products or models, there have been changes in the mapping techniques. If we look at the past, the basic scientific way of mapping involved was the interpretation of aerial photographs and more use of Survey of India Toposheets. The toposheets used to be at two common scales (1:250 k and 1:50 k) and such data are still relevant for any area, even for inaccessible Thar Desert. Recent publications of Survey of India however, does not include contour information as it used to provide. Such information can be obtained with special requisition and purpose. The aerial photographs at a scale ~33000 used to be kept under stereoprets in which one can view 3D (Three dimensional) surfaces of the landscape. To identify the objects, the images are analyzed and classified on the basis of their 'texture', 'structure', 'form' and their correlations (edge-gradient study). Once the photographic images are correlated with the type of land in the field, the sequence of events, which form a particular land unit, can often be delineated and many aspects of land, e.g. the land form, the soil, vegetation and the land-use settlement, can be inferred (Sen 1972). The mapping used to follow first, second and third order / stages. The first order maps are the individual maps, e.g. those of soil, geomorphology, vegetation, and geology and water resource. The second order of mapping dealt with eliminating the smaller ones with further photo interpretation or field checking. This is based on the principle of Minimum Mapping Unit or MMU. In the mapping of 1:50,000 Scale mapping, a minimum area of 0.5 sq. km or a minimum width of 0.25 km is the MMU. In the mapping in a GIS environment, at 1:50,000 scale, as per NRSC guidelines, a mapping unit about 2.5 × 2.5 ha are termed as the MMU. In the Bikaner District mapping by CAZRI, such land units were termed as biophysical units. The third order of mapping considers assessing the resource potential of the units, the land use, the water resources, the climate, the socio-economic conditions and the existing plants, if any, for the development. Thus using such procedures, Abichandani et al. (1975) delineated the MLRU of Bikaner district of western Rajasthan.

Use of remote sensing platforms for Natural resources assessment

Identification of any resource, their mapping and assessment is becoming easier with time and may be every year. The reason is the quick availability of data thanks to the advancement in the field of remote sensing technology, earth observation satellites, communication, physical infrastructure and computing environment. Let us see some of the current IRS (Indian Remote sensing) satellite Missions. There are 10 operational satellites in orbit - Resourcesat - 2, Cartsat - 2B, Oceansat - 2, Risat - 2, Cartosat - 2A, IMS - 1, Cartosat - 2, Cartosat - 1, Resourcesat-1 and TES. Like any other satellites like Landsat of USA, SPOT from France etc, the IRS series of satellites provide data in a variety of spatial, spectral and temporal resolutions. There are thematic series of satellites like Cartosat-3, Megha Tropiques, SARAL and Insat-3D (http://bhuvan.nrsc.gov.in/bhuvan/content/indian-remote-sensing-satellites).

Most commonly used EOS (Earth Observation Systems) are based on remote sensing satellites launched by ISRO (Indian Space Research Organization). IRS series of 1990s (IRS IC, ID) till 1997, enabled us to view the synoptic view of the area with scales equivalent to 50k. Oceansat was launched in 1999. Resourcesat 1 launched in 2003 (IRS P6) had 3 cameras (LISS 4 with 5.8 m spatial resolution, LISS III having 23.5 m resolution and AWiFS with 56 m resolution). All these are MSS or Multispectral images. Its later version; Resourcesat 2 launched in 2011, added LISS 4 (MX) with 5.8 m resolution with a scene width of 70 km. In 2005, there was a revolution in the form of Cartosat-1 with 2.5 m resolution and swath of 30 km. This was unique in the sense that it possessed stereo imaging capability which enabled generation of DEMs (Digital Elevation Models). Now we have Carto DEMS from our own organization in addition to Japanese ASTER (Advanced Space borne Thermal Emission and Reflection Radiometer) at 30m ground solution and NASA's SRTM (Space Shuttle Radar Topography Mission) at 90 m ground resolution. In addition to above, there are utility satellites like RISAT, which stands for all weather Radar Imaging Satellite launched in 2012. This is useful for facilitating agriculture and disaster management. SARAL is another satellite which is launched as an Indo French satellite mission. Apart from Indian satellites, there have been satellites like LANDSAT, SPOT etc. which have very high spatial and radiometric resolution and some of them can be downloaded free from US websites.

The objective of mentioning all these remote sensing satellites is to inform the readers about the availability of data and information these days. Depending upon the requirement and use, one has to select the right kind of satellite data products. In India, National Remote Sensing Centre at Hyderabad are the nodal agency to provide remote sensing data products.

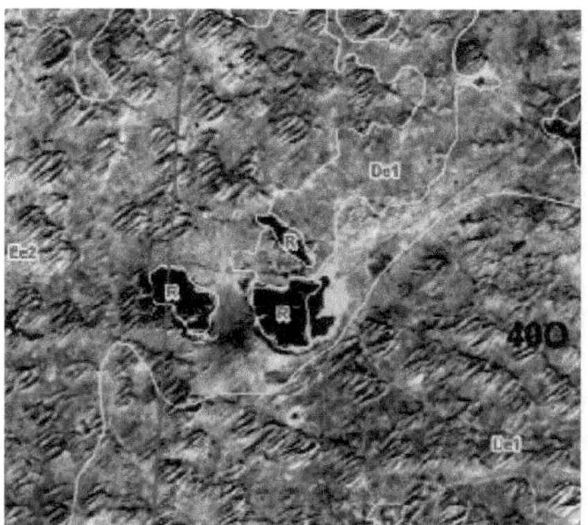

A barchan dune field at Chohtan in Barmer district) viewed in IRS LISS III FCC image (Resolution 23.5 m)

Same landform (barchans field) viewed in high resolution Google Earth images (Resolution < 1 m)

The field photograph of the dune field at same location

Fig. 18. 2: Terrain features as viewed by different satellite sensors.

The way forward

MLRUs can be made more effective depending upon its specific use. The size of the MLRU would also depend on the requirement for a purpose as it is scale dependent. For example, for district level planning, the satellite data like IRS-LISS-III images with 23.5 m spatial resolution was equivalent to 50,000 scale Survey of India toposheets. However, if a planning of agricultural operation is required for a micro-watershed level (less than 1000 ha area), the satellite images like LISS-IV with 5.8 m resolution will more helpful. Further down, if agricultural operation at plot level needs to be carried out, there are also satellite images which have <1 m spatial resolution. There are many other user friendly satellite data providers like Google Earth images which can improve the interpretation because of micro-level information. In that case the scope for micro level Land Resources Units will be interesting option. This will be helped by GIS (Geographic Information System). The analysis of individual layers in reference to each, have been greatly improved using different modules of GIS like spatial analyst and Geostatistical module etc. apart from ARC/MAP of ESRI'S GIS. The advantages of software and computing environment can be used to compute the MLRU much faster for timely decision and accuracy of measurement.

References

Anonymous 1992. Integrated Natural and human resources appraisal of Jaisalmer district (Eds. P.C.Chatterjee & Amal Kar), Central Arid Zone Research Institute, Jodhpur. 156p.

Anonymous 1995. Integrated Natural and human resources for sustainable development of Jalor district (Surendra Singh, K.D. Sharma and D.C. Joshi, Eds.), Central Arid Zone Research Institute, Jodhpur. 108 p.

Abichandani, C.T. and Sen, A.K. 1977. MLRU mapping - a concept of composite mapping unit for integrated survey. Annals of Arid Zone 16: 263 270.

Abichandani, C.T., Singh, S., Saxena, S.K., Kolarkar, A.S. and Sen, A.K. 1975. Assessment and management of the biophysical resources in district Bikaner, Western Rajasthan. Annals of Arid Zone 14(4): 292-301.

Christian, C.S. and Stewart, G.A. 1953. General Report on Survey of Katherine Darwin Region. C.S.I.R.O. Aust, Land Res. Ser. No. 1, Melbourne, Australia.

Mabbutt, J.A. and Mcalpine, J.R. 1965. Introduction to lands of the Port Moresby Karaiku area. C.S.I.R.O. Aust, Land Res. Ser. No. 14, Melbourne, Australia.

Sen, A.K. 1967. Documentation and cartography of the base map for coordinated land survey, based on aerial photographs. Annals for Arid Zone 6(2): 170-77.

Sen, A.K. 1972. Land-utilization mapping to estimate the waste lands of arid zone in Rajasthan by photo interpretation technique. Indian Natural Science Academy Bulletin No. 44: 66-71.

Sen, A.K. 1974a. Categorization of land utilization units in arid zone. Dec. Geog. 12(1): 61-72.

Sen, A.K. 1974b. Land-use mapping for arid zone by aerial photo-interpretation technique. Proceedings Summer Institute-Desert Ecosystem and its Improvement. ICAR, Jodhpur, October, 1974.

Kar, A. 2004. Framework of Integrated Natural Resources Assessment : Concept, methodology and contribution of CAZRI (In. NAIP publication on Natural Resources appraisal for land use planning in arid agro ecosystem, Eds. Balak Ram, Pratap Narain and D.C.Joshi), Central Arid Zone Research Institute, Jodhpur. 106p.

http://bhuvan.nrsc.gov.in/bhuvan/content/indian-remote-sensing-satellites

http://www.nrcs.usda.gov/wps/portal/nrcs/detail/tx/home/?cid=nrcs142p2_053625

www.nrcs.usda.gov/Internet/FSE.../nrcs142p2_052855.doc

Fenneman, N.M and Johnson, D.W. 1946. Physiographic divisions of the conterminous U.S. Geospatial_Data_Presentation_Form: map, http://water.usgs.gov/lookup/getspatial?physio

19

Variability in Soil Hydro-physical Properties in Field Conditions

R.K. Singh, Priyabrata Santra, H.M. Meena and R.K. Goyal

ICAR- Central Arid Zone Research Institute, Jodhpur, Rajasthan, India

Introduction

Soil is a dynamic natural body, characterized by high degree of spatial variability due to combined effect of physical, chemical or biological processes that operate with different intensities at different scales (Goovaerts 1998). Reports have shown that there is large variability in soil properties, crop yield, disease, weed etc., not only in large-sized fields (Godwin and Miller 2003,Vrindts et al. 2005), but also in small-sized fields (Bhattacharya et al. 2008). Therefore, knowledge of hydro-physical properties of soils is crucial for management practices for all kinds of agricultural practices for efficient utilization of input resources. Plant growth and development is highly related to soil hydro-physical properties, which not only depend on soil intrinsic textural and chemical characteristics (Hamza and Anderson 2005) but also on soil management.

Soil physical properties which change with the variation in soil structure include bulk density, total porosity, pore geometry (size distribution, shape, continuity and tortuosity), penetration resistance and aggregate stability, consequently affecting soil hydraulic properties such as water retention characteristics, plant available water capacity, infiltration capacity and saturated hydraulic conductivity. These soil properties are dependent on seasonal climatic conditions, management practices, crop development and biological activity (Reynolds et al. 2007). The hydro-physical properties of soils, i.e., water retention and water permeability in both saturated and unsaturated zones not only shape soil water balance but also decide the conditions for plant growth, development and yield. They also determine water availability for the plant root system and the transfer of water with chemical compounds dissolved in it into deeper layers.

Soil hydraulic properties

Water retention behaviour of soil i.e. the pressure head-water content (*h*-q) relationship can be best described by van Genuchten (VG) water retention model (van Genuchten 1980):

$$S_e = \left(\frac{\theta - \theta_r}{\theta_s - \theta_r}\right) = \left[\frac{1}{1+(\alpha h)^n}\right]^{1-\frac{1}{n}} \quad (1)$$

where S_e is the relative saturation, θ_r is the residual soil water content (cm^3 cm^{-3}), θ_s is the saturated soil water content (cm^3 cm^{-3}), h is the matric potential head (cm) and α (cm^{-1}) and *n* are shape parameters of the water retention curve. The parameter α changes with different soil types for the same value of *n* and *vice versa*. Therefore, the ordered-pair of α and *n* uniquely characterizes water retentive capacity of soil. Similarly, the ordered-pair α and *n* is also used for describing the unsaturated hydraulic conductivity, $K(\theta)$ (van Genuchten 1980):

$$K(S_e) = K_s S_e^{\,L}\left[1-(1-S_e^{\,n/(n-1)})^{1-1/n}\right]^2 \quad (2)$$

where, *L* is an empirical pore-connectivity parameter equals to 0.5 (Mualem 1976). While water retention capacity describes soil's ability to store water, hydraulic conductivity describes soil's capacity to allow water to flow through soil.

From laboratory measurement of soil water retention and the parameters of VG model it has been found that hydraulic properties varies largely across different soil texture. Specifically the variation of *k*(*h*) in unsaturated soil follows 6 to 7 order (Fig. 19.1a and b).

Steady state infiltration rate is one such hydraulic property, which governs the water balance equations at different landscape scales. Infiltration is defined as the process of entry of rain water or irrigation water into the soil and, therefore,

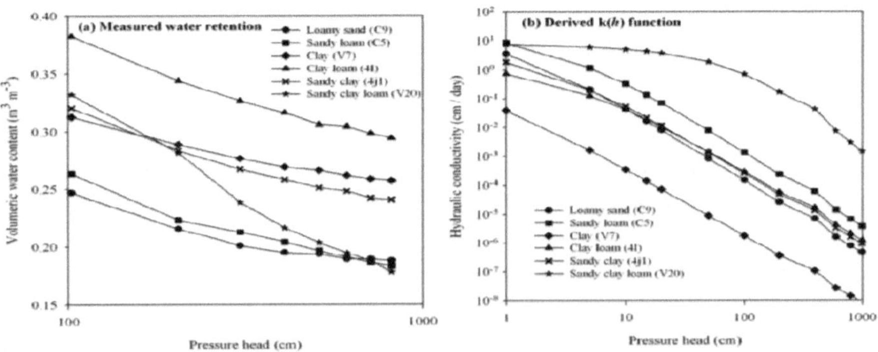

Fig. 19.1. a) Measured values of water retention and [θ (*h*)] and b) derived unsaturated hydraulic conductivity [*K*(*h*)] for some selected soil samples of Chilika database

decides the partitioning of rainwater into soil moisture and runoff. Infiltration characteristics of soil can be described by different type of models i.e., Kostiakov model, Horton model, Philip model, and Green &Ampt model.

Green &Ampt model $i = i_c + \frac{B}{I}$ (1)

Horton model $i = i_c + (i_0 - i_c)e^{-kt}$ (2)

where, i = instantaneous infiltration rate (mm/min) of soil, i_c = steady state infiltration rate (mm/min), B = constant, I = cumulative infiltration (mm), i_0 = initial infiltration rate (mm/min) at $t = 0$, k = constant that determines the rate at which i_0 reaches to i_c, t = time (min). Field measurements of infiltration characteristics over different soil surface in arid region reveal large variation (Fig.19.2) and also reported in literatures (Mallants et al. 1996, Sobieraj et al. 2004). It has been observed that infiltration rate quickly attained the steady state in most cases and within 10-15 minutes after initiation of the experiment from unsaturated state (dry soil).

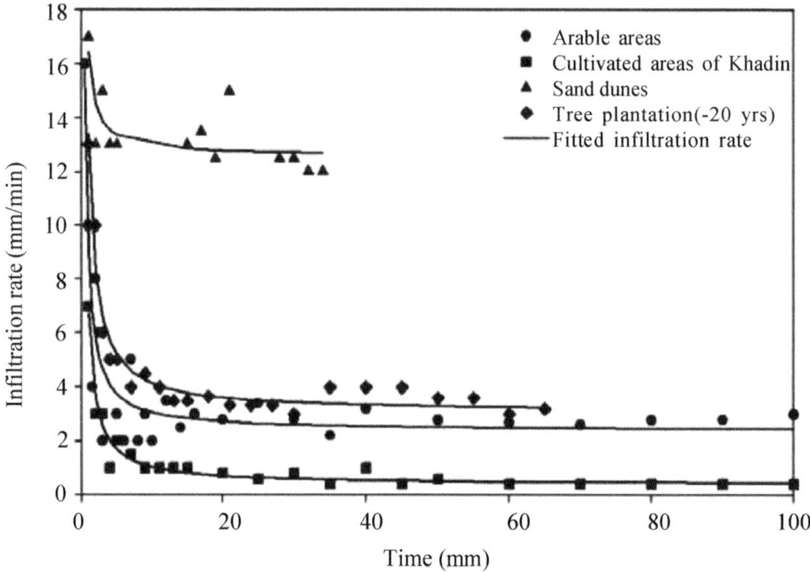

Fig. 19.2: Infiltration characteristics at different soil surfaces in Indian arid zone

Variation of soil properties at different scales

Spatial variability of soil physical properties at a large-scale, like catchment-scale, is mainly due to geological, geomorophological and pedological soil forming factors that could be altered and induced by other factors such as different land use managements. Among various soil properties, hydraulic conductivity and water retention are reported to have highest statistical variability (Biggar and Nielsen, 1976).

Stockton and Warrick (1971) indicated that variability in soil hydro-physical properties is a function of both soil depth and position in the landscape, as well as depends on experimental error while measuring them in field or laboratory. Degree of variability of soil properties can be studied through identifying its spatial variance structure, which can further be used in defining the sampling interval (Iqbal et al. 2005). Here, the degree of variation of soil properties at different scale has been presented in Table 19.1. Coefficient of variability (COV) was found lower for bulk density (7-12%) and pH (10-19%), whereas it was higher for soil hydraulic parameters (22-57%), soil organic carbon contents (45-100%) at different scales. Even, particle size distribution was found to vary greatly at different scale with COV for sand content ranging from 36 to 77% and for clay content ranging from 43 to 62%. Similar magnitude of variation of soil properties was also reported in literatures (Upchurch et al. 1988, Farkas et al. 2007). It is noted here that in Table 19.1, variation of soil properties is presented in different scales starting from a micro-watershed scale to a national scale. Larger variation of soil properties is expected whenever we are moving from a small scale to a larger scale. However, sufficient number of samples is required for true representation of variations in a particular scale.

Table 19.1: Variation of soil properties at different scales

Soil properties	Soil database from a watershed (42 km^2) in Chilika catchment (N = 100)		SOTER-IGP database (N = 176)		Benchmark soils of India (N = 78)	
	Mean	$^{\dagger}\sigma$	Mean	σ	Mean	σ
Bulk density (ρ_b) (Mg m^{-3})	1.50	0.18	1.44	0.10	1.51	0.14
OC (%)	0.92	0.41	0.31	0.31	0.29	0.21
Sand (%)	48.71	17.80	35.57	24.51	35.52	27.38
Silt (%)	16.51	6.34	36.00	14.87	27.93	16.23
Clay (%)	34.99	14.93	28.43	14.75	25.44	15.62
pH	6.83	1.20	7.67	1.49	7.56	0.80
ln (K_s) (cm day^{-1})	3.09	1.77	-	-		
θ_{FC} (cm^3 cm^{-3})	0.257	0.058	-	-		
θ_{PWP} (cm^3 cm^{-3})	-	-	-	-		
θ_{AWC} (cm^3 cm^{-3})	-	-	0.118	0.037		

where, q_{FC} = Soil water content at field capacity; q_{PWP} = Soil water content at permanent wilting point and $_{AWC}$ = Available water content.

Apart from the variation of soil properties explained above, another soil database from arid western India (SAWI) is presented in Table 19.2 for experiencing the variation. Sand content of the soil samples were high with a mean value of 63.87%. The average silt and clay content of soil samples were 14.22% and 22.92%, respectively. Wide variation in sand, silt, and clay content was observed in the SAWI database. Soil textural class was sandy for most of the soil samples. Organic carbon content of soils in the SAWI database was low with an average value of 2.55 g kg^{-1} of soil. The average $\theta_{1/3bar}$ was 17.49% (kg kg^{-1}) and varied widely from a very low value of 2.50% (kg kg^{-1}) to as high as 47.70% (kg kg^{-1}). The average θ_{15bar} was 7.04% (kg kg^{-1}). Wide variation in θ_{15bar} was also observed with a minimum value of 1.20% (kg kg^{-1}) and a maximum value of 19.40% (kg kg^{-1}).

Table 19.2: Descriptive statistics of soil properties from the database on soils from arid western India (SAWI)

Soil properties	Total SAWI database (N = 380)		
	Mean	Standard deviation	Range
Sand (%)	63.87	22.64	3.50 – 97.00
Silt (%)	14.22	11.15	0.40 – 67.90
Clay (%)	22.92	14.37	1.00 – 68.40
OC (g kg^{-1})	2.55	1.94	0.1 – 11.00
$^{\S}\theta_{1/3bar}$ (%, kg kg^{-1})	17.49	9.39	2.50 – 47.70
$^{\S}\theta_{15bar}$ (%, kg kg^{-1})	7.04	4.17	1.20 – 19.40

$^{\S}\theta_{1/3bar}$ and θ_{15bar} represents the soil water content (%, kg kg^{-1}) of soil at 1/3 bar and 15 bar, respectively.

Steady state infiltration rate was found 1.5-3 mm min^{-1} on land surface dominated with coarse sands and limited soil depth (<50 cm). For land surface dominated with deposited aeolian sands and soil depth of ~100 cm, steady state infiltration rate was found very high, 12.6 mm min^{-1}. Steady state infiltration rate in bed area of runoff farming system (*khadin*) in desert was found low, 0.39 mm min^{-1}. In contrast, steady state infiltration rate of the catchment area of *khadin* was found high, 3.28 mm min^{-1} (Fig. 19.3).

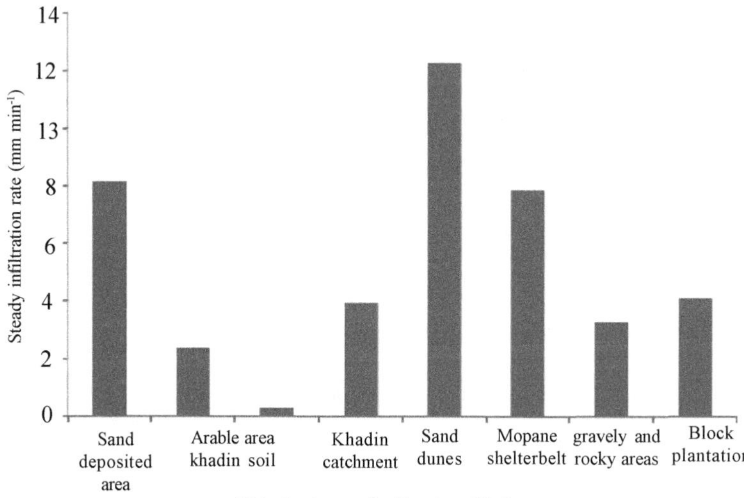

Fig. 19.3: Steady state infiltration characteristics at different land use situations in arid zone

Variation of soil properties at temporal scale and across different management practices

Many soil properties are strongly dependent on the dynamics of soil structure with the management systems being primary agents for changing the soil structural conditions (Alletto and Coquet, 2009). Temporal variability of soil physical and hydraulic properties can be even greater than spatial variability in agriculturally managed soils. Soil physical and hydraulic properties are expected to vary even during a crop cycle especially after tillage and irrigation or rain. Spatial variability of soil properties may appear in yield variation within a single field even in areas considered homogeneous from the soil survey point of view. Soil spatial distribution is represented by the soil patch pattern on soil maps. The level of heterogeneity within the soil pattern depends on the examined soil property.

Factors affecting soil physical and hydraulic properties

Changes of soil physical and hydraulic properties have been the subject of much research to quantify the effects of different management systems within growing seasons, over cropping systems and between years. Agricultural management practices are applied as events with their transient effects realized over different time scales. The resulting variability in soil physical and hydraulic properties is dynamic in space and time. Many soil properties are strongly dependent on the dynamics of soil structure with the management systems being primary agents for changing the soil structural conditions (Alletto and Coquet 2009). Tillage and irrigation are the most influential soil manipulations having short-term effects on soil physical and hydraulic properties. Tillage practices manipulate the soil and potentially alter its structure differently resulting in variability in temporal dynamics of the soil physical and hydraulic properties which are unstable (Strudley et al. 2008). Prominent soil properties get affected due to tillage and irrigation management practices have been described below:

Soil bulk density and total porosity

Soil bulk density and total porosity are the most commonly measured soil physical properties affected by tillage. They are primarily used as measures of soil compaction which may adversely affect soil hydraulic properties and processes. The bulk density and penetration resistance have been found to increase with the number of traffic passes while air permeability decreases with increased traffic intensity (Ohu et al. 2006).

Soil structure and aggregate stability

Maintenance of stable soil structure is an important issue from both agricultural and environmental aspects. Tillage directly affects soil structure by mechanically breaking macro-aggregates into smaller units and indirectly through its effect on root growth, biological processes and soil organic matter content. Soil properties such as soil structure, aggregate stability and porosity are often used as the criteria to differentiate tillage system effects on soil physical and hydraulic properties (Mielke et al. 1986, Voorhees 1983).

Soil pore geometry

In agricultural soils, management and tillage practices have a strong impact on the physical properties of the topsoil and the characteristics of the macropore system (Frede et al. 1994, Shipitalo et al. 2000). The pore geometry, which is the size distribution, shape, continuity and tortuosity of the soil pores, would be expected to vary between tilled and untilled cropping systems due to different methods of creation of pores. Schjonning and Rasmussen (2000) reported that direct drilling compared to mouldboard ploughing resulted in lower volume of macro pores (>30 μm) on coarse sandy soil and silt loam, whereas the opposite effect was found on sandy loam at the time of plant emergence.

Water flow in structured soils is mainly conducted by macro pores (>75 μm) and larger meso pores (30-75 μm) even though they constitute only a small fraction of the total porosity (Moret and Arrúe 2007). The main impact of different tillage techniques on soil hydraulic properties is expected to occur in these structural pores due to changes in different groups of macro and meso pores.

Soil water retention characteristics

Soil water retention is an important soil property controlled by the soil pores. Soil pore size distribution facilitates the characterization of soil structure (Pagliai et al. 1998), which can be estimated by soil water retention curves (Arya and Paris 1981). Changes in pore geometry resulting from tillage, as well as other disturbances, cause a change in water retention (Azooz et al. 1996). Tillage typically decreases the bulk density and increases total porosity, increasing the amount of water held at high soil water potentials and decreasing the amount held at lower potentials (Unger and Cassel, 1991).

Infiltration capacity and hydraulic conductivity

The mechanical action of tillage implements modifies soil structure, porosity, crop residue distribution and surface roughness. Changes in soil hydraulic properties of the surface soil induced by tillage can alter the infiltration rate of rain or irrigation and significantly affect the subsurface drainage discharge. When a soil is tilled, the infiltration rate may increase due to an increase in total porosity, but it may decrease if continuity of the large pores is disrupted.

References

Alletto, L. and Coquet, Y. 2009. Temporal and spatial variability of soil bulk density and near-saturated hydraulic conductivity under two contrasted tillage management systems. *Geoderma* 152: 85-94.

Arya, L.M. and Paris, J.F. 1981. A physicoempirical model to predict the soil moisture characteristic from particle-size distribution and bulk density data. Soil Science Society of America Journal 45: 1023-1030.

Azooz, R.H., Arshad, M.A. and Franzluebbers, A.J. 1996. Pore size distribution and hydraulic conductivity affected by tillage in northwestern Canada. Soil Science Society of America Journal 60: 1197-1201.

Bhattacharya, P., Tiwari, A.K. and Bhatt, V.K. 2008. Spatial variation of soil strength in small hilly watershed of Shivalik-Himalayan region in India. Indian Journal of Soil Conservation 36: 16-21.

Bigger, J.W. and Nielsen, D.R. 1976. Spatial variability of the leaching characteristics of a field soil. water Resour-Research. 12:78-84

Ersahin, S. 2003. Comparing ordinary kriging and cokriging to estimate infiltration rate. Soil Science Society of America Journal 67: 1848-1855.

Farkas, C.S., Rajkai, K., Kertesz, M. and VanMeirvenne, M. 2007. Spatial variability of soil hydro-physical properties: The Herceghalom case-study. In: Geostatistics and soil geography. Moscow:Nauka. p. 43-66. (In Russian)

Frede, H.G., Beisecker, R. and Gäth, S.1994. Long-term impacts of tillage on the soil ecosystem. Journal of Plant Nutrition and Soil Science 157: 197-203.

Godwin, R.J. and Miller, P.C.H. 2003. A review of the technologies for mapping within-field variability. Biosystems Engineering 84: 393-407.

Goovaerts, P. 1998. Geo-statistical tools for characterizing the spatial variability of microbiological and physic-chemical soil properties. Biology and Fertility of Soils, 27: 315-334.

Hamza, M.A. and Anderson, W.K. 2005. Soil compaction in cropping systems: A review of the nature, causes and possible solutions. Soil & Tillage Research 82: 121-145.

Iqbal, J., Thomasson, A., Jenkins, J.N., Owens, P.R. and Whisler, F.D. 2005. Spatial variability analysis of soil physical properties of alluvial soils. Soil Science Society of America Journal 69: 1338-1350.

Mallants, D., Binayak, P.M., Diederik, J. and Feyen, J. 1996. Spatial variability of hydraulic properties in a multi-layered soil profile. Soil Science 161: 167-181.

McBratney, A.B. 1998. Some considerations on methods for spatially aggregating and is aggregating soil information. Nutrient Cycling Agroecosystem 50: 51-62.

Mielke, L.N., Doran, J.W. and Richards, K.A. 1986. Physical environment near the surface of ploughed and no-tilled soils. Soil & Tillage Research 7: 355-366.

Mohanty, B.P., Ankeny, M.D., Horton R. and Kanwar, R.S. 1994. Spatial analysis of hydraulic conductivity measured using disc infiltrometers. Water Resources Research 30: 2489-2498.

Moret, D. and Arrúe,J.L.2007. Dynamics of soil hydraulic properties during fallow as affected by tillage. Soil & Tillage Research 96: 103-113.

Mualem, Y 1976. A new model for predicting the hydraulic conductivity of unsaturated porous media. Water Resources Research 12: 513-522.

Ohu, J.O., Mamman, E. and Muni, U.B. 2006. Influence of vehicular traffic on air permeability and groundnut production in a semi-arid sandy loam soil. International Agrophysics 20: 309-315.

Pagliai, M., S. RoussevaVignozzi, N., Piovanelli, C., Pellegrini, S. and Miclaus, N. 1998.Tillage impact on soil quality. Italian Journal of Agronomy 2: 11-20.

Reynolds, W.D., Drury, C.F., Yang, X.M., Fox, C.A., Tan, C.S. and Zhang, T.Q. 2007. Land management effects on the near-surface physical quality of a clay loam soil. Soil & Tillage Research 96: 316-330.

Schjonning, P. and Rasmussen, K.J. 2000. Soil strength and soil pore characteristics for direct drilled and ploughed soils. Soil & Tillage Research 57: 69-82.

Shipitalo, M.J., Dick,W.A. and Edwards, W.M. 2000.Conservation tillage and macropore factors that affect water movement and the fate of chemicals. Soil & Tillage Research 53: 167-183.

Sobieraj, J.A., Elsenbeer, H. and Cameron, G. 2004. Scale dependency in spatial patterns of saturated hydraulic conductivity. Catena 55: 49-77.

Stockton, J.G. and Warrick, A.W. 1971. Spatial Variability of unsaturated hydraulic conductivity. Soil Science Soiciety of America Journal 35: 847-848.

Strudley, M.W., Green,T.R. and Ascough, J.C. 2008. Tillage effects on soil hydraulic properties in space and time: state of the science. Soil & Tillage Research 99: 4-48.

Unger, P.W. and Cassel, D.K. 1991. Tillage implement disturbance effects on soil properties related to soil and water conservation: a literature review. Soil & Tillage Research 19: 363-382.

Upchurch, D.L., Wilding, L.P. andHarteld, J.L. 1988. Methods to evaluate spatial variability.In:Hossner, L.R. (Ed.) Reclamation of Disturbed Lands. CRC Press, Boca Raton, Fl., p. 201-229.

Van Genuchten, M.Th. 1980. A closed-form equation for predicting the hydraulic conductivity of unsaturated soils, 44:892-898.

Voorhees, W.B. 1983. Relative effectiveness of tillage and natural forces in alleviating wheel-induced soil compaction. Soil Science Society of America Journal 47: 129-133.

Vrindts, E., Mouazen, A.M., Reyniers, M., Martens, K., Maleki, M.R., Ramon, H. and De Baerdemaeker, J. 2005.Management zones based on correlation between soil compaction, yield and crop data. Biosystems Engineering 9: 419-428.

20

Soil and Water Conservation Measures for Enhancing Crop Productivity in Arid Regions

R.K. Goyal and R.K. Singh

ICAR-Central Arid Zone Research Institute, Jodhpur, Rajasthan, India

Introduction

The natural resources of arid regions particularly soil and water are limited and is often in a delicate environmental balance. Desert encroachment due to lack of conservation planning, and the dangers of destroying or depleting beyond recovery these productive resources, are evident at present time and may be disastrous if development is based on short term expediency rather than long term stability.

The arid zone of India is spread over 38.7 million ha area, out of which 31.7 m ha in under hot arid region and 7 m ha under cold region. The hot arid region occupies major part of north-western India (28.57 m ha) and occurs in small pockets (3.13 m ha) in south India. The north-western arid region occurs between 22°30' and 32°05' N latitudes and from 68°05' to 75°45' E longitudes, covering western part of Rajasthan (19.6 m ha, 69%), north-western Gujarat (6.22 m ha, 21%) and 2.75 m ha (10%) in south-western part of Haryana and Punjab (Faroda et al. 1999). Rainfall distribution is highly uneven over space and time (CV>60%). The region receives low rainfall (<100 mm to 500 mm), has high evapotranspiration and high temperature regime (Rao and Singh 1998). Groundwater is deep and often brackish. The western-central area is devoid of drainage system and surface water resources are meager (Fig. 20.1). Due to low and erratic rainfall, replenishment of water resources is also very poor. The entire Rajasthan state is being categorized as the driest state and water scarce (having per capita water availability below 1000 m^3 $year^{-1}$) since 1991 in the country. Increasing pollution by industrial units, big and small, unregulated mining and even over-extraction of water from deep wells also add to the water quality problem in number of districts. Rapid urbanization and industrialization make such existing differences even more

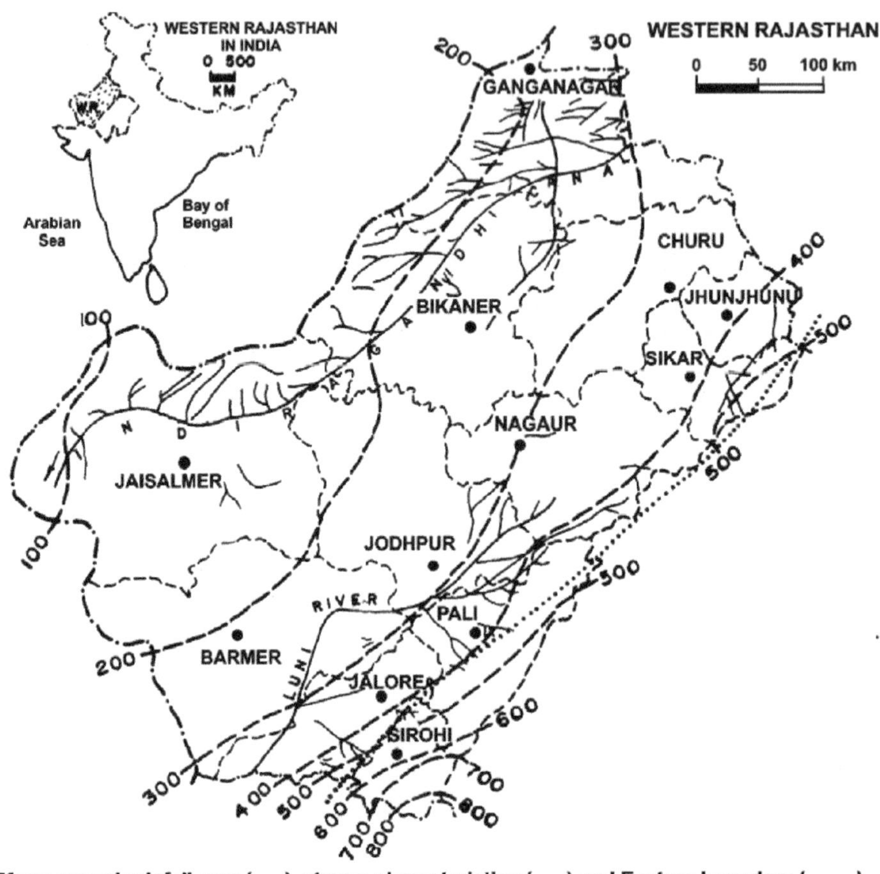

Fig. 20.1: Drainage system and surface water resources in arid western Rajasthan

glaring. During the twentieth century, the region experienced agricultural drought once in three years to every alternate year in one or the other part of the region. The overall probability of drought for the state is 47%. Every alternate year is drought year for the state. The weather condition, even in average years, for most part of the year remains too dry and inhospitable for successful growth of crops. Under such conditions of uncertainty, conventional cropping is risky and is essentially for sustenance only (Goyal and Vittal 2009). The above scenario leads to question of risk in arid agriculture. The main cause of risk in arid agriculture is the variability of rainfall.

Rainfall characteristics of the Indian arid zone

The mean annual rainfall over the Indian arid region varies from more than 500 mm in the southeastern parts to less than 100 mm in the northwestern and western part of the arid region. More than 85% of the total annual rainfall is received during the southwest monsoon season (July to September) mainly under the influence of depressions passing across the Rajasthan. The eastern parts of Rajasthan get rains by the last week of June and gradually covers the entire arid region by middle of July. The withdrawal phase of monsoon again start in the extreme western part by middle of September and retreats by the end of September.

The rainy season varies from 50 days in the western part to 80 days in the eastern part of arid Rajasthan. A small quantum of rainfall of about 7-10 per cent of the annual is received during the winter season under the influence of western disturbances. Rainfall is low and erratic and the coefficient of variation of annual rainfall varies from 42 per cent to more than 64 per cent (Table 20.1).

Table 20.1: Normal annual rainfall and its coefficient of variation in the Indian arid region

Station	Rainfall (mm)	C.V. (%)	Station	Rainfall (mm)	C.V. (%)
Barmer	267	63	Nagaur	340	53
Bikaner	287	47	Sikar	457	42
Sri Ganganagar	245	53	Hisar	446	45
Jaisalmer	188	64	Bhuj	342	65
Jodhpur	366	52	Anantpur	562	50

Rainfall distribution models of Jodhpur indicate that out of 97 years (1901-97), 52 years at Jodhpur recorded average to above average (350 to more than 400 mm) rainfall which indicate that one in two years, Jodhpur region receives substantial rainfall. About 19 years recorded a rainfall of 250-350-mm (Ramakrishna et al. 1988). Appropriate crop production technology can stabilize yield levels in such years. The rest 26 years received less than 250 mm. This would mean specific technology to over come deficit rainfall situations.

Surface water resources

The surface water resources of the arid region is scarce and because of low and erratic rainfall, replenishment of these water resources is also very poor. Due to high atmospheric temperature and low humidity, a large part of the rainwater is lost as evapotranspiration. Surface water potential, except in canal command area, is very low in the central, western and southern parts. In central and western parts, the run-off generated in response to some high-magnitude rainstorms gets lost in sandy terrain. Sharma and Vangani (1992) estimated the surface water potential of this region as 1360×10^6 m^3 out of which 47 per cent was utilized till 1988. Large numbers of tanks, reservoirs, minor irrigation dams and check dams have been constructed at different locations in Luni basin and other areas to store runoff water during monsoon period. In western Rajasthan, 550 storage tanks in the capacity ranging from less than 1.51 to 208×10^6 m^3 are functional with total utilizable capacity of nearly 1169.28×10^6 m^3 for providing irrigation in 0.102×10^6 ha land (Khan 1997). Out of these, six reservoirs viz. Jaswantsagar, Sardar Samand, Jawai, Hemawas, Ora and Bankali are the major irrigation tanks with capacity of irrigation of more than 4000 ha each. Jawai is the main source of drinking water supply to many towns and villages (Source: Irrigation Department, Govt. of Rajasthan). Hydrologically the western Rajasthan can be divided into three broad zones (CAZRI 1990).

Zone – I : Region with major input of surface water from more humid region, frequently with extensive irrigated agriculture. About 60 per cent area of Ganganagar district in the north and 50 per cent area of Bikaner district and 25 per cent area of Jaisalmer district in the northwest lie in this zone. This is the main canal irrigated zone in arid Rajasthan.

Zone –II: Plain lands with a primitive or no stream network. The region has a system of repetitive micro-hydrology. Churu, Jhunjhunun, Sikar, Nagaur, Jodhpur and parts of Bikaner, Jaisalmer and Barmer districts come under this category. This zone occupies 52% area of arid Rajasthan.

Zone-III: Sloping region with an integrated stream network. The Luni basin, occupying the districts of Pali, Jalore, part of Jodhpur and Barmer districts, lie in this zone.

Table 20.2 and 20.3 presents surface water resources and the estimated water demand for the arid Rajasthan respectively.

Table 20.2: Surface water resources (mcm) in arid Rajasthan

District	Zone-I	Zone -II	Zone-III	Total
Ganganagar *	6.49	11.19	-	17.68
Bikaner	7.14	26.16	-	33.3
Churu	-	22.82	-	22.82
Jhunjhunun	-	8.04	96	104.04
Sikar	-	10.48	96	106.48
Jaisalmer	5.03	38.38	-	43.41
Jodhpur	-	20.66	-	20.66
Nagaur	-	53.18	-	53.18
Pali	-	-	869	869
Barmer	-	19.64	-	19.64
Jalore	-	-	71	71
Total	18.66	210.55	1132	1361.21

* including Hanumangarh

Table 20.3: Estimated present water demand (mcm) of arid Rajasthan (CAZRI, 1990)

Demand for	Year				
	1981	1991	1995	2001	2011
Human Consumption @ 40 lpd*	196.85	236.06	261.82	289.23	349.02
Livestock Consumption @ 30 lpd	249.00	290.00	308.00	332.50	376.00
Irrigation @ 0.30 m $ha^{-1}year^{-1}$	5178.00	5696.00	5900.00	6265.00	6892.00
Industry	16.00	17.00	17.50	18.00	21.00

*lpd: liter per day

Since the water is limited, such trend is forcing people to use even 'marginal' quality water in some areas.

Groundwater resources of arid Rajasthan

The quantity and quality of groundwater in this region is not sufficient even for drinking purposes. Over and above insufficient quantity, the ground water is moderately to highly saline over large area. A dominantly sandy terrain and disorganized drainage network (drainage density is as low as 0.3 km^{-2}), and recurring droughts constantly exert pressure on already meager groundwater resources. The stage of groundwater development has exceeded 100% in Barmer, Jalore, Jhunjhunun, Jodhpur, Nagaur, and Sikar districts. Number of Safe blocks has been significantly reduced because of meager rainfall and over exploitation of groundwater resources mainly for irrigation (Fig. 20.2).

Factors responsible for resource depletion/degradation in drylands

- Low, erratic and intense rainfall leading to meager surface water resources and soil erosion.
- High wind velocity causing wind erosion.
- High solar regime resulting in high evapotranspiration losses.
- Low water holding capacity of soil consequently low crop production.
- Cultivation on marginal lands causing severe water and wind erosion.
- Inappropriate selection of crops (high water demanding); depletion of ground water.
- Faulty agricultural practices particularly on sloping lands.
- High human and animal pressure per unit land.
- Excessive cutting of tree leaving extensive soil surface barren for action of wind and water.

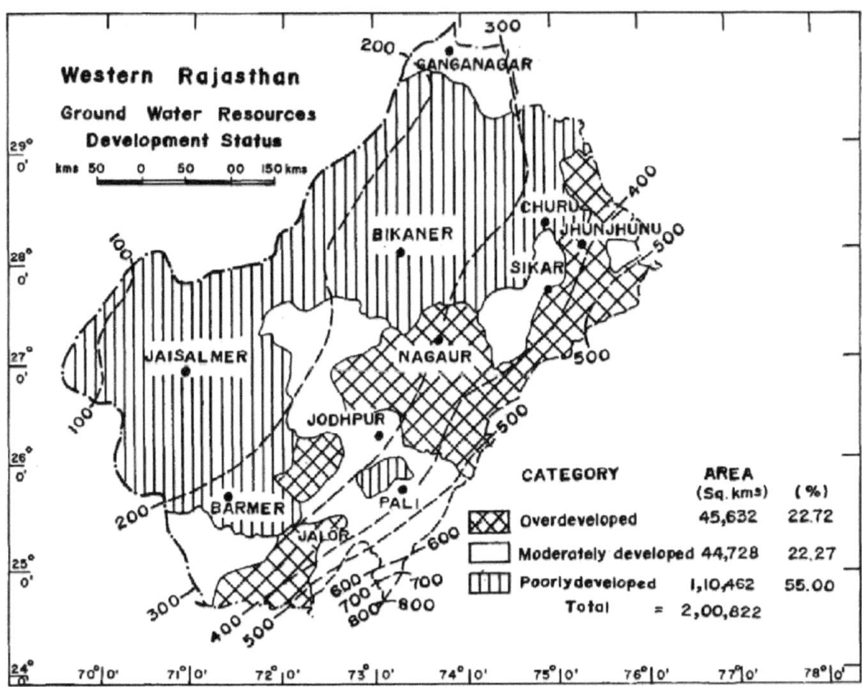

Fig. 20.2: Ground water resources development status in arid western Rajasthan
Source: G.W.D. Rajasthan

Strategies for soil water conservation in arid Rajasthan

Resource conservation

Soil and water are the two basic resources, needed for crop production. Measures for soil and water conservation can be divided in two main categories i.e. physical/engineering and biological. While the physical measures are the first line of defense and are very necessary to prevent immediate damage, biological measures are the offensive attack made from a position of good defensive strength. Physical measures consist of construction of mechanical barriers across the direction of flow of water to retard/retain the runoff and thereby reduce soil and water losses. Contour bunds, contour cultivation, trenching, sub-surface barrier etc. are some of the physical measures, which are found to be useful in arid areas. Biological measures like vegetative barrier, windbreak/shelter belt, stubble mulching are very effective for moisture conservation and controlling soil erosion due to wind and water.

Resource utilization

Efficient utilization of available resources is of paramount importance for successful crop production. Improved agronomic practices i.e. inter cropping, improved crops and seed varieties, crop rotation, minimum tillage etc. plays vital role in ensuring risk distribution. For aberrant weather situation contingency planning like life saving irrigation, integrated insect-pest management, differential fertilizer application etc. are crucial to minimize loss of seed, crop and fertilizer. The land which is unfit for conventional farming as per land capability classification may be used for alternative land uses, dairying, agro-forestry and agro-horticulture, cut flowers and fisheries, where water is stored for longtime. Value addition should be in built with the program for income generation and livelihood.

Conservation measures for arable lands

Contour farming

Contour farming is beneficial on all slopes where line sown is adopted. All ridges and row of plants place across the slope form continual series of miniature barriers to water that moves over the soil surface and offers maximum opportunity for infiltration. Moreover, contour operations reduce the power of the water to erode, suspend and carry away soil particles and increase the moisture storage. Increased uniform moisture storage can boost up the yield above 10%. Contour farming alone cannot control runoff volume from higher slope lands which may need bunding and grass waterways in natural drainage. For small fields with uniform slopes, one contour guideline is enough and where slopes are irregular two to three guidelines may be enough for farming operation. Contour farming is most effective on to the moderate slope of 2 to 7%. Smith and Wischmeier (1962) have reported that contour farming reduces soil loss upto 60% in comparison to up and down farming on 1% slope.

Contour bunding

Contour bunds act as barrier to the flow of water and at the same time impounding water increases soil moisture. Contour bunding is suitable for low rainfall area (< 600 mm) and for permeable soils having slope less than 6% to serve both as water and soil conservation measures. Contour bunding consists of constructing narrow trapezoidal bunds at contour to impound runoff water behind them so that all the impounded water is absorbed gradually in to the soil profile for crop use. The design of a contour bunding system involves the determination of the spacing between the bunds and the cross-sectional area of the bund and thc type and dimensions of the system. The main criteria for spacing between two bunds is to intercept the water before it attains erosive velocity. It is always desirable to remove local ridges and depression by leveling prior to alignment of contour bunding to avoid uneven impounding and danger of breaching. On average, contour bunds had 27 per cent higher soil moisture and 14 to 181 per cent higher fodder yield than flat surfaces on grasslands of western Rajasthan (Wasi-Ullah et al. 1972).

Graded bunding

Graded bunds are found suitable in areas where an annual rainfall is more than 600 mm to dispose off safely the excess water from the agricultural fields to avoid water stagnation. Graded bunds may also be adopted in areas of rainfall 500 mm, if the soils are highly impermeable. Graded bunds usually have wide and shallow channels and earthen bund laid along a predetermined longitudinal slope. As graded bunds are essentially means for the safe disposal of excess water from cropped lands, suitable outlets are required to be constructed on graded bund. Draining of excess water from one plot to another through outlets provided in the bund require special attention since considerable amount of soil may be lost through these outlets. Provision should be made to arrest the silt and allow only clear water to flow away. The vegetated watercourse strengthens the system. In areas with high rainfall where the volume of water to be disposed from one plot to the next lower level plot is large and the vertical interval between the plots is reasonably high, a site specific outlet design like a pipe outlet, drop structure etc. must be provided for safe disposal of the excess water. Graded bund is reported to reduce the run-off from 20 to 4.8 per cent and soil loss from 24 to 4.12 t ha^{-1} yr^{-1}. Besides other benefits intercropping on contour resulted in 48 per cent higher grain yield (Singh et al. 1997).

Vegetative barrier

Vegetative barriers of suitable grass species are grown along contours at suitable vertical interval to intercept part of runoff and to control erosion in agricultural fields having flat to slight undulating topography. The vegetative barrier moderates the velocity of overland flow and traps silt at low cost, and augment production of food, fuel and fodder or fiber from lands by growing suitable vegetation species. In recent years, vegetative barrier has found acceptability among the farmers, as it is cheaper over mechanical measures.

Vegetative barriers can be easily established across a wide spectrum of soil-climatic conditions. Selection of species depends upon purpose of barrier, site specific conditions, particularly soil and climatic variables. The spacing between plant to plant and row to row are governed by vegetation species to be planted as barrier. In general the plant to plant spacing of 0.20 to 0.30 m at predetermined or 0.50 to 1 m vertical interval between the barriers have been found effective for soil and water conservation. Generally, paired row of barrier planted in staggered form across the slope proves more effective. Dominant grass or shrub species of the region should be preferred for vegetative barrier.

Grass waterways

Grass waterways are developed for safe disposal of excess water from agricultural fields. These may be natural or man-made courses protected against erosion by suitable grass cover. Grass waterways are also used for channelizing and regulating runoff flows for water harvesting purposes. The best location for waterways is a natural depression or along valley line. These may also be constructed along field boundaries for safe disposal of excess rainfall from agricultural fields. Vegetative waterways may be located in all classes of lands except hard rocks, where construction may be difficult. The cross section of waterways may be trapezoidal, triangular or parabolic with shallow depth and flat side slope to facilitate easy movement of man, animal and machinery. The depth of waterways may be kept within 20 to 50 cm and side slope more than 4:1. The channel should have free board of 15 cm. The channel cross-section and bed slope should be such that the computed velocity is within permissible limit (Singh et al. 1990). Cost of construction of grass waterways depends upon type of soil, channel cross-section, length of channel and grass plantation technique.

Conservation measures for non-arable lands

Contour furrowing

Contour furrowing is most effective measure to reduce runoff and soil loss, increase in yield and is commonly adopted in grasslands and forestlands. However, in very sandy soils or soils with heavy clay pan area, their benefit is limited. Contour furrows varying from 30-60 cm wide and 10-25 cm deep can be used. The shape varies from "V" to square, rectangular, or parabolic. The cross section and depth of furrows mainly depend on soil and equipment used for making them. Furrows spaced at 8-10 m give better distribution of runoff water and higher yield of fodder. The effectiveness of contour furrows to hold water depends upon the degree of slope smoothness of the surface and accuracy in following contours and its life depends upon stability of soil and water storage capacity of the furrow. Layout of contour lines with the help of dumpy level, leveling staff and stacking is usual. The demarcation of contour lines can easily be done more economically with the help of tractor attached with sub-soiler tractor. After completing the demarcation work, in place of sub-soiler, disc plough with two discs set at 25° may be attached for digging of soil. The loose excavated earth will directly deposit on one edge of furrow and the regular shape of mound can be made manually. In a study conducted in arid part of Iran it is found that contour furrow and pitting has significantly helped in controlling soil erosion, increasing water penetration and soil moisture content and promoted propagation of *Hammada saliconica* species, a desirable plant species for both soil conservation and livestock grazing in the region (Jahantigh and Pessarakli, 2009).

Contour trenches

Contour trenches are useful practice in forestry areas. In pasturelands, contour trenches become great obstruction to grazing animals. This practice can be adopted in area, which are unsuitable for cultivation but suitable for forestry. Normal size of a trench is 1 m × 0.30 m × 0.60 m with an unexcavated portion of 1.5 m between two trenches. Length spacing or vertical interval depends on the slope of land. Spacing may vary from 30-60 m. After the trenches are excavated the correct size, they are refill partially, and stocking the remaining excavated material as a small bund on the downstream side. Mane et al. (2009) reported effectiveness of continuous contour trench for runoff control and recommended as best soil conservation practice on area having 7 to 8 per cent slope in konkan region of India.

Gully control measures

Check dams

Check dams are masonry overflow barriers (weirs) constructed across seasonal streams. A check dam as such has a relatively limited storage capacity but a large volume of water can still be pumped from such storage as the stream continues to flow and the check dam serves the purpose of an ideal intake structure. A check dam, by storing the base flow, maintains a supply of water for recharge as well as for direct use beyond the monsoon period (Goyal and Narain 2006). It creates flooding of upstream area which requires surplusing arrangements at suitable intervals to drain water. Check dam should be avoided in isolation. The number of check dams primarily depends upon the slope of the gully and the quantity of runoff. It may not be advisable to construct check dams on bigger streams with high gradient and where runoff is very high. The bigger streams should be treated with drainage line treatment like gabionic structures and boulder checks with masonry work to curtail the runoff.

Gabionic cheek dams

Gabionic check dams are useful in a locality where stones are readily available and their irregular shape makes them unsuitable for making loose stone cheek dams. If the expected water velocity is very high, gabion is recommended in place of loose rock dams. A gabion is rectangular shaped cage made of galvanized wire, which is filled with locally available boulders, rocks or stones. The gabion may be conveyed flat and are folded to shape at the construction site. Usually, gabions are 1 m wide and 0.75-12 m high with varying lengths ranging between 2-10 m. The gabionic check dams are constructed by connecting several gabions in horizontal and vertical direction. The gabionic check dams are very stable and semi-permanent in nature. These structures are flexible; they may even change shape automatically according to the streambed, even when the bed shape changes due to erosion, without losing stability.

Loose stone/dry stone masonry check dams

These structures are effective for checking runoff velocity in steep and broad gullies. These are suitable at upper reaches of the catchment. They have a relatively longer life and, usually require less maintenance. The bed of the gully is excavated to a uniform depth of about 0.3 m. Stones are then hand packed from the foundation level. Flat stones of size 20-30 cm are the best for construction and laid in such a way that all the stones are keyed together. Large size stones are placed at the center of the dam and gaps between stones may be filled with small piece stones. The dam should go up to 0.3 to 0.6 m into the stable portion of the sides of the gully to prevent end cutting. In the center of the dam, sufficient spillway is provided to allow maximum runoff to discharge. Vangani et al. (1998) reported sediment deposition of 3.86 t ha^{-1} yr^{-1} against loose stone check dam in Osian-Bigmi watershed (Distt. Jodhpur).

Brushwood check dams

These check dams are constructed by using locally available brushwood and supported by wooden stakes and used in the small gully heads not deeper than 1 m. These check dam are of two types; single row post brush dam and double row post brush dam. Brushwood check dams are constructed in areas where wooden posts, brushwood etc., are available in plenty. These check dams can only be used in the small gully heads not deeper than 1 m. Single row post brush dam are made of single row of wood stakes to which long branches of trees are tied length wise along gully with their butt ends facing upstream while in double row post brush dam, the straw and brushwood are laid across the gully between two rows of wooden posts, the distance between the rows being not more than 0.9 m. The longest branches are laid at the bottom and the shorter length branches are laid above it till the required dam height (0.3 to 0.7 m) is obtained.

Earthen gully plugs

An Earthen gully plug is constructed out of local soil across the stream to check soil erosion and flow of water. It is suitable in the upper catchment areas having scope of water storage and where the soil for the embankment is close by Depth of the gully should be less than 2 meter and gully bed slope is less than 10%. The site should have scope for side spillway. The bed of the gully is excavated to a uniform depth of about 0.3 m. Foundation trench is filled with compacted earth and earthen embankment is constructed by using local soil.

Agronomic practices

All operations carried out in the field, from land preparation to crop harvesting, with the aim of increasing the crop yield are included under agronomical practices. Certain simple agronomical practices like optimum tillage, administration of organic manure, suitable cropping pattern, and strip cropping have been found to be effective in retaining soil fertility as well as giving satisfactory crop yield.

Tillage operation

Tillage (ploughing) is the practice of breaking and working the soil to the desired depth prior to sowing. Tillage makes soil loose and hence prone to erosion. Timing and depth of tillage are the two important factors, which need special attention. Tillage should be done immediately before the crop season to take advantage of one or two early showers for land preparation. In arid region, land tilled into ridges and furrows across the wind direction has been found to reduce the effects of wind erosion during the summer months (Gupta 1993).

Cropping pattern

When one particular crop is grown in the same plot year after year, the land is likely to develop certain deficiencies. If the crop is erosion permitting in nature, the land is likely to get degraded. Similarly, due to the practice of monoculture, the land may develop serious deficiency in certain plant nutrients. Sparsely spaced crops like Sorghum, pearl millet, corn are erosion permitting crops. Densely grown crops like grasses, legumes and other deep-rooted crops are erosion resistant crops. Erosion resistant crops should therefore, be grown at least once in two years in the plots where such crops am not being grown and the land shows visible signs of degradation. Weeding at appropriate time significantly improves crop yields (Singh and Singh 1988, Gupta and Gupta 1982). Therefore, elimination of weed canopy early in the season is one of the important practices to reduce water use per unit of yield.

Land leveling

Farm lands should be leveled for efficient management of water. Level lands allow more infiltration, thus increasing soil moisture and leaching. This in turn reduces run-off and hence soil erosion. Leveling of irregular land is done by the cut and fill method. Soil from the elevated portions is removed and placed in the depressed portions to obtain a level land.

Strip cropping

In strip cropping two or more crops are grown in alternate strips preferably across the slope. One strip of erosion permitting crop should be alternated with another strip of relatively more erosion-resisting crop. In strip cropping, the entire land remains in a kind of balance. Soil eroded from one strip is retained by the next strip and the overall fertility of the land is maintained. Narrow strips are more effective in reducing wind erosion in lighter soils. The width of the strip varies from 6 m in sand to 30 m in sandy loam. Reduction in sand drift due to protective strips of grass at Bikaner and Hingoli (Jodhpur) was also reported by Singh (1989). Another advantage of strip cropping is that it helps in the prevention of pest attack on the crops. Since pests are mostly crop specific, one particular strip affected by one particular pest remains confined within that strip itself and does not spread to the next strip, thus preventing the spreading of the diseases to the entire field.

Mulching

Mulching of open land surface is achieved by spreading stubble, trash or any other vegetation. The objectives of mulching are to minimize splash influence of rain drops on base surface; reduce evaporation; increase absorption of the rainfall; obstruct surface flow thereby retarding erosion and allow microbiological changes to occur at optimum temperature. Sometimes, spreading of organic residues, instead of mixing can help in reduction of soil and water loss to a considerable extent. Polyethylene mulches have also been utilized for water harvesting and control of seepage. Trash farming, in which crop remains are cut, chopped and partly mixed in ground and partly left on land surface is also a form of mulching. Mulching reduces the evapotranspiration losses from soil surface and helps in promoting better plant establishment and results in higher yields. Application of grass mulch at the rate of 6 t ha^{-1} resulted in reduced mean maximum soil temperature, reduced evapotranspiration and consequently increase of 40 per cent yield of green gram (Gupta 1978, 1980).

Rainwater harvesting structures

Nadi

Nadi is a local name of dugout pond, used for storing limited water quantities available from an adjoining natural catchment during the rainy season. In and Rajasthan, *nadi* system of water harvesting is the oldest practice and still the principal source of water supplies for human and livestock consumption. The size and number of *nadis* depend upon demand, physiographic conditions and rainfall pattern. A *nadi* has the limitation of high evaporation losses due to a large exposed surface area, high seepage losses through porous sides and bottom, heavy sedimentation due to biotic interference in the catchment and water contamination causing health hazard. Evaporation losses ranged from 55 to 80% of the total losses in various environments. The highest evaporative losses in a *nadi* occur during March to June, the driest season, when the demand of water is the highest. High evaporation also occurs in October and November after the rainy season. Seepage losses are greatest during the rainy season (July-September) when *nadi* is completely filled. To overcome this problem CAZRI has developed designed *Nadis* with LDPE lining on sides and bottom keeping surface to volume ratio 0.28 and provision of silt trap at inlet. *Nadis* also help to recharge ground-water aquifers although their effect varies depending on the underlying soils and rocks

Selection of site for construction of a *nadi* is primarily made on the basis of an available natural depression (valley formation) with its catchment and its water yield potential. Rocky hills are excellent natural rainwater catchments. Catchment area of rock outcrop or stony/gravelly pediments with steep slope also have high water yielding efficiency. Construction of *nadi* at the lowest elevation has the benefit of natural drainage and need for minimum excavation of earth. Preference should be given to locations with a low soil permeability or side having impermeable murram (hard pan).

Tanka

Tanka is an underground tank for collection and storage of surface runoff from the natural or artificially prepared catchments or from roof top. The stored water is used mainly for drinking and cooking purposes and also to raise nursery for plantation. *Tanka* is constructed for an individual family or for a community depending upon the requirements. However, an individual family *tanka* is better managed and conserved water is used judiciously than community *tanka* (Vangani et al. 1988). Community *tanka* is generally constructed for a school, panchayat ghar or large settlements/villlage. Settlements have one or more *tankas*. In the traditional method, 5 mm thick lime mortar is used to plaster the bare horizontal and vertical soil surfaces to a thickness. In some cases, a second layer of plaster of cement mortar is also applied to a thickness of about half of lime mortar. The top is covered with *Zizyphus nummularia* bushes. CAZRI, Jodhpur has developed improved design of *tanka* for capacity ranging from 10 to 600 m^3 which has been replicated in large numbers in western Rajasthan for conjunctive use. These *tankas* were successfully constructed in Jhanwar, Sar, and Baorali-Bambore watersheds (Bhati et al. 1997). Harvested water of these tankas was used to provide life saving irrigation to plants. The Benefit cost ratio of tanka ranged from 1.25 to 1.40 under different uses (Goyal et al. 1995, 1997).The catchment of a *tanka* is made by spreading the excavated material around the structure. A second layer of 5-6 cm thick pond silt on bentonite is compacted to make the surface smooth and semi- impervious. A uniform slope of 2-3 per cent towards *tanka* is provided for harvesting maximum possible runoff.

Khadin

Khadin is a unique practice of water harvesting and moisture conservation in suitable deep soil surrounded by natural catchment zone. In this system, runoff from upland and rocky surfaces is collected in the adjoining valley by enclosing a segment with an earthen bund. Any excess water in *khadin* bed is passed out through a spillway provided in the bund. The plots are rigorously built and managed to make the entire system a self-contained unit for winter cultivation. Under the condition of intense evaporation, the moisture threshold and soil fertility are maintained. The total energy input of rainwater, sand-silt-clay accumulation and cultivators' own activities are interwoven into a complete production system of winter crops. There is progressive increase in crop yield every year as more and more fresh silt and clay accumulate in the *khadin* bed. Since the rainy season crops are not grown in *khadin*, the losses by transpiration are nearly absent. The ratio of farmland and catchment areas is regulated to be about 1:10 so that a suitable moisture supply is uniformly maintained. The water holding capacity and water infiltration, rates are balanced by the shallow depth of soils in these plots. The basement of *khadin* is invariably a hard surface upon which sand-silt-clay is made to accumulate just to the depth of a few meters. This maintains a convenient supply of natural moisture and nutrients within the crop root zone. Few wells are generally kept on down side of earthen bund. These are recharged through accumulated in *khadin* and the well is utilized during summer months when *khadin* dries up (Narain and Goyal 2005). Studies of groundwater recharge through *khadins* in different morphological settings suggest that 11 to 48 per cent of the stored water contributed to groundwater in a single season. This replenishment of aquifers means that sub-surface water can be extracted through bore wells dug downstream from the *khadin.* The average water-level rise in wells bored into sandstone and deep alluvium was 0.8 m and 2.2 m, respectively (Khan, 1996).

Conclusions

A successful application of any measures for improving production requires an integrated approach. Management of resources for enhancing crop productivity in arid areas is real challenge. For management of scares water resources, multiple point strategies are needed. On one hand technological advancement is needed for the better and early forecast of drought and on other hand technologies of rainwater harvesting and conservation needs to be popularize and percolated at extreme down end. On cropping fronts appropriate technology is needed for development of drought tolerant early maturing crops to combat drought. Traditional rainwater harvesting structures like *nadi, baori, talab* etc needs renovation on continuous basis. Efforts special efforts are needed to harvest flash floodwater for the lean period by construction of large storage structures at appropriate sites. Efforts are also needed to control the indiscriminate extraction of groundwater by the private tube well owners by law and recharge of groundwater should be made mandatory. All the technologies discussed above are essentially site specific and different components need to be integrated as a holistic approach to conserve rainwater to provide water for different needs on sustainable basis. These technologies are time tested and are of proven soundness for extreme conditions such as of arid.

References

Bhati, T.K., Goyal, R.K. and Daulay, H.S. 1997. Development of Dryland Agriculture on Watershed basis in Hot Arid Tropics of India - A Case Study. Annals of Arid Zone 36(2): 115-121.

CAZRI. 1990. WATER: 2000AD, The Scenario for Arid Rajasthan. Central Arid Zone Research Institute, Jodhpur.

Faroda, A.S., Joshi, D.C. and Ram, Balak. 1999. Agro-ecological Zones of North-western Hot Arid Region of India. Central Arid Zone Research Institute- Jodhpur.

Goyal, R.K. and Vittal, K.P.R. 2008. Water Resource Management in Hot Arid Zone of India. Journal of Hydrological Research and Development, 23: 37-54.

Goyal, R.K. and Narain, P. 2006. Impact evaluation of watershed conservation measures in hot arid zone of India. In Conference on Natural Resources Management for Sustainable Development in Western India (NRMSD-2006), Indian Association of Soil & Water Conservationist, Dehradun - Pune.

Goyal, R.K., Ojasvi, P.R. and Gupta, J.P. 1997. Rainwater Management for Sustainable Production in Indian Arid Zone. In 8th International Conference on Rainwater Catchment (Aminipouri, B and Ghoddoussi, J, Eds.), Tehran.

Goyal, R.K., Ojasvi, P.R. and Bhati, T.K. 1995. Economic evaluation of water harvesting pond under arid conditions. Indian Journal of Soil Conservation 23(1): 74-76.

Gupta, J.P. 1978. Evaporation from a sandy soil under mulches. Annals of Arid Zone 17(3): 287-290.

Gupta, J.P. 1980. Effect of mulches on moisture and thermal regimes of soil and yield of pearl millet. Annals of Arid Zone 19: 132-138.

Gupta, J.P. 1993. Wind erosion of soil in drought prone areas. In Desertification and Its Control in the Thar, Sahara and Sahel Regions (Sen, AK and Kar, A., Eds.), Scientific Publishers, Jodhpur.

Gupta, J.P. and Gupta, G.N. 1982. Effect of post emergence cultivation on weed growth, nutrient uptake and yield of pearl millet. Annals of Arid Zone 21: 241-247.

Jahantigh, M. and Mohammad, P. 2009. Utilization of Contour Furrow and Pitting Techniques on Desert Rangelands: Evaluation of Runoff, Sediment, Soil Water Content and Vegetation Cover. Journal of Food, Agriculture & Environment 7: 736-739.

Khan, M.A. 1996. Inducement of Groundwater Recharge for Sustainable Development. In 28th Annual Convention, Indian Water Works Association, Jodhpur, India.

Khan, M.A. 1997. Water resources of Indian arid zone. In Desertification Control in the Arid Ecosystems of India for Sustainable Development (Singh, S and Kar, A, Eds.). Agro Botanical Publishers (India), Bikaner.

Mane, M.S., Mahadkar, U.V., Ayare, B.L. and Thorat, T.N. 2009. Performance of Mechanical Soil Conservation Measures in Cashew Plantation Grown on Steep Slopes of Konkan. Indian Journal of Soil Conservation 37: 81-184.

Narain, P. and Goyal, R.K. 2005. Lead paper on 'Rainwater harvesting for increasing productivity in arid zones'. National Symposium on "Efficient water management for eco-friendly sustainable and profitable agriculture". Symposium Abstract. Organized by Indian society of water management and Indian agriculture research institute from Dec. 1-3, 2005 at Water Technology Center, New Delhi: 141-142.

Ramakrishna, Y.S., Rao, A.S. and Joshi, N.L. 1988. Adjustments to weather variations for efficient agricultural production system in arid western plains of India. Fertilizer News 4: 29-34.

Rao, A.S. and Singh, R.S. 1998.Climatic features and crop production. In Fifty Years of Arid Zone Research in India (Faroda, AS and Singh, M, Eds). Central Arid Zone Research Institute, Jodhpur.

Sharma, K.D. and Vangani, N.S. 1992. Surface water resources in the Thar desert and adjoining arid areas of India. In Perspectives on the Thar and the Karakum. Department of Science and Technology, New Delhi.

Singh, A.K., Kumar, A.K., Katiyar, V.S., Singh, K.D. and Singh, U.S. 1997. Soil and Water Conservation Measures in Semi-arid Regions of South-Eastern Rajasthan. Indian Journal of Soil Conservation 25(3): 186-189.

Singh, G.C., Venkatramanan, G., Sastry and Joshi, B.P. 1990. Manual of Soil and Water Conservation Practices. Oxford and IBH Publishing Co. Pvt. Ltd., New Delhi.

Singh, K. C. 1989. Importance of grasses in management of sandy soil-strip cropping of grasses. In Proceedings of the International Symposium on Managing Sandy Soils, CAZRI, Jodhpur.

Singh, S.D. and Singh, M. 1988. Integration of planting geometry. In Water Harvesting in Arid Tropics (Singh, SD, Ed). CAZRI, Jodhpur.

Smith, D.D. and Wischmeier, W.H. 1962. Rainfall erosion. Advances in Agronomy 14: 109-148.

Vangani, N.S., Sharma, K.D. and Chatterji, P.C. 1988. Tanka - a reliable system of rainwater harvesting in the Indian desert. CAZRJ Bulletin No. 33, CAZRI, Jodhpur.

Vangani, N.S., Singh, S. and Sharma, R.P. 1998. Index catchment - a new concept for sustainable integrated development. Annals of Arid Zone 37(2): 133-137.

Wasi-Ullah, Chakravarty, A.K., Mathur, C.P. and Vangani, N.S. 1972. Effect of contour furrows and contour bunds on water conservation in grasslands of western Rajasthan. Annals of Arid Zone 11(3/4): 170-182.

21

Soil Resources in Forests of Rajasthan

N. Bala

Arid Forest Research Institute, Jodhpur, Rajasthan, India

Introduction

The life support systems of a country and socioeconomic development of its people largely depend on soils. The soils are the most valuable natural resources for the society. It caters the basic needs of mankind by producing food, fibre and timber. It is therefore essential to know the distribution and extent of different soils and their qualities, potentialities and constraints for sustainable utilization of this valuable natural resource (Shyampura and Sehgal 1995). Information on distribution, potential and constraints of major forest soils are essential to design most appropriate soil management systems in order to increase productivity of forests as forests also offer excellent potential for poverty reduction and rural economic growth in India. However, unlike agriculture lands forest soils usually draw less attention and are ignored in spite of the fact that they nourish the lungs of the Earth. Soil quality is the most important factor in forest management decisions. Soils will determine productivity of a particular forest and management strategy. Knowing about forest soils can serve as a basis for forest management decisions, including land acquisition, species selection for planting, site preparation requirements, watershed development, fertilization prescriptions, stand density/ composition, and harvest timing, as well as decisions affecting land ownership and use.

Soils are important component of the forest ecosystem. It is a complex and variable medium comprising mineral particles, organic matter, water, air and living organisms. The characteristics of soil largely determine the nature of the flora and fauna that sustains the world's terrestrial biodiversity and its productive potential. Geology, topography and climate all play a part in creating the many different soil types, which often vary within short distances. The physical, chemical and biological properties of soils are continually modified by a number of natural processes, which include leaching, waterlogging and the addition and decomposition of organic matter.

When seen in detail at the village level, the soils of Rajasthan are complex and highly variable, reflecting a variety of differing parent materials, physiographic land features, range of distribution of rainfall and its effects etc. However, broadly, based on the soil texture they have been grouped into five major types. As per USDA classification soils of Rajasthan have been grouped into 5 orders, 8 suborders, 16 great groups, 32 subgroups and 86 families (Shyampura *et al.*, 2002).

However, all these efforts are mainly concentrated on agricultural lands and wastelands. Though forest soils and the biodiversity contained within the forest ecosystem produce a range of resources useful to people very few have addressed forest soils. Forest soils provide a wide range of ecosystem services.

Keeping these in view a study was done to characterize and classify forest soils of Rajasthan. The idea was to generate scientific understanding of the soil resources in forest areas of Rajasthan for conservation, improvement and efficient management of the precious resource base. Stratified sampling method was adopted for the study. Survey was conducted in different forest blocks, district wise, to select representative forest/vegetation types including topographic variations prevailing in the district. Soil profiles were then studied at 541 places by opening pits of 6 × 6 sq. ft up to a depth of 6 feet or to a lithic contact/soil depth in the 33 districts of Rajasthan covering 478 forest blocks and 243 ranges.

Physico-chemical characterization of the soils was done in the field as well as in laboratory. Soil structure, consistency, colour, soil profile descriptions, physiographical parameters were recorded, as per standard format, in the field. Soil pH, electrical conductivity, organic carbon, inorganic carbon, NO_3^--N and NH_4^+-N, PO_4^{-3}-P, potassium, cation exchange capacity (CEC), soil texture (sand, silt and clay content) have been estimated following standard procedure. Ecological study in an area of 0.1 ha near each of the soil profile pits was also conducted.

Soil organic carbon (SOC) was calculated after applying corrections both for gravel content and rock out crops. Soil organic carbon stock has been estimated for each layer of soil profile as well as for the whole soil profile. The soils have been classified as per the USDA classification.

Forest Scenario in Rajasthan

As per legal status the total forest area is 32737 km^2 (FSI, 2015) which constitute 9.57% of geographic area of the State. However, based on the satellite data (Oct 2013 to Feb 2014) the forest cover is 16171 km^2 which is only 4.73% of the geographical area of the state (FSI, Fig. 21.1). There are two major forest types namely tropical dry deciduous and tropical thorn forests. In terms of forest canopy density classes, the state has 76 km^2 area under very dense forest, 4426 km^2 area under moderately dense forest and 11669 km^2 area under open forest. Comparison of the current assessment with the previous assessment shows a gain of 85 km^2 of forest cover. The state has 20 forest types which belong to two forest type groups viz. Tropical dry deciduous and Tropical Thorn Forests. Majority (88.3%) of the forest fall under Tropical dry deciduous type whereas Tropical thorn forests occupy 6.18% of the forest cover. Tree outside forest (TOF) and/or plantation also contribute significantly (5.8%) towards total forest cover of the state. The estimated tree cover in the state is 8269 km^2 which is 2.42% of its geographical area.

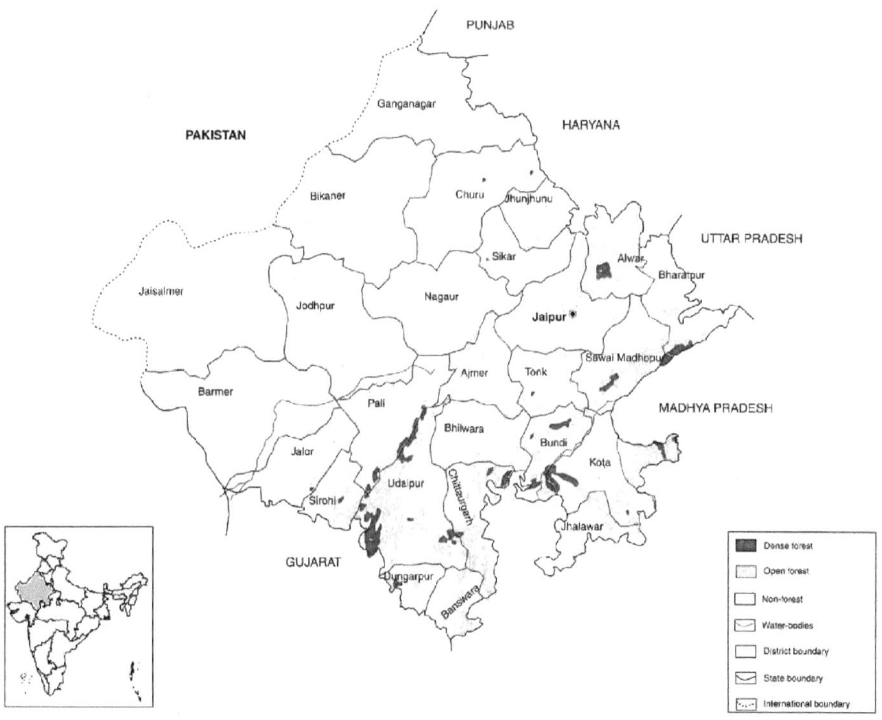

Fig. 21.1: Forest cover map of Rajasthan *Source:* (FSI, 2015)

Compared to 1987 assessment there has been an increase of 3413 km^2 total forest cover in the state as per the latest survey. Gain in the forest cover can be ascribed to massive plantation activities undertaken by the state forest department. Natural re-growth of *Prosopis juliflora* at several places has also contributed towards improvement of forest cover.

Vegetation and forest type

The vegetation of the study sites are categorized into 31 forest sub types as per Champion and Seth (1968). *Anogeissus sericea*, *Acacia leucophloea*, *Heteropogon-Acacia leucophloea* grassland also observed at few places as dominant vegetation type. *Prosopis juliflora* and *Anogeissus pendula* were observed as the most dominant tree species. Among the study sites *Anogeissus pendula* forest was observed as the most prevalent forest type which was found at 56 sites. This was followed by Tropical dry deciduous scrub (49 sites), *Prosopis juliflora* scrub (44 sites), *Acacia senegal* forest (42 sites), N. Tropical Dry Mix. Deci. Forest (39 sites), Desert thorn forest (27 sites), *Butea* forest (26 sites), *Boswellia serrata* forest (26 sites), *Anogeissus pendula* scrub (21 sites), Dry teak forest (20 sites). Plantation of different species was observed in 13% of the study sites, *Acacia tortilis* being the most dominant planting species covering 52 study sites. Two most important invasive species *P. juliflora* and *Lanata camara* were recorded in 36.7% and 7.2% forest blocks distributed in 32 and 15 distric, respectively in Rajasthan.

Regeneration of different tree and shrub species was observed in as many as 437 forest blocks/study sites. A total of 62 species were observed to regenerate at different sites in varied numbers. *Prosopis juliflora*, *Anogeissus pendula*, *Butea monosperma*, *Acacia senegal*, *Diospyros melanoxylon*, *Acacia leucophloea* and *Acacia tortilis* were observed as major regenerating species. Among the species, regeneration of *Prosopis juliflora* was observed in 28% sites followed by *Anogeissus pendula* in 20% sites, *Butea monosperma* in 15% sites and *Acacia senegal* in 8% sites.

Soils in different forest blocks of Rajasthan experience a wide variation in physiographic, relief and drainage condition. Over 15% study sites are situated on steep slope (>30%), 6.5% on 20-30% slope, 10.5% on 10-20% slope and the rest are on 0-10% slope. Most of the forest area in Udaipur are situated on steep slope (72% study sites were situated on >30% slope). This is followed by Jaipur with 67.74% sites on >30% slope, Rajsamand with 38.46% sites, Bundi with 26.32% and Bhilwara with 26.09% sites on >30% slope.

Soil characteristics

Soil physical characteristics like gravel content, structure, consistency, colour and soil bulk density varied widely in different sites. Western arid districts are dominated by coarse textured soils as compared to soils of eastern districts. At 258 sites (47.7% of the total sites) soils were characterized as skeletal (having >35% gravels or rock fragments). A large number of soil profiles studied have Lithic contact at shallow soil depth. Lithic contact was observed in 232 study sites constituting 42.88% of the total study sites.

Nitrogen availability is low throughout the forests of Rajasthan. NH_4^+-N in soils varied from a low of 0.05 ppm (1410 RD, IGNP, Jaisalmer) to 76 ppm in Ghanerao forest block, Pali district. Higher NH_4^+-N content was observed in soils of Pali, Alwar and Dungarpur districts whereas, it was low in Jaisalmer and Churu districts. NO^{-3}-N varied from a low of 0.12 ppm (Ghadsisar, Churu) to 55.07 ppm in Dhanni forest block, Bhilwara. NO^{-3}-N was comparatively high in soils of Bhilwara, Pali, Jhalawar, Pratapgarh districts and low in Churu, Barmer, Hanumangarh districts. High soil PO_4^{-3}-P was observed in Dausa, Pratapgarh, Dungarpur districts, maximum (41.5 ppm) being in Balaji block, Dausa. Low PO_4^{-3}-P was recorded from forest blocks of Barmer and Ajmer districts. Lowest Potassium was observed in soils of Mishroli forest block Jhalawar and highest Potassium (92.5 ppm) was recorded in Ashapura oran and Barli of Pali district.

As per the index of soil fertility (GoI 2011), 58% soil samples were low in PO_4^{-3}-P content and 34.7% were medium leaving only 7.3% samples with high PO_4^{-3}-P. All the soil samples were found low in available nitrogen. The soil samples were mostly (65%) medium to high in available K content. Thirty five percent samples were found having low available K content. Soils of Chittorgarh, Alwar, Dholpur, Jaipur, Sikar, Jhunjhunu and Hanumangarh forest blocks showed high available K content. Similar observation was reported by Meena et al. (2012) for some typical pedons representing major landforms of Malwa plateau transect of Banswara district viz., hill top, side slope, foot slope, undulating pediment, moderately sloping pediment, gently sloping pediment and very gently sloping alluvial plain, developed from basalt and occurring at different elevations under varying land use. The soils were low in available nitrogen while soils of lithic ustorthents and typic haplusterts were low in available phosphorus whereas, soils of vertic haplustepts and lithic haplustepts were medium in available phosphorus. Singh and Rathore (2013) also reported low available nitrogen and phosphorus and medium to high in available potassium. Most districts in Rajasthan have been categorized as medium in K status. Districts Barmer, Churu, Sri Ganganagar, Jhunjhunu, Jodhpur and Sikar are in High K status. Districts Dholpur, Rajsamand and Udaipur are in Low Medium status. Rehanul Hasan (2012) has placed 24 districts of Rajasthan under medium and 3 districts having high potassium fertility status.

Comparatively greater number of soil samples showing low available K in the present study may be because of the sampling process. In our study we explored greater soil depth (up to 180 cm) compared to soil depth (15/30 cm) usually considered for agricultural soils.

Low (1.85 cmol (ρ+) kg^{-1}) cation exchange capacity (CEC) was recorded in soils of Taragarh forest block of Ajmer district and high CEC of 77.82 cmol (ρ+) kg^{-1} was recorded in soils of Naua forest block, Udaipur. Cation exchange capacity reported in the present study is higher than that reported by Lavti and Paliwal (1980) but in line with the CEC reported by Somasundaram *et al.* (2013) for some soils of Chambal region of Rajasthan. Higher cation exchange capacity was observed in natural vegetation compared to cropland and plantation (Somasundaram *et al.*, 2013). Lower CEC has also been reported by Gill et al. (2012) for cultivated soils of Phalasia block of Udaipur district, Rajasthan. Comparatively higher soil organic carbon in forest soils may the reason for higher CEC.

Soil organic carbon stock

Soil organic carbon was estimated using wet digestion method. It is well established that about 60-86% of soil organic carbon is oxidized in the Walkley and Black method. Therefore, a correction factor is sometime used to arrive at correct SOC value. However, since the recovery percentage of carbon varies due to land-use, texture etc. the correction factor was not used here. As a result the reported SOC% and SOC stock may be slightly underestimated. We have calculated soil organic carbon (SOC) stock after applying corrections both for gravel content and rock out crops. Soil organic carbon stock has been estimated for each layer of soil profile as well as for the whole soil profile.

Percent soil organic carbon was high in Panund block of Udaipur district and low in 2 LKD, Bikaner district. Soil organic carbon stock varied between 237.7 Mg ha^{-1} (Mg = 10^6 g) in Sagbari block, Banswara and 0.546 Mg ha^{-1} in Runawasa (hill top) block, Sikar district. High soil organic carbon stock was observed in Banswara, Alwar, Kota, Udaipur, Baran and Churu districts and very low soil organic carbon stock was recorded in soils of Barmer, Sikar and Ajmer districts. Santra et al. (2012) reported 3.57 tonnes ha^{-1} (Mg ha^{-1}) for an agricultural farm located in hot arid region of western Rajasthan (Jaisalmer). However, this estimate is for the top 15 cm soil layer. In comparison to this, the SOC stock calculated for Jaisalmer varied between 11.6 to 77.5 Mg ha^{-1} in different forest blocks computed for the entire soil profile. A huge depletion in soil organic carbon due to cultivation practice has been reported by Santra et al. (2012). Little higher SOC stock in forest blocks of Jaisalmer compared to the report of Santra et al. (2012) may be well justified considering the land use.

There is always a possibility of either underestimate or overestimate the SOC stock at a small scale if regional scale estimates of SOC stock based on global mean is applied. SOC stock of 78.68 Mg ha^{-1} for 0-30 cm soil layer in hot arid ecoregion of India (Punjab, Haryana and Rajasthan) has been reported by Velayutham et al. (2009), which is higher than that reported in the present study. Difference in methods of SOC determination and SOC stock calculation may have lead to these differences. On the contrary, the report of SOC stock of 1550 kg km^{-2} for 0-25 cm soil layer of Typic Torripsamments soil profile in hot arid region of Rajasthan (Singh et al. 2007) is much lower than the SOC stock estimated in the present study. Soil organic carbon stock depended on percent organic carbon, soil depth, gravel content in soils and presence of rock outcrop.

Soil classification

Soils of Rajasthan forests have been classified into 5 orders (*viz.* Inceptisols, Entisols, Aridisols, Alfisols and Vertisols), 9 sub orders (Psamments, Ustepts, Orthents, Ustalfs, Salids, Calcids, Cambids, Gypsids and Ustarts), 14 Great groups and 37 Subgroups. The Inceptisols are dominantly observed in forests of Rajasthan covering 67.65 percent study sites. This was followed by Entisols, Alfisols and Aridisols covering 19.60, 7.76 and 4.44% of the study sites respectively. Vertisols are least represented covering only 0.55 of the study sites. Shyampura et al. (2002) has classified the soils of Rajasthan in to 5 orders, 8 suborders, 16 great groups, 32 sub groups and 86 families. The Entisols are dominant order observed in 36%, followed by Inceptisols, Aridisols, Vertisols, covering 22.8, 19.5 and 2.3% of the total geographical area, respectively. The Alfisols are least represented covering 0.73% area. The rock out crops constitute 4.8% of the total geographical area of the state.

Soil profile Desert thorn forest(6B/C1)

Fig. 21.2: Sandy, hyperthermic, Typic Torripsamments, Jaisalmer

Soil profile Boswellia forest (5/E2)

Fig. 21.3: Loamy-skeletal, hyperthermic, Lithic Haplustepts, Jaipur

Soil profile Dry Teak forests (5A/C1)

Fig. 21.4: Fine, smecti tie, hyperthermic, Typic Haplusterts, Banswara

Soil profile *Acacia senegal* forest (6/E2)

Fig. 21.5: Sandy, hyperthermic, TypicTorripsamments, Jaisalmer

References

Champion, H.G. and Seth, S.K. 1968. A Revised Survey of the Forest Types of India. New Delhi India: The Manager of Publications.

FSI, 2015. India State of Forest Report. Forest Survey of India, Dehradun.

Gill, R.S., Kanthaliya, P.C. and Giri, J.D. 2012. Characterization and classification of soils of Phalasis block of Udaipur district, Rajasthan. Agropedology 22: 61-65.

GoI. 2011. Methods Manual-Soil Testing in India. Department of Agriculture & Cooperation, Ministry of Agriculture, New Delhi, 208 pp.

Hasan, R. 2012. Potassium Status of Soils in India. Better Crops International 16: 3-5.

Lavti and Paliwal. 1980. Evaluation of soil structure of some arid and semi arid soils of Rajasthan. Proceeding of Indian National Science Academy B46(2): 236-242

Meena, R.H., Giri, J.D. and Sharma, S.K. 2012. Soil-site suitability evaluation for chick pea in Malwa plateau of Banswara district, Rajasthan. International Journal of Scientific and Research Publications 2: 1-6.

Santra, P., Kumawat , R.N., Mertia, R.S., Mahla, H.R. and Sinha, N.K. 2012. Spatial variation of soil organic carbon stock in a typical agricultural farm of hot arid ecosystem of India. Current Science 102: 1303-1309.

Shyampura, R.L. and Sehgal, J. 1995. Soils of Rajasthan for optimizing land use. NBSS publication (Soils of India series). National Bureau of Soil Survey and Land Use Planning, Nagpur, India, 76+6 pp.

Shyampura, R.L., Singh, S.K., Singh, R.S., Jain, B.L. and Gajbhiye, K.S. 2002. Soil series of Rajasthan. NBSS Publ. No. 95, NBSS & LUP, Nagpur, 364 p.

Singh, D.P. and Rathore, M.S. 2013. Available nutrient status and their relationship with soil properties of Aravalli mountain ranges and Malwa Plateau of Pratapgarh, Rajasthan, India. African Journal of Agricultural Research 41: 5096-5103.

Singh, S.K., Singh, A.K., Sharma, B.K. and Tarafdar, J.C. 2007. Carbon stock and organic carbon dynamics in soils of Rajasthan, India. Journal of Arid Environment 68: 408-421.

Somasundaram, J., Singh, R.K., Parandiyal, A.K., Ali, S., Chauhan, V., Sinha, N.K., Lakaria, B. L., Saha, R., Chaudhary, R.S., Coumar, M.V., Singh, R.K. and Simaiya, R.R. 2013. Soil properties under different land use systems in parts of Chambal region of Rajasthan. Journal of Agricultural Physics 13: 139-147.

Velayutham, M., Pal, D.K. and Bhattacharya, T. 2009. Organic carbon stocks in soils of India. 71-96. In Global Climate Change and Tropical Ecosystems, Lal, R., Kimble, J. M. and Stewart, B. A. (eds.) Lewis Publishers, Boca Raton, FL, USA.

Walkley, A. and Black, I. 1934. An examination of Degtjareff method for determining soil organic matter, and a proposed modification of the chromic acid titration method. Soil Science 37: 29-38.

22

Soil Microbial Diversity in the Drylands

Ramesh C. Kasana and N.R. Panwar

ICAR-Central Arid Zone Research Institute, Jodhpur, Rajasthan, India

Introduction

Microorganisms generally present in all the environments are the most abundant and diverse living organisms present on earth. Microorganisms have great impact on the biosphere due to their inbuilt ability to transform the key elements of life into forms usable by other organisms and plants. Because of heterogeneity in nature of soils from variations in physical, chemical and biological properties the various natural habitat are microbially most diverse environments on the earth maintaining the stability between complex microbial communities and their function (Daniel 2005). One gram of soil contains about 10 billion microorganisms of possibly thousands of different species but microbial diversity in general, and particularly in extreme environments, and its role in ecology is poorly understood (Rossello-Mora and Amann 2001).

Deserts present a unique environment with very little precipitation and large variations in day and night temperatures, making it inhospitable environment for living organisms. Still a number of microbial species do survive in those harsh environments. The populations of aerobic bacteria in deserts across the world are reported to vary from < 10 in Atacama desert to 1.6×10^7 cfu g^{-1} in soils of Nevada (Skujnis 1984), and India is not exception to that with the bacterial population in the range of 1.0×10^3 to 8.35×10^6 cfu g^{-1} sediment or ml^{-1} water in the samples collected from cold deserts of north western Himalayas (Yadav et al. 2015).

Drylands which occupy ~41% of Earth´s surface are expected to expand in global area by 11–23% by 2100 (Huang et al. 2015, Maestre et al. 2015). Arid region in India is spread over in 38.7 million hectare area. Out of the total, 31.7 m ha lies in hot and remaining 7 m ha lies in cold region. About 62% area of arid region falls in western Rajasthan followed by 20% in Gujarat and 7% in Haryana and Punjab.

Andhra Pradesh, Karnataka and Maharashtra together constitute about 11% area of arid region. The tropical desert of Asia extends to India through Rajasthan and Gujarat where it is called the Thar. The soil in the desert characteristically has a high percentage of soluble salts, high pH, low loss on ignition, varying percentage of calcium carbonate and poor organic matter content. The major limiting factor is water and these soils can be reclaimed if proper facilities for irrigation are available. The region is subjected to intense winds and a wide variation in temperature ranging from a minimum of 5°C to a maximum of 45°C (Gupta 1993). Photographs for hot desert in Rajasthan and cold desert in Himachal Pradesh are shown in Fig. 22.1.

Fig. 22.1: Hot desert in Rajasthan (A) and cold desert in Himachal Pradesh (B)

Microbial diversity in hot deserts

Microbial diversity in desert soils is highly influenced by various factors such as temperature, moisture and the availability of organic carbon. Of these, moisture availability is the major constraint affecting microbial diversity, community structure and activity. Sand dunes from Thar desert are reported to have relatively smaller population ($1.5 \times 10^2 - 5 \times 10^4$ g^{-1} soil) (Venkateswarulu and Rao 1981, Kathuria, 1998). Gram-positive spore formers were dominant and the populations do not decline significantly even during summers (Rao and Venkateswarlu 1983). Actinomycetes may constitute ~ 50% of the total microbial bacterial population in desert soils. Dominant microflora of desert soils is made up of coryneforms, i.e. *Archangium*, *Cystobacter*, *Myxococcus*, *Polyangium*, *Sorangium* and *Stigmatella*; sub-dominant forms comprise *Acinetobacter*, *Bacillus*, *Micrococcus*, *Proteus* and *Pseudomonas*. Cyanobacteria also contribute significantly (0.02 to 2.63×10^4 g^{-1} soil) to the biota of hot arid regions in terms of primary productivity and nitrogen fixation (Bhatnagar et al. 2003). The dominant cyanobacterial forms of Thar Desert are *Chroococcus minutus*, *Oscillatoria pseudogeminata* and *Phormidium tenue*, while *Nostoc* sp. dominates amongst heterocystous forms. Fungal populations range from 0.5 to 14.7×10^3 cfu with the dominant genera as *Aspergillus*,

Curvularia, *Fusarium*, *Mucor*, *Penicillium*, *Paecilomyces*, *Phoma* and *Stemphylium*. Xeric mushrooms such as *Coprinus*, *Fomes*, *Terfezia* and *Teramania* and arbuscular mycorrhizal fungi such as *Glomus*, *Gigaspora* and *Sclerocystis* have also been reported from desert (Pande and Tarafdar 2004). Lower organic content of sandy soils of Jodhpur reduced the fungi and actinomycetes (10^4-10^5 and 10^3-10^4 cfu g^{-1} dry soil, respectively) than those (about 10^8-10^6 cfu g^{-1} soil) of the aerobic and non-sandy soils because moisture content has negative and organic content has positive relation with those organisms. *Bacillus thuringiensis* being a saprophytic and aerobic soil bacteria (Joung and Cote 2000), its population would be lower in the nutritionally poor sandy soils (Kaur and Singh, 2000, Danger et al. 2010).

Rhizosphere is an important site of microbial activity in desert soils, since it provides ample carbon substrate in an otherwise organic matter poor arid soil. Generally R:S (rhizosphere : soil) ratio is high in arid soils for nearly all metabolic types of bacteria (viz. heterotrophs, diazotrophs, and nitrifiers) and fungi in most plants studied (Khathuria 1998, Singh and Tarafdar 2002). The soil zone penetrated by fine roots and held together by mucilage is called the rhizosheath. *Bacillus polymyxa* and the fungus *Olpidium* were found associated with rhizosheaths. *Ancalomicrobium* and *Hyphomicrobium*- like organisms were also present (Wullstein and Pratt 1981) Rhizosheaths are important because of the associated diazotrophs and enhanced water retention and nutrient uptake (Watt et al. 1994). In the rhizosphere of desert plants mycorrhiza play a very significant role in plant nutrition and stabilization (Tarafdar and Rao, 1997, Tarafdar and Rao 2002, Tarafdar and Gharu 2006). Most species in families of *Asteraceae*, *Fabaceae*, *Poaceae*, *Rosaceae* and *Solanaceae* usually form endomycorrhizal associations in arid habitats (Skujins 1984). Trappe (1981) listed 264 plant species from arid and semi-arid environments that had mycorrhizal-based root colonization and approximately 25% species exhibited specific associations with endomycorrhizal fungi. *Glomus deserticola* is indigenous to many desert soils (Bhatnagar et al. 2003). Kiran Bala et al. (1989) reported >50% infection by vesicular-arbuscular mycorrhizal (VAM) fungi in 17 tree species of the Indian desert, with genera *Glomus* and *Gigaspora* being dominant. *Opunitia* and *Euphorbia* showed considerable root infection. In desert, incidence of arbuscular mycorrhiza (AM) infection varies with the availability of water (Tarafdar 1995) and with composition of plant community (Kiran Bala et al. 1989). Mycorrhiza also helps in desert reclamation and soil stabilization. They link soil particles to each other and to the roots in part by producing glomalin, important glue that holds aggregates together. Beneficial effects of *Acaulospora mellea*, *Gigaspora margarita*, *G. gigantean*, *Glomus deserticola*, *G. fasciculatum*, *Sclerocystis rubiformis*, *Scutellospora calospora* and *S. nigra* on *Moringa concanensis* and use of such AM fungi in conservation of this endangered multipurpose tree species in Indian desert have been documented (Panwar and Vyas 2002).

Crusts on soils are formed by microorganisms and microphytes. These are variously called as cryptogannic / cryptobiotic / biological / cynobacterial / microphytic crusts (West 1990). A consortium of green algae, cyanobacteria, lichen, fungi, bacteria, diatoms, mosses and liverworts form cryptobiotic crusts. Mosses and liverworts seem to favour the comparatively more mesic sites. Cyanobacteria favour the harshest sites, and lichens dominate in intermediate sites. Cyanobacterial crusts generally dominate poor sandy soils. Lichens increase proportionately with carbonates, gypsum and silt content of the substrate (Garcia -Pichel et al. 2001, Budel 2002). *Microcoleus chthonoplastes* is dominant in saline sand crusts (Garcia -Pichel et al. 2001). The most common genera are *Microcoleus* (*M. chthonoplastes*, *M. paludosus*, *M. sociatus* and *M. vaginatus*) and *Nostoc* sp. Other forms are *Calothrix*, *Lyngbya*, *Oscillatoria*, *Phormidium*, *Scytonema* and *Tolypothrix*. Common green algae of soil crusts are *Chlorella*, *Chlorococcum*, *Coccomyxa* and *Klebsormidium*. Dominant fungal genera in crusts are *Alternaria*, *Fusarium* and *Phialomyces* whereas in non-crusted soils, *Alternaria* and *Penicillium* dominate followed by *Fusarium*. Approximately 90% of the aerobic bacterial population of cryptobiotic crusts consists of coryneform bacteria (Skujins 1984). Diversity status of cryptobiotic crusts at Thar Desert in India showed 43 morphotypes of diazotrophs in BG11-N enrichment and 71 of algae and cyanobacteria in the same medium supplemented with nitrogen (Bhatnagar et al. 2003). Most frequent form was *Phormidium tenue*. In case of diazotrophs the most frequent forms were *Nostoc punctiforme*, *Nostoc commune* and *Nostoc polludosum*. Presence of Cyanobacteria was influenced by their plant partners. *Alternaria* sp. was found to be the dominant fungus in these crusts. A photograph of fungal diversity from hot desert soil on rose Bengal agar plate is shown in Fig. 22.2A.

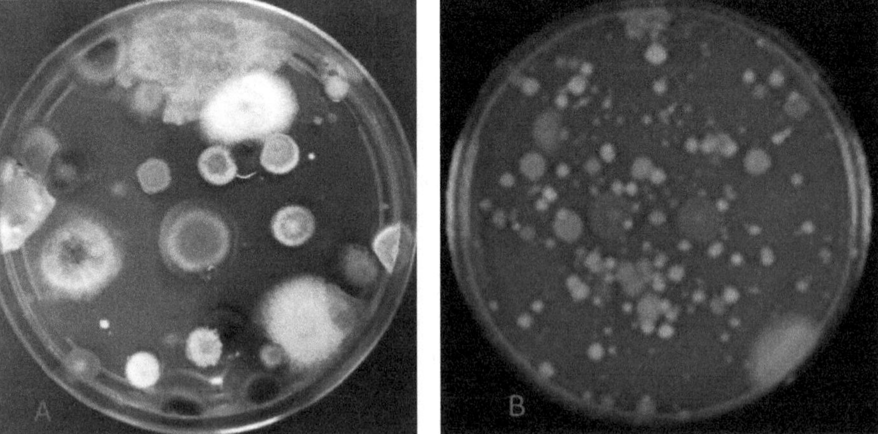

Fig. 22.2: Diversity of fungi (A) and bacteria (B) from hot and cold desert soils on agar plates

Microbiological properties of these soils also varied depending on the type of land use patterns. Grasslands, in general, supported higher number of microorganisms than tree plantations, cultivated fields or barren land. Stabilization of shifting and sand dunes introduction of vegetation has markedly increased the soil microflora. In general, the low organic matter content and poor moisture availability of desert soils were the major factors limiting optimum microbial activity.

Microbial diversity in cold deserts

There is imposition of environmental extremes in the cold deserts which, for the Antarctic Dry Valleys, include low temperatures, wide temperature fluctuations, low nutrient status, low water availability, high incident radiation, and physical disturbance. Studies have shown that most Antarctic microorganisms belong to a restricted number of cosmopolitan texa and were largely aerobic, with only few reported anaerobic isolates. Large numbers of coryneform-related bacteria such as *Arthrobacter, Brevibacterium*, *Cellulomonas*, and *Corynebacterium* were reported together with *Pseudomonas* and *Flavobacterium*. Firmicutes bacteria belonging to *Bacillus, Micrococcus, Nocardia,* and *Streptomyces* were isolated. A number of less common genera such as *Beijerinckia*, which rarely occur outside tropical soils, *Xanthomonas*, a pathogen associated with higher plants, and *Planococcus*, a marine genus, have also been isolated from Antarctic soils (Friedmann 1993). The comparison of microbial diversity by culture independent method for mineral soil samples from three dry mineral soil sites: (i) underneath acrabeater seal carcass on Bratina Island (BIS1), (ii) the mid slopes of Miers Valley (MVG), and (iii) fine gravels from Penance Pass, a high-altitude site between the Miers and Shangri La Valleys (PENP was carried out). Overall, Cyanobacteria (13%), Actinobacteria (26%), and Acidobacteria (16%) represented the majority of the identified phylotypes. Cyanobacteria appeared to be restricted to only the high-altitude PENP sample site. Actinobacteria were present in all three sites, with the most highly populated clade in the nutrient-rich BIS sample. Eighteen percent of all phylotypes obtained were assigned as so-called uncultured and were prevalent in all three sites (Smith et al. 2006). Twenty strains isolated from the arid region within the Atacama Desert, Chile were identified based on analysis of 16S rRNA gene sequences, and belonged to *Rhodopseudomonas*, *Sphingomonas* sp., *Mesorhizobium* sp., *Asticcacaulis* sp., *Bradyrhizobium* sp., *Bacillus subtilis*, *Bacillus pumilus*, and *Burkholderia* sp. (Lester et al. 2007). The 16S rRNA phylogenetic analysis of alkaline protease-producing psychrotrophic bacteria from glacier and cold environments showed that isolates belonged to three classes i.e. Actinobacteria, Gammaproteobacteria and Alphaproteobacteria, and were affiliated with the genera *Acinetobacter*, *Arthrobacter*, *Mycoplana*, *Pseudomonas*, *Pseudoxanthomonas*, *Serratia* and *Stenotrophomonas* (Salwan et al. 2010).

The bacterial diversity analysis of arid soils of the Atacama Desert revealed the high abundances of novel Actinobacteria and Chloroflexi and low levels of Acidobacteria and Proteobacteria, phyla that are dominant in many biomes (Neilson et al. 2012). Microbial diversity analysis carried out in three soil types collected from drylands of Moab, UT, USA showed the presence of six bacterial phyla (Acidobacteria, Actinobacteria, Bacteroidetes, Chloroflexi, Cyanobacteria, and Proteobacteria), as well as Archaea, comprising the majority of the community in the biocrusts and below-crust soils (Steven et al. 2013). The population of heterotrophic bacteria was enumerated in different samples collected from cold deserts of north western Himalayas and variations were observed among the culturable bacterial population of each sample on eleven different growth media. In the phylogenetic analysis based on 16S rRNA gene sequencing the bacteria belonging to phylum Firmicutes (*Bacillus psychrosaccharolyticus, Bacillus amyloliquefaciens, Bacillus muralis, Bacillus simplex, Bacillus thuringiensis, Bacillus altitudinis, Bacillus megaterium, Bacillus subtilis, Virgibacillus halodenitrificans, Virgibacillus* sp., *Lysinibacillus fusiformis* and *Lysinibacillus sphaericus*), phylum Actinobacteria (*Aeromicrobium, Arthrobacter, Citricoccus, Janibacter, Kocuria, Microbacterium, Plantibacter, Rhodococcus and Sanguibacter*), phylum Proteobacteria (*Brevundimonas terrae, Bosea* sp. *and Methylobacterium* sp., *Burkholderia* sp., *Burkholderia cepacia, Variovorax ginsengisoli, Janthinobacterium lividum, Janthinobacterium* sp., *Aeromonas, Pantoea, Providencia, Pseudomonas, Psychrobacter* and *Yersinia*) and phylum Bacteroidetes (*Flavobacterium antarcticum, Flavobacterium* sp. and *Sphingobacterium* sp.) were reported recently by Yadav et al. (2015). So there is a great diversity in the microbial population of the samples collected from various cold deserts. A photograph of bacterial diversity from cold environment soil on nutrient agar plate is shown in Fig. 22.2 (B).

Global field study conducted across 80 dryland sites from all continents showed that bacterial communities were dominated by Actinobacteria, Proteobacteria, Acidobacteria and Planctomycetes, while the most abundant soil fungi were those from the phylum Ascomycota, Basidiomycota, Chytridiomycota and Zygomycetous (Maestre et al. 2015).

In vitro studies for plant growth promoting attributes

Studies on stress tolerance of 19 phosphate-solubilizing fluorescent *Pseudomonas* from the cold deserts of the trans- Himalayas reported that majority of the strains showed growth in temperature range of 15-35°C, up to pH 12 and 2.5% $CaCl_2$, growth was restricted at 40°C to ten strains, pH 13 to six strains, 45°C to three strains, and 10% NaCl to one strain (Vyas et al. 2009).

Acinetobacter rhizosphaerae strain BIHB 723 from the cold deserts of the Himalayas exhibited the plant growth-promoting attributes of inorganic and organic phosphate solubilization, auxin production, 1-aminocyclopropane-1-carboxylate deaminase activity, ammonia generation, and siderophore production (Gulati et al. 2009). Among 164 bacterial strains belonging to various genera, 93% were able to grow at 42°C, 13% at 50°C and 71% at 4°C, while twenty strains were able to grow in temperature between 4 and 50°C. In case of salinity, 99% and 92% of the bacterial strains grew at lower (5%) or higher (20%) salinity and 74% were unable to grow in the absence of NaCl in the medium. Bacteria able to grow in presence of 5, 10 and 20% polyethylene Glycol (PEG) and survivals were 90, 87 and 81% respectively. 96% of the isolates showed the ability to produce indole-3-acetic acid (IAA), where as only (2%) showed ACC-deaminase activity, The phosphate solubilisation activity was present in 65% of the whole collection, Ammonia production was also a common PGP trait shown by 93% of the isolates (Mapelli et al. 2013). Of the 82 representative strains isolated from cold deserts of north western Himalayas solubilisation of phosphorus (40), production of ammonia (62), gibberellic acid (20) and indole-3-acetic acid (54) strains, ACC deaminase activity (21) production of siderophores (39) and HCN production (8) and antagonistic activity against *Rhizoctonia solani* and *Macrophomina phaseolina* (10) was reported (Yadav et al. 2015).

Diversity of bacteria associated with plants in desert

The diversity of diazotrophs associated with plant of north Sinai deserts study showed that among 51 nitrogen-fixing isolates majority belonged to genus *Bacillus* and other isolates having affiliation to genera *Enterobacter*, *Klebsiella*, *Pseudomonas*, *Agrobacterium* and *Azospirillum* (Othman et al. 2003). A total of 60 new bacterial isolates belonging to genera *Rhizobium, Sinorhizobium, Mesorhizobium* and *Bradyrhizobium* were obtained from root nodules of 19 legume species growing in the infra-arid climatic zone. *Rhizobium galegae*, a species generally an endosymbiont of *Galega officinalis* and *G. orientalis* was found in nodules of *Astragalus cruciatus, Lotus creticus* and *Anthyllis henoniana* (Zakhia et al. 2004). Studies on diazotrophic culturable bacterial community associated with roots of *Lasiurus sindicus*, a perennial grass of Thar desert, India showed that the bacteria belonging to *Staphylococcus warneri*, *Bacillus* sp., *Micrococcus luteus, Microbacterium* sp., *Azospirillum* sp., *Rhizobium* sp., *Agrobacterium tumefaciens, Inquilinus limosus, Ralstonia* sp., *Variovorax paradoxus, Bordetella petrii, Pseudomonas pseudoalcaligenes, Stenotrophomonas* sp. and *Chryseobacterium defluvii* (Chowdhury et al. 2007). Eighteen bacterial isolates showing acetylene reduction activity (ARA) ranging (1658 n moles d^{-1}) to (8303 n moles d^{-1}) were isolated from the rhizospheric soil of (*Calligonum polygonoides* and *Lasiurus sindicus*) two hot arid zone plants (Gothwal et al. 2008). Studies on diversity of rhizobial bacteria associated with nodules of the gypsophyte *Ononis tridentata* L. growing in semiarid mediterranean areas in Spanish soils showed that bacteria belonged to genera *Rhizobium, Mesorhizobium, Phylobacterium* and *Bosea* (Rincon et al. 2008).

The diversity analysis and structure of the bacterial communities in the rhizosphere and non rhizosphere of three cactus species from semi-arid highlands in central Mexico: *Mammillaria carnea, Opuntia pilifera,* and *Stenocereus stellatus,* showed that bacteria belonged to five major bacterial groups: alpha, beta, and gamma-

Proteobacteria, Actinobacteria and Firmicutes with representative genera; *Acinetobacter, Arthrobacter, Bacillus, Burkholderia, Enterobacter, Herbaspirillum, Leifsonia, Massilia, Microbacterium, Micrococcus, Ochrobactrum, Paenibacillus, Pseudomonas, Rhizobium* and *Sinomonas* (Aguirre-Garrido et al. 2012). Based on FAME profile and/or 16S rRNA gene sequencing the cacti-associated bacteria isolated from the Brazilian semi-arid region belonged to different species of genera *Bacillus, Brevibacillus, Virgibacillus, Paenibacillus, Enterobacter, Pantoea Arthrobacter, Gordonia, Cellulosimicrobium* and *Nocardia* with *Bacillus* as dominant genus (Kavamura et al. 2013). The diversity of bacteria present on the roots and shoots of 43 plant species growing in north Sinai deserts, Egypt showed that majority of nitrogen-fixing bacilli isolates belonged to genus *Bacillus*, while bacteria belonging to other genera like *Enterobacter, Serratia, Klebsiella, Pantoae Agrobacterium, Pseudomonas, Stenotrophomonas, Ochrobactrum, Sphingomonas* and *Chrysemonas* were also reported. Many tested plants showed the presence of *Gluconacetobacter diazotrophicus* inside root and shoot tissues (Hanna et al. 2013). One hundred twenty two (122) haloalkaliphilic bacteria growing optimally in media with 10% salt and at pH 10 were isolated from 5 distinct arid saline systems of southern Tunisia. Phylogenetic analysis based on 16S rRNA gene showed that the isolates belonged to thirteen different genera: *Halomonas, Salinicoccus, Nesterenkonia, Oceanobacillus, Virgibacillus, Halobacillus, Salimicrobium, Bacillus, Piscibacillus, Marinococcus, Brevibacillus, Leucobacter,* and *Arthrobacter* (EI Hidri et al. 2013). Bacterial diversity associated with *Salicornia* growing in Tunisian hypersaline soils was carried out by culture dependant and culture in-dependant approach. Based on culturing approach the bacteria belonging to genera *Chromohalobacter, Halobacillus, Halomonas, Kushneria, Marinococcus, Nesterenkonia, Oceanobacillus* and *Virgibacillus* were reported, while in culture independant approach using DGGE analysis of the bacterial microbiome inhabiting *Salicornia* rhizosphere and surrounding bulk soil showed the presence of prevalent taxonomic groups as Alpha-, Beta- and Gammaproteobacteria, Bacilli, and Actinobacteria (Mapelli et al. 2013).

Plant growth promotion

In contrast to desertification, which has been recognized as a major threat to biodiversity, to convert deserts into arable, green landscapes is a global vision as well as an answer to world hunger and climate change. Studies conducted on effect of plant growth-promoting bacteria (*Azospirillum brasilense* and *Bacillus pumilus*), arbuscular mycorrhizal (AM) fungi (mainly *Glomus* sp.) on three plants *Prosopis articulate, Parkinsonia microphylla* and *Parkinsonia florida* for desert reforestation and urban gardening in the Sonoran Desert of Northwestern Mexico and the Southwestern USA showed that two plant species had different positive responses to several of the parameters tested while third plant did not respond. Moreover inoculation with these growth-promoting microorganisms induced significant effects on the leaf gas exchange of these trees when they were cultivated without water restrictions (Bashan et al. 2009). Inoculation with *Acinetobacter rhizosphaerae* (isolated from cold deserts) resulted in a significant increase in the growth of pea, chickpea, maize, and barley under controlled conditions. Field testing with the pea also showed a significant increment in plant growth and yield (Gulati et al. 2009). The effect of local N -fixing bacteria; *Azospirillum lipoferum* and *Azotobacter chroococcum* as biofertilizers on the growth and yield of three species of *Mentha* plants were studied in field experiments carried out under desert conditions in, Desert research center (DRC), Egypt. Application of mixture of *Azospirillum lipoferum* and *Azotobacter chroococcum* on *Mentha piperita* L. give the highest growth parameters, while *Mentha arvensis* L give the best essential oil percentage (Abd El-Hadi Nadia et al. 2009). Based on screening results of bacterial isolates for in vitro plant growth promotion seven strains (six belonging to genus *Bacillus* and one to *Pantoea*) were checked for growth promotion of *Zea mays* L. under water stress (30% of field capacity) and it was found that one strain of *Bacillus* increased all the three analyzed plant parameters *viz*. leaf area, stalk length and shoot dry biomass and also helps to protect the plant against the negative effects of desiccation (Kavamura et al. 2013).

Conclusion

There appears no embellishment in saying that the desert though they appear to devoid of life, possesses a great wealth of microorganisms. The microbial diversity in these environments is adapted to the prevailing conditions and plays an important role in these ecosystems. The rich microbial diversity in the deserts is under explored and unexploited to its potential and needs attention. The microorganisms present in these environments can help in the plant adaptation and restoration of degraded environments of the deserts.

References

Abd El-Hadi Nadia, I.M., Abo El-Ala, H.K, and Abd El-Azim, W.M. 2009. Response of some *Mentha* species to plant growth promoting bacteria (PGPB) isolated from soil rhizosphere. Australian Journal of Basic and Applied Sciences 3: 4437-4448.

Aguirre-Garrido, J.F., Montiel-Lugo, D., Hernández-Rodríguez, C.C., Torres-Cortes, G., Millán, V., Toro, N., Martínez-Abarca, F. and Ramírez-Saad, H.C. 2012. Bacterial community structure in the rhizosphere of three cactus species from semi-arid highlands in central Mexico. Antonie Van Leeuwenhoek 101: 891–904.

Bashan, Y., Salazar, B. and Puente, M.E. 2009. Responses of native legume desert trees used for reforestation in the Sonoran Desert to plant Growth-promoting microorganisms in screen house. Biology and Fertility of Soils 45: 655–662.

Bhatnagar, A., Bhatnagar, M., Makandar, M.B. and Garg, M.K. 2003. Satellite centre for microalgal biodiversity in arid zones of Rajasthan. Project completion report, funded by Department of Biotechnology, New Delhi.

Budel, B. 2002. Diversity and ecology of biological crusts. In Progress in Botany (Eds. K. Esser, U. Luttge, W. Beyschlag, and F. Hellwig), pp. 386-404. Springer-Verlag, Berlin.

Chowdhury, S.P., Schmid, M., Hartmann, A. and Tripathi, A.K. 2007. Identification of diazotrophs in the culturable bacterial community associated with roots of *Lasiurus sindicus*, a perennial grass of Thar Desert, India. Microbial Ecology 54: 82-90.

Danger, T.K., Babu, Y.K. and Das, J. 2010. Population dynamics of soil microbes and diversity of *Bacillus thuringiensis* in agricultural and botanic garden soils of India. African Journal of Biotechnology 9: 496-501.

Daniel, R. 2005. The metagenomics of soil. Nature Reviews Microbiology 3: 470-478.

El Hidri, D., Guesmi, A., Najjari, A., Cherif, H., Ettoumi, B., Hamdi, C., Boudabous, A. and Cherif, A. 2013. Cultivation-dependant assessment, diversity, and ecology of haloalkaliphilic bacteria in arid saline systems of southern Tunisia. BioMed Research International, ID 648141, 15 pp.

Friedmann, E.I. 1993. Antarctic Microbiology. Wiley-Liss. New York, pp 634.

Garcia-Pichel, F., Lopez-Cortes, A. and Nubel, U. 2001. Phylogenetic and morphological diversity of cyanobacteria in soil desert crusts from the Colorado plateau. Applied Environmental Microbiology 67: 1902-1910.

Gothwal, R.K., Nigam, V.K., Mohan, M.K., Sasmal, D. and Ghosh, P. 2008. Screening of nitrogen fixers from rhizospheric Bacterial isolates associated with important Desert plants. Applied Ecology and Environmental Research 6: 101-109.

Gulati, A., Vyas, P., Rahi, P. and Kasana, R.C. 2009. Plant growth-promoting and rhizosphere-competent *Acinetobacter rhizosphaerae* strain BIHB 723 from the cold deserts of the Himalayas. Current Microbiology 58: 371–377.

Gupta, G.N. 1993. Research priorities in arid zone forestry. In: Aforestation of arid lands. Dwevedi, AP and Gupta GN (eds). Scientific Publsisher, Jodhpur, India. pp 87-101.

Hanna, A.L., Youssef, H.H., Amer, W.M., Monib, M., Fayez, M. and Hegazi, N.A. 2013. Diversity of bacteria nesting the plant cover of north Sinai deserts, Egypt. Journal of Advanced Research 4: 13–26.

Huang, J., Yu, H., Guan, X., Wang, G. and Guo, R. 2015. Accelerated dryland expansion under climate change. Nature Climate Change, 10.1038/nclimate2837.

Joung, K.B. and Cote, J.C. 2000. A review of the environmental impacts of the microbial insecticide *Bacillus thuringiensis*, Technical Bulletin no. 29. Horticulture Research and Development Center, Canada, p. 16.

Kaur, S. and Singh, A. 2000. Distribution of *Bacillus thuringiensis* strains in different soil types from North India. Indian Journal of Ecology 27: 52-60.

Kavamura, V.N., Santos, S.N., Silva, J.L., Parma, M.M., Avila, L.A., Visconti, A., Zucchi, T.D., Taketani. R.G., Andreote, F.D. and Melo, I.S. 2013. Screening of Brazilian cacti rhizobacteria for plant growth promotion under drought. Microbiology Research 168: 183-191.

Khathuria, N. 1998. Rhizosphere microbiology of desert, M.Sc. dissertation, pp. 52. Department of Microbiology, Maharshi Dayanand Saraswati University, Ajmer, Rajasthan.

Kiran Bala, Rao, A.V. and Tarafdar, J.C. 1989. Occurrence of VAM associations in different plant species of the Indian desert. Arid Soil Research and Rehabilitation 3: 391-396.

Lester, E.D., Satom, M. and Ponce, A. 2007. Microflora of extreme arid Atacama Desert soils. Soil Biology and Biochemistry 39: 704–708.

Maestre, F.T., Delgado-Baquerizo, M., Jeffries, T.C., Eldridge, D.J., Ochoa, V., Gozalo, B., Quero, J.L., García-Gómez, M., Gallardo, A., Ulrich, W., Bowker, M.A., Arredondo, T., Barraza-Zepeda, C., Bran, D., Florentino, A., Gaitán, J., Gutiérrez, J.R., Huber-Sannwald, E., Jankju, M., Mau, R.L., Miriti, M., Naseri, K., Ospina, A., Stavi, I., Wang, D., Woods, N.N., Yuan, X., Zaady, E. and Singh, B.K. 2015. Increasing aridity reduces soil microbial diversity and abundance in global drylands. Proc National Academy of Science USA 112: 15684-15689.

Mapelli, F., Marasco, R., Rolli, E., Barbato, M., Cherif, H., Guesmi, A., Ouzari, I., Daffonchio, D. and Borin, S. 2013. Potential for plant growth promotion of rhizobacteria associated with S*alicornia* growing in Tunisian hypersaline soils. BioMed Research International, doi.org/10.1155/2013/248078.

Neilson, J.W., Quade, J., Ortiz, M., Nelson, W.M., Legatzki, A., Tian, F., LaComb, M., Betancourt, J.L., Wing, R.A., Soderlund, C.A. and Maier, R.M. 2012. Life at the hyper arid margin: novel bacterial diversity in arid soils of the Atacama Desert, Chile. Extremophiles 16: 553-566.

Othman, A.A., Amer, W.A., Fayez, M., Monib, M. and Hegazi, N.A. 2003. Biodiversity of diazotrophs associated to the plant cover of north Sinai deserts. Archives of Agronomy and Soil Science 49: 683 – 705.

Pande, M. and Tarafdar, J.C. 2004. Arbuscular mycorrhizal fungal diversity in neem-based agroforestry systems in Rajasthan. Applied Soil Ecology 26: 233-241.

Panwar, J. and Vyas, A. 2002. A biological approach towards conservation of endangered plants in Thar desert, India. Current Science 82: 576-578.

Rao, A.V. and Venkateswarlu, B. 1983. Microbial ecology of the soils of Indian desert. Agriculture Ecosystem Environment 10: 361–369.

Rincon, A., Arenal, F., Gonzalez, I., Manrique, E,, Lucas, M.M. and Pueyo, J.J. 2008. Diversity of rhizobial bacteria isolated from nodules of the gypsophyte *Ononis tridentata* L. growing in Spanish soils. Microbial Ecology 56: 223-233.

Rossello-Mora, R. and Amann, R. 2001. The species concept for prokaryotes. FEMS Microbiology Reviews 25: 39-67.

Salwan, R., Gulati, A. and Kasana, R.C. 2010. Phylogenetic diversity of alkaline protease-producing psychrotrophic bacteria from glacier and cold environments of Lahaul and Spiti, India. Journal of Basic Microbiology 50: 150–159.

Singh, J.P. and Tarafdar, J.C. 2002. Rhizospheric microflora as influenced by sulphur application, herbicide and rhizobium inoculation in summer mung bean (*Vigna radiata* L.). Journal of the Indian Society of Soil Science 50: 127-129.

Skujins, J. 1984. Microbial ecology of desert soils. Advances in Microbial Ecology 7: 49-91.

Smith, J.J., Tow, L.A., Stafford, W., Cary, C. and Cowan, D.A. 2006. Bacterial diversity in three different Antarctic Cold Desert mineral soils. Microbial Ecology 51: 413-421.

Steven, B., Gallegos-Graves, L.V., Belnap, J. and Kuske, C.R. 2013. Dryland soil microbial communities display spatial biogeographic patterns associated with soil depth and soil parent material. FEMS Microbiology Ecology 86: 101-113.

Tarafdar, J.C. 1995. Role of VA mycorrhizal fungus on growth and water relations in wheat in presence of organic and inorganic phosphates. Journal of the Indian Society of Soil Science 43: 200-204.

Tarafdar, J.C. and Gharu, A. 2006. Mobilization of organic and poorly soluble phosphates by *Chaetomium globosum*. Applied Soil Ecology 32: 273-283.

Tarafdar, J.C. and Rao, A.V. 1997. Mycorrhizal colonization and nutrient concentration of naturally grown plants on gypsum mine-spoils in India. Agriculture Ecosystems and Environments 61: 13-18.

Tarafdar, J.C. and Rao, A.V. 2002. Possible role of arbuscular mycorrhizal fungi in development of soil structure. In, Arbuscular Mycorrhizae: Interactions in Plant, Rhizosphere and Soils (Eds. A.K. Sharma and B.N. Johri), pp. 189-206. Oxford & IBH Publishing Co. Pvt. Ltd.

Trappe, J.M. 1981. Mycorrhizae and productivity of arid and semi arid range lands. In Advances in Food Production System for Arid and Semi Arid Lands (Eds Manassah, J.J. and Briskey, E.J.), Academic Press, New York, pp. 581-599.

Venkateswarulu, B. and Rao, A.V. 1981. Distribution of microorganisms in stabilized and unstabilised sand dunes of Indian desert. Journal of Arid Environment 4: 203-208.

Vyas, P., Rahi, P. and Gulati, A. 2009. Stress tolerance and genetic variability of phosphate-solubilizing fluorescent *Pseudomonas* from the cold deserts of the trans-Himalayas. Microbial Ecology 58: 425–434.

Watt, M., McCully, M.E. and Canny, M.J. 1994. Formation and stabilization of rhizosheaths of *Zea mays* L. (Effect of soil water content). Plant Physiology 106: 179-186.

West, N.E. 1990. Structure and function of microphytic soil crusts in wild wind ecosystems of arid to semi-arid regions. Advanced Ecological Research 20: 179-223.

Wullstein, L.H. and Pratt, S. 1981. Scanning electron microscopy of rhizosheaths of ryzopsis hymenoides. American Journal of Botany 68: 408–419.

Yadav, A.N., Sachan, S.G., Verma, P. and Saxena, A.K. 2015. Prospecting cold deserts of north western Himalayas for microbial diversity and plant growth promoting attributes. Journal of Bioscience and Bioengineering 119: 683-693.

Zakhia, F., Jeder, H., Domergue, O., Willems, A., Cleyet-Marel, J.C., Gillis, M., Dreyfus, B. and de Lajudie, P. 2004. Characterization of wild legume nodulating bacteria (LNB) in the infra-arid zone of Tunisia. Systematic Applied Microbiology 27: 380-395.

23

Geostatistical Application for Assessing the Soil Water Availability in Farm Scale- A Case Study

Priyabrata Santra[1], U.K. Chopra[2] and Debashis Chakraborty[2]

[1]*ICAR-Central Arid Zone Research Institute, Jodhpur, Rajasthan, India*
[2]*ICAR-Indian Agricultural Research Institute, New Delhi, India*

Introduction

High degree of spatial variability in soil properties is observed at different scales. This variability is generally originated from the combined effect of physical, chemical or biological processes operating with different intensities at different scales (Goovaerts 1998). Knowledge of this spatial variationof soil properties is important in several land management applications e.g. variable rate application of nutrient and water in agricultural production system, land suitability classifications, modeling water balance components at landscape scales, precision farming *etc*. The concept of 'management zone' was also evolved in response to this large variability with the main purpose in efficient utilization of agricultural inputs with respect to spatial variation of soils and its properties (Franzluebbers et al. 1996, Atherton et al. 1999, Malhi et al. 2001). Therefore, an appropriate understanding of spatial variation of soil properties is essential. The most important way to gather knowledge in this aspect is to prepare soil maps through spatial interpolation of point-based measurements of soil properties. Here, a case study is discussed where soil water retention at two critical levels e.g. field capacity (FC) and permanent wilting point (PWP) was spatially characterized through geostatistical approaches. Because these two soil moisture contents guide farmers to apply right amount of irrigation water at right time. The study was carried out at the experimental farm of ICAR-Indian Agricultural Research Institute, New Delhi, India (28°37′-28°39′ N, 77°8′ 30″-77°10′30″ E, 217-241 m above mean sea level). A detailed discussion of the study is available in Santra et al. (2008)

Spatial data of soil properties in IARI farm

Soils of ICAR-IARI experimental farm falls under two broad soil subgroups – Typic Haplustepts and Typic Ustifluvent (NBSS&LUP 1976). Total six soil series are observed in the farm and these are Mehrauli, Palam, Holambi, Daryapur, Nagar and Jagat. Sandy loam is the dominant soil textural class of the farm. Soil samples were collected from 50 sites of ICAR-IARI farm (243 ha) covering six soil series under two soil subgroups (Fig. 23.1). From each site, disturbed and undisturbed soil samples were collected from surface soil layers: 0-15 cm and 15-30 cm. Core samplers with cylindrical cores of 7.1 cm in diameter and 7.1 cm in length were used for undisturbed soil sample collection, while disturbed soil samples were collected in polythene bags and air dried in the laboratory. Disturbed samples were air-dried, ground and passed through 2-mm sieve, and used to determine soil texture (Bouyoucos 1962), organic carbon content (Walkley and Black 1934) and water content at FC (33 kPa) and PWP (1500 kPa) (Klute and Dirkson 1986). Undisturbed soil cores were oven-dried at 105°C and used to determine bulk density (ρ_b) (Blake and Hartge 1986).

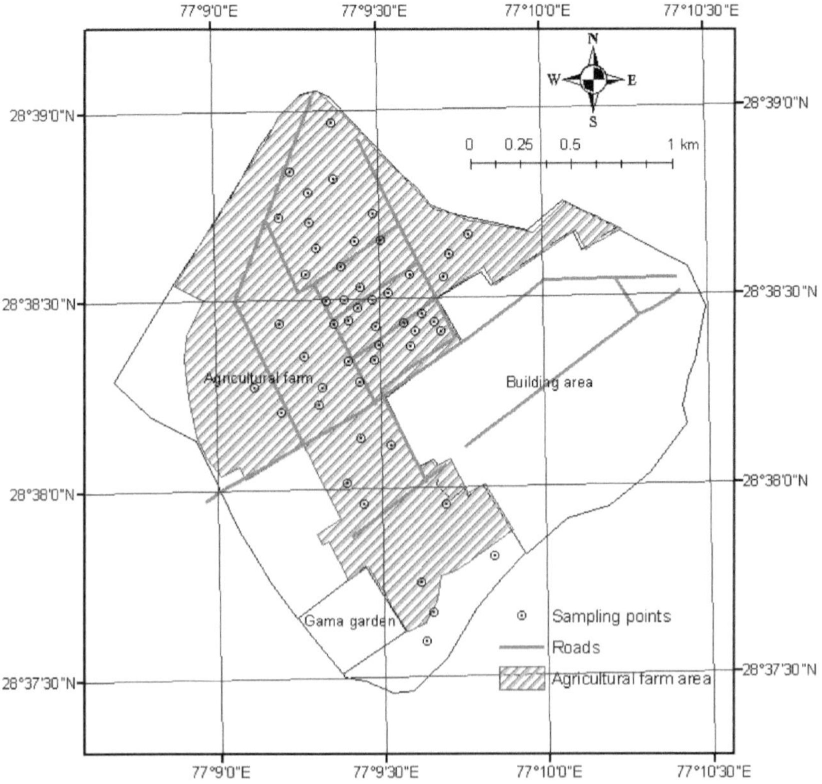

Fig. 23.1: Location of sampling points in experimental farm of ICAR-Indian Agricultural Research Institute (IARI), New Delhi.

Mean (μ), and standard deviation (σ) of determined soil properties are presented in Table 23.1. Before calculating the descriptive statistics, each soil property was checked for normality using the Kolmogorov-Smirnov (K-S) test statistics at 5% level of significance. Logarithmic transformations were made for silt content to fit it in normal distribution. Except silt content, other soil properties were fitted in normal distribution. Average bulk density for surface (0-15 cm) and subsurface layer (15-30 cm) was 1.54 and 1.62 Mg m^{-3}. Intensive use of heavy tillage implements resulted in compact layer at 15-30 cm soil depth. Organic carbon content in surface is twice that of subsurface, which may be due to continuous addition of crop residues on surface of cropped field. Among primary soil particles, silt content is slightly lower in subsurface soil layer than surface layer, whereas clay content is higher in subsurface layer. Eluviation-illuviation process due to downward movement of water through soil might have resulted in deposition of fine size particles at greater depth. In general, the variation of bulk density and organic carbon content is higher in surface than subsurface soil layers whereas the trend is reverse for silt content and clay content. Varieties of crops grown at different time and different blocks of the farm and the modified soil physical environment for each crop resulted in high variation of bulk density at surface layer than subsurface layer. Moreover, differential growth behaviour of each crop results into different quantities of shoot and root biomass, some of which were incorporated during tillage. For example, the amount of crop residue added over soil surface is higher for cereal crops like rice, wheat and maize in comparison to pulse and oil-seed crops. Hence, the variation of organic carbon content is higher in surface layer than subsurface layer. Particle size distribution does not vary significantly in surface soil layer. Mixing of soil during tillage operation resulted in less variation of particle size distribution at surface layer than subsurface layer.

Table 23.1: Descriptive statistics of soil properties

Soil properties	Depth	Mean	Standard deviation
Bulk density (Mg m^{-3})	0-15 cm	1.54	0.09
	15-30 cm	1.62	0.08
Organic carbon content (%)	0-15 cm	0.62	0.22
	15-30 cm	0.29	0.17
ln (silt content) (%)	0-15 cm	2.51	0.27
	15-30 cm	2.50	0.31
Clay content (%)	0-15 cm	31.16	5.27
	15-30 cm	33.42	6.59

Spatial variation of soil properties

Experimental semivariograms for each soil property was computed using GEOEAS (Geostatistical Environmental Assessment Software) and plotted with lag distance *h* (Isaaks and Srivastava 1989). During pair calculation for computing semivariogram, maximum lag distance was taken as the half of the minimum extent of sampling area to minimize the border effect. Lag increment was fixed as 100 m. In this study, omnidirectional semivariogram was computed for each soil property because no significant directional trend was observed. The computed semivariogram values [$\hat{\gamma}(h)$] for corresponding lag (*h*) were fitted with available theoretical semivariogram models using weighted least square technique using solver function of MS Excel spreadsheet (Jian *et al.* 1996). Weight for each lag was assigned according to the number of pairs for that particular lag. Best fit model with lowest value of residual sum of square was selected for each soil property and each soil depth. Three commonly used semivariogram models were fitted for each soil property. These are spherical, exponential and Gaussian models. In case of fitting semivariogram for organic carbon content, hole-effect model was also included. Expressions for different semivariogram models used in this study are given below.

Spherical model: $$\gamma(h) = C_0 + C\left[1.5\frac{h}{a} - 0.5\left(\frac{h}{a}\right)^3\right] \quad \text{if} \le \text{h} \le \text{a} \tag{1}$$

$$= \text{C}_0\text{+C otherwise}$$

Exponential model: $$\gamma(h) = C_0 + C_1\left[1 - \exp\left\{-\frac{h}{a}\right\}\right] \quad \text{for} \ge 0 \tag{2}$$

Gaussian model: $$\gamma(h) = C_0 + C\left[1 - \exp\left\{\frac{-h^2}{a^2}\right\}\right] \quad \text{for h} \ge 0 \tag{3}$$

Hole-effect model: $$\gamma(h) = C_0 + C_1\left[1 - \frac{\sin(\pi h / a)}{\pi h / a}\right] \quad \text{for h} \ge 0 \tag{4}$$

In all these semivariogram models, nugget, sill and range were expressed by C_0, $(C+C_0)$ and *a* respectively. In case of exponential and Gaussian models, *a* represents the theoretical range. Practical range for these two semivariogram models was calculated as the lag distance for which semivariogram value was 95% of sill.

Best fitted semivariogram models for each soil property are given in Table 23.2 along with the parameters. Among different theoretical models tested, Gaussian model was found as the best fit in most cases. In case of organic carbon content, spatial variation for both soil depths was best described by hole-effect model. It indicates the periodic appearance of homogeneous patches of organic carbon content in two dimensional spaces. Spherical model was best fitted in only one case i.e. for fitting the experimental semivariogram values for *ln*(silt) at 0-15 cm depth. For bulk density and major textural separates, range varies from 900 to 1200 m. It indicates that bulk density, silt and clay content of two locations separated with lag distances below 1 km are spatially correlated with each other.

Beyond this lag distance, existing variation is defined as random variation. Organic carbon content is spatially correlated for a short lag distance i.e. around 500 m. Nugget (C_0) defines the micro-scale variability and measurement error for the respective soil property whereas partial sill (C) indicates the amount of variation which can be defined by spatial correlation structure. Out of total variation, nugget component is 50% for bulk density which shows that micro-scale variation of this property is relatively high. Moreover, support area for measuring the bulk density is too small (7.1 cm in diameter) to average the micro-scale variations in the field. Minimum sampling distance in this study was 57.8 m, which might not be to capture the variability at small lag. For organic carbon content, nugget component is around 75% of the total variance for 0-15 cm soil depth. For particle size distribution, nugget component is less than 50% for both silt, and clay content. Silt content at 0-15 cm soil depth was found the best in terms of spatial correlation structure. Clay content is also highly spatially correlated and spatial correlation structure is better than bulk density and organic carbon content. Magnitude of spatial variance is higher in surface than subsurface layer for bulk density and organic carbon content and reverse is the trend for silt and clay content. Same findings were also pointed out from the classical statistical variance.

Table 23.2: Semivariogram parameters of soil properties of IARI farm

Soil property	Depth (cm)	Semivariogram model	Range (m)	Nugget (C_0)	Partial sill (C)
Bulk density (Mg m^{-3})	0-15 cm	Gaussian	1053	0.005	0.005
	15-30 cm	Gaussian	1201	0.004	0.004
Organic carbon content (%)	0-15 cm	Hole-effect	450	0.035	0.012
	15-30 cm	Hole-effect	550	0.014	0.022
ln (% silt content)	0-15 cm	Spherical	902	0.004	0.090
	15-30 cm	Gaussian	1224	0.064	0.069
Clay content (%)	0-15 cm	Gaussian	994	5.421	32.163
	15-30 cm	Gaussian	1179	15.086	49.690

Ordinary kriging and surface map of soil properties

Surface map of basic soil properties were prepared using semivariogram parameter through ordinary kriging. Accuracy of the soil maps were evaluated through cross-validation approach (Davis 1987). Three evaluation indices were used in this study to check the accuracy: mean absolute error (MAE), mean squared error (MSE) and goodness-of-prediction (G) and are mentioned in Table 23.3.

Table 23.3: Evaluation performances of krigged map through cross-validation

Soil properties	Depth	MAE	MSE	G
Bulk density (Mg m^{-3})	0-15 cm	0.074	0.008	-7.16
	15-30 cm	0.053	0.005	18.90
Organic carbon content (%)	0-15 cm	0.177	0.043	6.44
	15-30 cm	0.132	0.032	-18.09
ln (silt content) (%)	0-15 cm	0.175	0.051	29.21
	15-30 cm	0.229	0.084	10.37
Clay content (%)	0-15 cm	3.761	27.609	-1.51
	15-30 cm	4.490	34.467	18.96

Spatial map of basic soil properties for 0-15 cm and 15-30 soil depth prepared through ordinary kriging are presented in Fig. 23.2 and 23.3, respectively. Southern part of the farm is higher in bulk density (>1.59 Mg m^{-3}) at surface but lower at subsurface than other parts of the farm [Fig. 23.2(a)]. For north-western part of the farm, bulk density is 1.47-1.52 Mg m^{-3} at surface and >1.67 Mg m^{-3} at subsurface. This shows the presence of compacted subsurface layer at north-western part of the farm possible due to continuous rice cultivation in these areas. Increase in bulk density in East to West direction was also observed. Except for southern part of the farm a general trend of increase in bulk density with depth was observed. Spatial map of organic carbon content (0-15 cm) shows that 80% of the farm has medium organic carbon content (0.5-0.75%) and some patches of organic carbon content were also observed from the map [Fig. 23.2(b)]. For subsurface layer [Fig. 23.3(b)], the same pattern is followed with organic carbon content in low category (<0.5%) for maximum part of the farm. In general, organic carbon content is higher in surface than subsurface layer for entire farm area. Spatial map of silt content (%) shows that it decreases in East to West direction for both surface [Fig. 23.2(c)] and subsurface layer [Fig. 23.3(c)]. Western part of the farm is slightly higher in silt content for surface layer than subsurface layer. For other parts of the farm, there is no significant difference in silt content between surface layer and subsurface layer. Fig. 23.2(d) shows that except some few patches, clay content is <34% for surface layer. For subsurface layer [Fig. 23.3(d)], it is found that clay content is higher than surface layer. For western, north-western and south-western part of the farm, clay content is around 4-6% higher in surface layer than subsurface layer.

Surface map for water content at FC and PWP

Spatial maps on water content at field capacity (θ_{FC}) and permanent wilting point (θ_{PWP}) were prepared through linking soil maps on basic properties and PTFs. The PTFs for θ_{FC} and θ_{PWP} were developed from the available soil data in benchmark soils of India (Murthy et al. 1982).

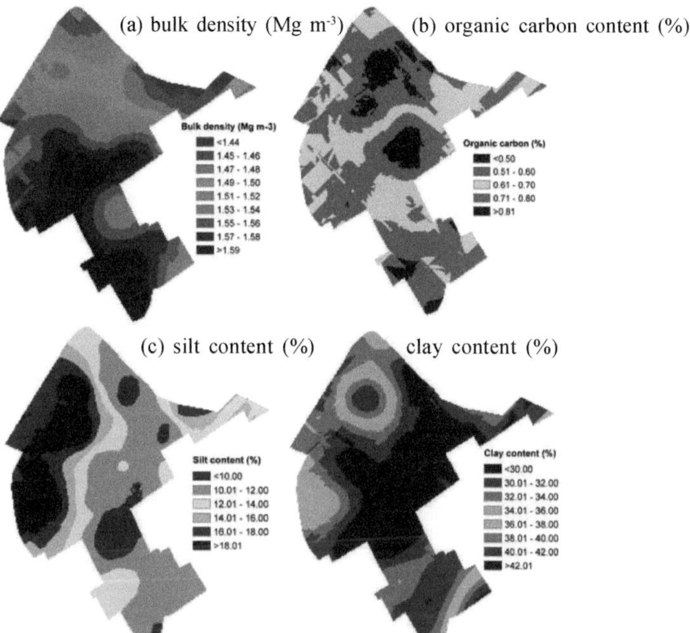

Fig. 23.2: Kriged maps of soil properties for 0-15 cm soil depth; (a) bulk density (Mg m^{-3}), (b) organic carbon content (%), (c) silt content (%), and (d) clay content (%). *Source*: Santra *et al.*, 2008

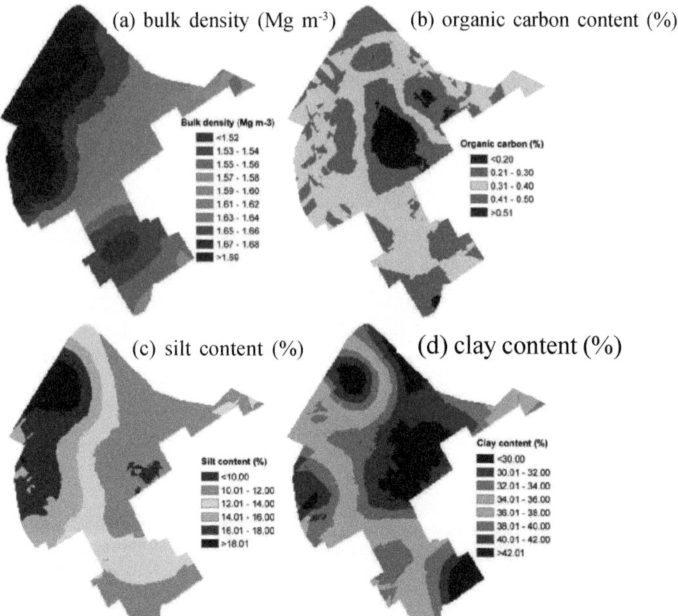

Fig. 23.3: Kriged maps of soil properties for 15-30 cm soil depth; (a) bulk density (Mg m^{-3}), (b) organic carbon content (%), (c) silt content (%), and (d) clay content (%). *Source*: Santra *et al.*, 2008

Performance of the developed PTFs was evaluated and compared with the established PTFs using the measured water retention data from the study area and it was found that the developed PTFs from Benchamark soils of India performed better. The set of PTFs developed here for estimating θ_{FC} and θ_{PWP} are given below:

$$\theta_{FC}(\%, w/w) = 21.931 - 0.20564 \times sand + 0.175 \times clay + 4.673 \times OC \quad (R^2 = 0.89) \tag{5}$$

$$\theta_{PWP}(\%, w/w) = 8.7255 - 0.092946 \times sand + 0.15944 \times clay \; (R^2 = 0.78) \tag{6}$$

In the above PTFs, sand refers to the per cent sand content of soil with particle size range of 0.05-2 mm, clay refers to the per cent clay content with particle size <0.002 mm and OC refers to the per cent organic carbon content in the soils. These PTFs were used to convert the maps of basic soil properties (c.g. sand, clay, OC etc) to soil water retention at FC and PWP. Uncertainty in spatial estimates of θ_{FC} and θ_{PWP}in developed maps was calculated by comparing their observed values at 50 locations of the study area with the estimated one.

The other way to generate the map of θ_{FC} and θ_{PWP} is by predicting the values of θ_{FC} and θ_{PWP}at measured point locations and then interpolating the point predicted values. Both these approaches of preparing soil maps by (i) interpolating predictor soil properties first followed by joining component maps through model and (ii) predicting first at point locations through modeling followed by interpolation has been discussed by Bechini et al. (2003). In the first approach, uncertainty associated with maps of predictor soil properties will be propagated through PTF models, however it may be quantified to know the risks while applying them in field. But the advantage of first approach is that the spatial maps of predictor soil properties may also be used for farm level planning or modeling environmental processes.

Spatial maps of θ_{FC} and θ_{PWP} generated through the first approach as discussed above are presented in Fig. 23.4. Water content at field capacity increases in East to West direction of the farm for both surface and subsurface layer. The value of θ_{FC} (%, w/w) varies from 15.88 to 25.48% for surface layer and 15.00 to 23.78% for subsurface layer. Maximum water content is found at North-Western part of the farm where clay content was highest among all parts of the farm. Similar type of spatial trend is also observed for the map of θ_{PWP}. The value of θ_{PWP}(%, w/w) varies from 6.72 to 11.92% for surface layer and 7.52 to 12.21% for subsurface layer.

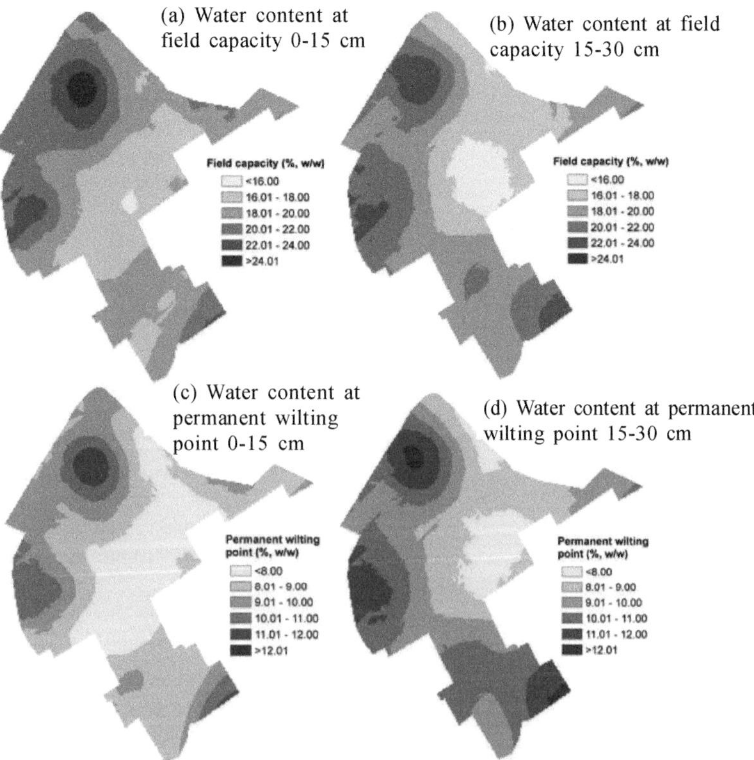

Fig. 23.4: Spatially predicted map of water content for IARI farm; (a) water content at field capacity for 0-15 cm soil depth, (b) water content at field capacity for 15-30 cm soil depth, (c) Water content at permanent wilting point for 0-15 cm soil depth. (d) Water content at permanent wilting point for 15-30 cm soil depth *Source*: Santra *et al.*, 2008

Conclusion

Spatial maps of soil water retention at FC and PWP were prepared through kriging and application of PTFs for an agricultural farm. These maps may serve the guideline for proper irrigation scheduling because water content at FC and PWP are considered as two important moisture constants for plant growth. Besides these, spatial maps of predictor soil properties *e.g.* sand content, clay content, OC content, bulk density etc. also helps in identifying specific crops to be grown in an area.Therefore, crop specific management zones in the farm may be planned based on these spatial soil maps. Although such type of blocking of large farm area based on field expertise is available in few cases but such maps may lead to segregating the farm in precise zones, which is the need for precision agriculture. Such type of study in farm scale may be extended for major nutrients (N, P, K) required for plant growth and thus may help in site-specific nutrient management.

References

Atherton, B.C., Morgan, M.T., Shearer, S.A., Stombaugh, T.S. and Ward, A.D. 1999. Site-specific farming: A perspective on information needs, benefits and limitations. Journal of Soil and Water Conservation 54: 455-461.

Bechini, L., Bocchi, S. and Maggiore, T. 2003. Spatial interpolation of soil physical properties for irrigation planning. A simulation study in northern Italy. European Journal of Agronomy 19: 1-14.

Blake, G.R. and Hartge, K.H. 1986. Bulk density. InMethods of Soil Analysis, Part 1. Physical and Mineralogical Methods (ed. Klute et.al.), 2nd ed. Agronomy Monograph9, ASA, Madison, WI, pp. 363-375.

Bouyoucos, G.J. 1962. Hydrometer method improved for making particle size analysis of soils. Agronomy Journal 54: 464-465.

Davis, B.M. 1987. Uses and abuses of cross-validation in geostatistics. Mathematical Geology 19: 241–248.

Franzluebbers, A.J. and Hons, F.M. 1996. Soil-profile distribution of primary and secondary plant-available nutrients under conventional and no tillage. Soil and Tillage Research 39: 229–239.

Goovaerts, P. 1998. Geostatistical tools for characterizing the spatial variability of microbiological and physico-chemical soil properties. Biology and Fertility of Soils 27: 315-334.

Isaaks, E.H. and Srivastava, R.M. 1989. An Introduction to Applied Geostatistics, Oxford Univ. Press, New York, pp. 1-561.

Jian, X., Olea, R.A., and Yu, Y. 1996. Semivariogram modeling by weighted least squares. Computer and Geoscience 22:387-397.

Klute, A. and Dirkson, C. 1986. Hydraulic conductivity and diffusivity: Laboratory methods. In (ed. Klute et.al.), Methods of soil analysis Part 1, 2nd ed., Agronomy Monograph 9, ASA, Madison, WI, pp. 687-734.

Malhi, S.S., Grant, C.A., Johnston, A.M. and Gill, K.S. 2001. Nitrogen fertilization management for no-till cereal production in the Canadian Great Plains: A review. Soil and Tillage Research 60: 101–122.

Murthy, R.S., Hirekerur, L.R., Deshpande, S.B., Venkata Rao, B.V. and Shankaranarayana, H.S. 1982. Benchmark Soils of India - Morphology, characteristics and classification for resource management, NBSS&LUP, Nagpur, India, pp. 373.

NBSS & LUP. 1976. Soil Survey and Land Use Planning of the IARI farm. National Bureau of Soil Survey and Land Use Planning New Delhi, India.

Santra, P., Chopra, U.K., and Chakraborty, D. 2008. Spatial variability of soil properties in agricultural farm and its application in predicting surface map of hydraulic property. Current Science 95(7): 937-945.

Walkley, A. and Black, I.A. 1934. An examination of Degtjareff method for determining soil organic matter, and a proposed modification of the chromic acid titration method. Soil Science 37: 29-38.

24

Geostatistical Modeling of Soil Organic Carbon Contents–A Case Study

Priyabrata Santra and R.N. Kumawat

ICAR-Central Arid Zone Research Institute, Jodhpur, Rajasthan, India

Introduction

Soil organic carbon (SOC) content plays a key role in maintaining soil fertility in agricultural farm. At present, there has been great interest to reduce the atmospheric CO_2 level through a chain of increasing vegetation carbon pools first and then to store them as soil carbon pool. To quickly assess the soil carbon restoration programmes, it is important to know how much soil carbon is stored in an agricultural farm through adoption of different land management practices. Average SOC content is considered in most soil carbon pool calculations in spite of its large spatial variation in landscape, and thus leads to inaccurate estimate of SOC stock for an area. Therefore, reliable estimates of SOC pools and their spatial variability are essential to establish the soil carbon sequestration programmes at different landscape scales. Here, a case example of assessing the spatial variation at a typical agricultural farm is discussed. The experimental field was located at Badoda village of Jaisalmer district, Rajasthan covering an area of about 76.12 ha. The farm lies between 265°2′30″-26°53′30″ N and 71°12′15″-71°13′15″ E (Fig. 24.1) and 20 km away from Jaisalmer city in eastward direction. Elevation of the farm ranged from 232 m to 248 m above mean sea level with an overall slope from south to north. Among total area of the farm, 85.38% area was under cultivation in four main blocks: block A, B, C, and D whereas rest area was under grassland.

Cultivation in these four blocks was started with digging of tube well at each block in successive stages. The first tube well in the farm was established in the year 1999 and cultivation was started in the block A. Afterwards cultivation was

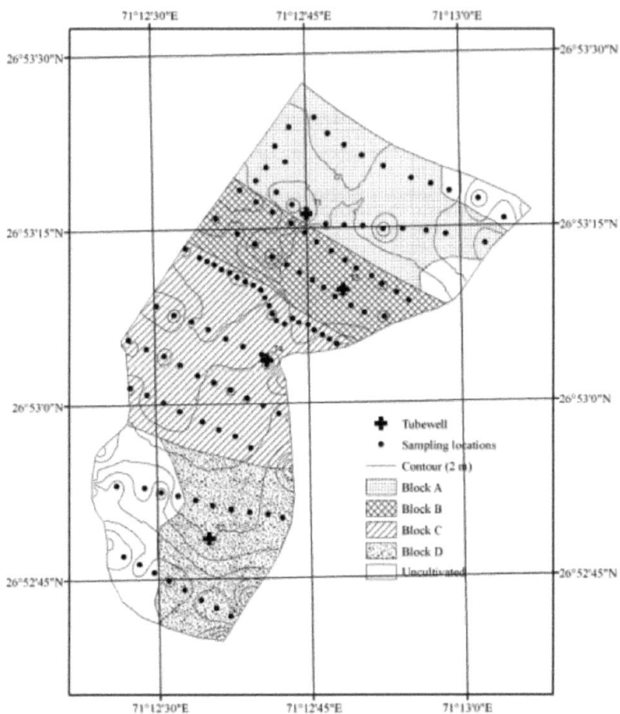

Fig. 24.1: Layout of the agricultural farm along with the location of sampling points within it from hot arid ecosystem of India located at Badoda village of Jaisalmer, Rajasthan

successively started in the block D, B, and C with the establishment of tube wells in the year 2003, 2005, and 2008, respectively. Average depth of groundwater in these four tube wells in the year 2009 was 95 m. Groundnut, mustard, moong bean, clusterbean and wheat were the major crops under cultivation in the farm. The cultivation in a particular plot of the farm has been practiced once in a year and kept the land fallow for remaining periods of the season since the year 1999. Groundout was the major crop in the block A and C. Mustard and moong bean were mostly cultivated at the eastern part of the block A, B, and C. Wheat was the major crop in the block D. Clusterbean was cultivated in scattered small areas of the farm. The soil of the farm was sandy to loamy sand in texture with abundant presence of freely available $CaCO_3$ (5-10%) and taxonomically defined as Typic Torripsaments (Singh et al. 2007). Further detailed description of the farm is available in Santra et al. (2012).

Spatial data of soil carbon in the farm

Spatial data on soil organic carbon content of the farm was generated by collecting 116 surface soil samples (0-15 cm) followed by their laboratory analysis using wet digestion method (Walkley and Black 1934). The georaphical coordinates of each sampling location was recorded using handheld global positioning system (GPS), Model Etrex H. Spatial distributions of the sampling locations are presented in Fig. 24.1. The average SOC content of the farm was 1.66 g kg^{-1} and ranged from as low as 0.11 g kg^{-1} to 6.34 g kg^{-1}. Soils of groundnut cultivated plots were comparatively higher in SOC content (1.88 g kg^{-1}) than rest cultivated plots of the farm. Among groundnut cultivated plots, the highest SOC content (2.51 g kg^{-1}) was observed in the block C and the lowest (0.86 g kg^{-1}) in the block A of the farm. Natural fallow lands at south eastern and southern portion of the farm, where deposition of aeolian sands was very common and rocky outcrops with presence of a few grass species of *Eleusine compressa, Lasiurus sindicus* etc are the dominant landscape features, had the average SOC content of 2 g kg^{-1}. Farm areas under clusterbean, mustard, and moong bean cultivation were comparatively lower in SOC content (1.58 g kg^{-1}, 1.54 g kg^{-1} and 0.77 g kg^{-1}, respectively) than rest of the plots. Comparatively higher SOC content in groundnut cultivated plots was mainly due to spreading nature of groundnut crop, which covered a major portion of soil surface and thus protected carbon stock of soil from its loss through wind eroded aeolian sediments. Overall, it was found that the soils under block A of the farm, where cultivation was started 10 years ago, had the lowest SOC content (1.06 g kg^{-1}). Reversely, the highest SOC content (2.13 g kg^{-1}) was observed in block C, where cultivation started just one year before. These clearly showed that cultivation of land even for once in a year had resulted in a significant reduction of SOC content over last 10 years. Disturbances of soil surface through cultivation resulted into loose soil surface and thus aggravated the rate of SOC depletion through wind erosion events, which are dominant land degradation processes in hot arid areas.

Spatial variation of SOC content

Spatial variation of SOC content within the farm was studied by calculating semivariogramat different lag distances, $\hat{\gamma}(h)$. During pair calculation for computing the semivariogram, maximum lag distance was taken as 500 m, which was about the half of the minimum extent of sampling area of the farm and thus avoided the border effect. During computation of experimental semivariogram, number of lag and the lag tolerance was taken as 10 and 25%, respectively. The directional trend of SOC content within the farm was found negligible and hence omni-directional semivariogram was computed. The computed semivariogram values [(h)] for corresponding lag (h) were fitted with available theoretical semivariogram models using weighed least square technique. Weight for each lag was directly proportional to the number of sampling pairs and inversely proportional to the standard deviation of experimental semivariogram values. Best-fit model with lowest value of Akaike Information Criteria (AIC) was selected for defining the spatial correlation parameters of SOC content:

$$\mathrm{AIC} = n\ln\left[\sum_{i=1}^{n}\frac{N_i(h)}{\sigma[\gamma_i(h)]}[\gamma_i(h)-\hat{\gamma}_i(h)]^2\right]+2p \qquad (1)$$

where n is the number of lag distance (n =10), $\sigma[\gamma_i(h)]$ is the standard deviation of $\gamma_i(h)$ at the lag h, $N_i(h)$ is the number of sample pairs at the lag h, $\gamma_i(h)$ is the measured value of experimental semivariogram at the lag h $\hat{\gamma}_i$, (h) is the estimated semivariogram value at the lag h, and p is the number of parameters in the model. The model with the lowest value of AIC was adjudged as the best model to describe the spatial variation of SOC content. Three commonly used semivariogram models were tried to fit the computed semivariogram values of SOC content. These are the spherical, exponential, and Gaussian models. Expressions of these three models are commonly available in literatures (Santra et al. 2008). After fitting the standard semivariogram model, three spatial correlation parameters of SOC content were identified: nugget (C_0), sill (C_0+C), and range (a). The spherical model was found with the lowest value of AIC and thus was considered as the best to fit the computed semivariogram values. The fitted semivariogram of SOC content of the farm is depicted in Fig. 24.2.

The nugget (C_0) and sill (C_0 + C) were 0.77 and 2.12, respectively. It was found that the nugget, which indicates the small-scale variation of a regionalized variable, was 36% of the sill. However, the contribution of nugget to sill may be reduced

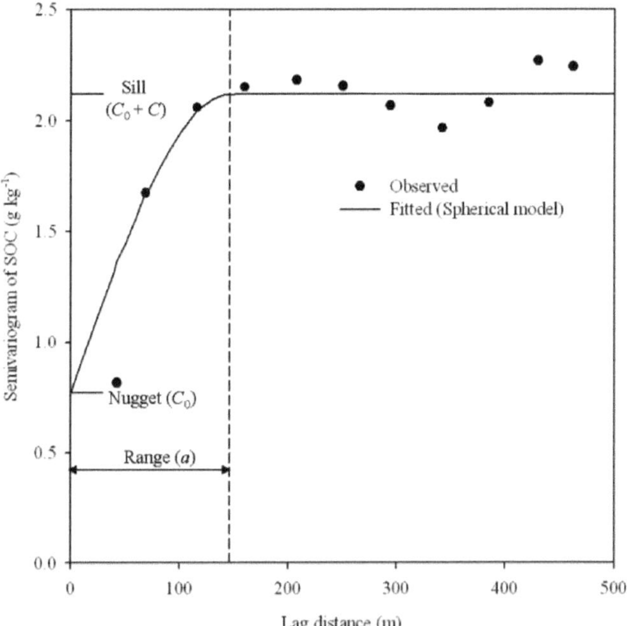

Fig. 24.2: Semivariogram of soil organic carbon content (g kg^{-1}) within the farm area

through adoption of intensive sampling efforts but depends on the budget for sampling. The range (a) parameter of SOC content was 146 m. These spatial correlation parameters indicated that SOC content were highly variable in 2-dimensional soil surface and such spatial variation might be captured only through sampling strategy with minimum separation distance $\leq$150 m.

Surface map of SOC contents

Surface map of SOC content of the farm was prepared using the spatial correlation parameters (C_0, C, and a) through ordinary kriging (OK). Kriged map of SOC content with the grid size of 5×5 m was prepared using geostatistical wizard of ArcGIS 9.1 and is presented in Fig. 24.3. At the centre of the farm covering Block C, SOC content was highest (>2 g kg^{-1}). Lowest SOC content (<0.5 g kg^{-1}) was observed at the top left side of the farm covering mostly block B. However the SOC content at right side of the block B was found slightly higher and in some pocketed location it was >2 g kg^{-1}. Major portion of block D had SOC content in the range 1-2 g kg^{-1}. A small central pocket of the block D had SOC content in the range 0.51-1 g kg^{-1}.

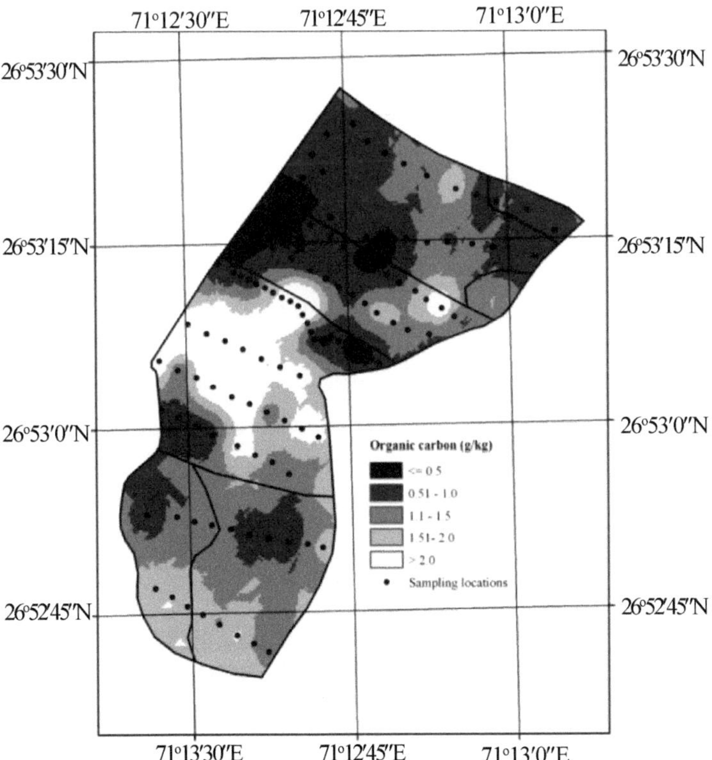

Fig. 24.3: Spatial distribution of soil organic carbon content (g kg^{-1}) of 0-15 cm soil layer in the agricultural farm at Jaisalmer, Rajasthan *Source*: Santra *et al.*, 2012

SOC stock of the farm

The surface map of SOC stock density of 0-15 cm soil layer of the farm was calculated following the stepwise methodology described in Batjes (1986). In the first step, SOC content (g kg^{-1}) was multiplied with the bulk density (kg m^{-3}) to obtain the SOC content on volumetric basis (g m^{-3}). The reported mean bulk density of 1.45×10^3 kg m^{-3} for sandy soils from hot arid zones was used here for the above conversion. In the second step, SOC content on volume basis (g m^{-3}) was multiplied with the thickness of soil layer (m) and the area (m^2) to obtain SOC stock (kg) of 0-15 cm soil layer for each 5×5 m grid of the farm. All the above calculations were performed using raster calculator option of ArcGIS 9.1. At the end of the second step, surface map of SOC stock density of the farm was obtained. In the third step, SOC stock of 0-15 cm soil layer within the total farm area or for different blocks of the farm were obtained by cumulating the SOC stock of grids lying within the desired polygon boundary. For block-wise calculation of SOC stock, grid data within the polygon of each block were extracted using extraction option of the spatial analyst tool of ArcGIS 9.1.

Total SOC stock of the farm was calculated as 272 t with an average density of 3.57 t ha^{-1}. Block-wise distribution of SOC stock of the farm is presented in Table 24.1. Block C covering 28.14% area contributed 41.41% SOC stock of the farm. In the contrary, SOC stock of block A was 52.38 t, which was 19.26% of the total SOC stock of the farm but covers 23.84% of the farm area. Here it is noted that block A has been cultivated since last 10 years from the year 1999 whereas in the block C cultivation was started in the year 2008. The results showed that continuous cultivation of loose sandy soils in hot arid areas may deplete the SOC stock in a faster rate. Among cultivated areas of the farm, the SOC stock density was higher in plots where cultivation has been just started (5.26 t ha^{-1}) than those plots where cultivation was started almost 10 years ago (2.89 t ha^{-1}). It was also observed that in areas with rocky outcrops of the farm had SOC stock density of 1.41 t ha^{-1}.

Table 24.1: Soil organic carbon stock in 0-15 cm soil layer of four blocks of the farm under cultivation for different time periods

Blocks of the farm	Year of cultivation	Area (ha)	SOC stock (t)	SOC stock density for 0-15 cm soil layer (t ha^{-1})
Block A	10 years	18.15 (23.84)	52.38 (19.26)	2.89
Block B	4 years	11.86 (15.58)	36.36 (13.34)	3.06
Block C	~1 years	21.42 (28.14)	112.68 (41.41)	5.26
Block D	6 years	13.56 (17.81)	54.97 (20.20)	4.05
Others	-	11.13 (14.62)	15.70 (5.77)	1.41

Comparison of SOC stock density

In the present study, SOC stock of a farm from the hot arid agro-ecological region of India was calculated considering its spatial variation within the farm. But most of the reported studies on SOC stock assessment in this region were based on mean SOC content of several soil profiles, which may be thought of as the global mean of SOC for a region. Following the approach of considering mean SOC content of unit area and then multiplying with the total area, SOC stock of the total farm area was also calculated. The SOC stock in such calculation was 275 t with a density of 3.61 t ha^{-1}, which was almost similar with the SOC stock calculated through consideration of spatial variation of SOC content. The present study showed that even though there was a little difference in estimated SOC stock of the farm between two approaches but wide variation in SOC stock density was observed at different portions of the farm. Till now most of the SOC stock assessment in India was carried out in regional scale where local variation is ignored. But for site-specific carbon management within a farm, knowledge on the spatial variation of SOC stock density is very important. Moreover, regional scale estimates of SOC stock based on global mean either underestimated or over estimated the SOC stock for a small farm scale. For example, in the present study, we obtained a SOC stock density of 3.57 t ha^{-1} for an agricultural farm located at extremely hot arid situation of western Rajasthan. SOC stock density of 78.68 t ha^{-1} for 0-30 cm soil layer in hot arid ecoregion of India (Punjab, Haryana, and Rajasthan) was reported in literature (Velayutham e al. 2009), which is quite higher than the present study. In contrary, reported SOC stock density of 0.0155 t ha^{-1} for 0-25 cm soil layer of Typic Torripsamments soil profile in hot arid region of Rajasthan, which is quite lower than the present study. Therefore, assessment of SOC stock at small farm scale is very essential for successful implementation of carbon sequestration programme.

Conclusion

In this chapter, spatial variation of SOC content in an agricultural farm has been discussed. Although the SOC stock of the farm might be calculated based on an average SOC content but the attention was made on exploring the spatial variation of SOC content and stock density within the farm. It was hypothesized that surface map of SOC stock density may guide the farmers for site-specific management of carbon in an agricultural farm. The present study showed that consideration of spatial variation of SOC content in a farm may lead to accurate assessment of SOC stock density. Surface map of SOC stock density of an agricultural farm may be an essential requirement for adoption of site-specific carbon sequestration strategies and for implementation of carbon credit programmes according to green productivity concepts.

References

Batjes, N.H. 1996. Total carbon and nitrogen in the soils of the world. European Journal of Soil Science 47: 151-163.

Santra, P., Chopra, U.K. and Chakraborty, D. 2008. Spatial variability of soil properties in agricultural farm and its application in predicting surface map of hydraulic property. Current Science 95(7): 937-945.

Santra, P., Kumwat, R.N., Mertia, R.S., Mahla, H.R. and Sinha, N.K. 2012. Soil organic carbon stock density and its spatial variation within a typical agricultural farm from hot arid ecosystem of India. Current Science 102(9): 1303-1309.

Singh, S.K., Singh, A.K., Sharma, B.K. and Tarafdar, J.C. 2007. Carbon stock and organic carbon dynamics in soils of Rajasthan,India. Journal of Arid Environment 68(3): 408-421.

Velayutham, M., Pal, D.K. and Bhattacharya, T. 2009. Organic carbon stocks in soils of India. In: *Global Climate Change and Tropical Ecosystems* (eds R. Lal, J.M. Kimble, and B.A. Stewart), pp. 71-96, Lewis Publishers, Boca Raton, FL.

Walkley, A. and Black, I.A. 1934. An examination of Degtjareff method for detrmining soil organic matter, and a proposed modification of the chromic acid titration method. Soil Science 37: 29-38.

25

Digital Soil Mapping of Organic Carbon in Fruit Tree Based Land Use System in Ar0id Region Using Geostatistical Approaches

Akath Singh, Priyabrata Santra, Mahesh Kumar and N.R. Panwar

ICAR-Central Arid Zone Research Institute, Jodhpur, Rajasthan, India

Introduction

Globally, the SOC pool of 1,550 Pg (1 Pg = 1015 g) to 1-m depth is about three times that of biotic and twice that of atmospheric pools (Post et al. 1982, Watson et al. 2000). In arid and semi-arid ecosystems, SOC contents show high degree of spatial variability due to patchiness of vegetation (Wiesmeier et al. 2009). The arid ecosystem might play a key role in mitigation of climate change effects through sequestering carbon in soils (United Nations 2011). Therefore, a suitable land use system having high C sequestration potential for arid ecosystem needs to be identified and promoted. However, the prevailing conditions of arid ecosystems normally does not support dense vegetation. Therefore, low carbon stock is expected in both above-ground vegetation and below-ground soil. Fruit trees are an integral part of arid farming system and played a significant role to improve the nutritional security and employment opportunity of farmers but also have potential role to increase carbon storage both in biomass and soil. To quickly assess the soil carbon restoration programmes, it is important to know how much soil carbon is stored in an agricultural farm through adoption of different land management practices. Therefore, reliable estimates of SOC pools are essential to establish the soil carbon sequestration programmes at different landscape scales.

Thomson and Kolka (2005) summarized two categories of techniques for assessing the SOC pool: (1) arithmetic averaging of SOC contents from multiple field measurements for a landscape unit followed by multiplying it with area and bulk density, (2) soil landscape modeling approach where spatial maps of SOC contents was prepared through geostatistical approaches. A major advantage of the second approach is that we can get an estimate at each possible location within an area and thus will be more useful for prioritization of suitable management zones as per its needs (Wiesmeier et al. 2011). Moreover, error or uncertainty of estimation in the second approach may be properly quantified, which may be quite useful in further modeling approach using these maps as input. Here, a study on spatial assessment of SOC content and stock density in a horticultural farm located at Jodhpur is discussed. Detailed description of the fruit orchard is available in Singh et al. (2016).

Fruit cultivation in arid western Rajasthan

Arid zone soils are low in organic matter (0.05 to 0.4%) and N on account of high temperature low rainfall, scanty vegetation cover and sandy texture of soil (Jenny and Rayachaudhari, 1960). Out of total geographical area of western Rajasthan 77.5% is affected by wind erosion/deposition hazard (Narain et al. 2000). Secondary salinization/sodicity in different part is also common. Cultivated land including fallow land occupies 62.94% of total geographical area of region. Agricultural lands are primarily rainfed with varying degree of cropping intensity but majority of which is less than 100%. Nearly 28.13% area is classified as wasteland (Ram and Lal, 1997). The vegetation of hot region is very sparse and consists of scattered thorny trees/shrubs and grasses. This region has 682 species, belonging to 352 genera and 87 families of flowering plants.

A vast land resource (39.54 m ha) in the country characterized as arid region is considered underutilized at present as it has good potential for quality production of several horticultural crops. The development of arid horticulture is comparatively recent. In order to achieve sustainable production, soil quality as assessed by physical, chemical and biological parameters need to be maintained. In this region, sustained production can be achieved by improved management practices that integrate soil conservation along with integrated nutrient management. Soil organic carbon (SOC) plays a major role in determining soil fertility at local farming situation and reserves major terrestrial carbon on global point of view.

Fruit trees in arid farming system have played a significant role to enhance land productivity and improve livelihoods. During last few decades, fruit cultivation in arid western Rajasthan has been increased significantly. Among different fruit crops grown in western Rajasthan, pomegranate (*Punica granatum*) is dominant covering about 5000 ha area. Other major fruits crops in western Rajasthan are ber (*Ziziphus* spp.), aonla (*Emblica officinalis*), date palm (*Phoenix dactylifera*) and fig (*Ficus carica*). Fruit tree based production system has not only improve the economic, nutritional security and employment opportunities but also have potential role in climate change mitigation strategies through increasing carbon storage in the above ground biomass and below ground soil.

Horticultural orchard at ICAR-CAZRI Jodhpur

The fruit orchard is located at horticulture research farm of ICAR-Central Arid Zone Research Institute (CAZRI), Jodhpur Rajasthan and lies at 26°14' 59.27" N and 72°59'19.54" E and at an elevation of about 240 m above mean sea level (Fig. 25.1).

The fruit orchard consisted of block plantations of different arid fruit tree species *e.g. Punica granatum* (pomegranate), *Cordia myxa* (*'gonda'*), *Emblica officinalis* (*'aonla'*), *Aegel marmelos* (*'bael'*), *Carisa carandas* (*'karonda'*), *Prosopis cineraria* ('khejri'), *Citrus limon* (lemon), *Opuntia ficus-indica* ('cactus pear') and *Ziziphus* sp. (*'ber'*). Area of the fruit orchard was 16 ha, out of which 14.33 ha area was planted with different common arid fruit trees 20-35 years before. A details of the different fruit tree species present in the orchard with different accessions along with the cultural management practices are presented in Table 25.1.

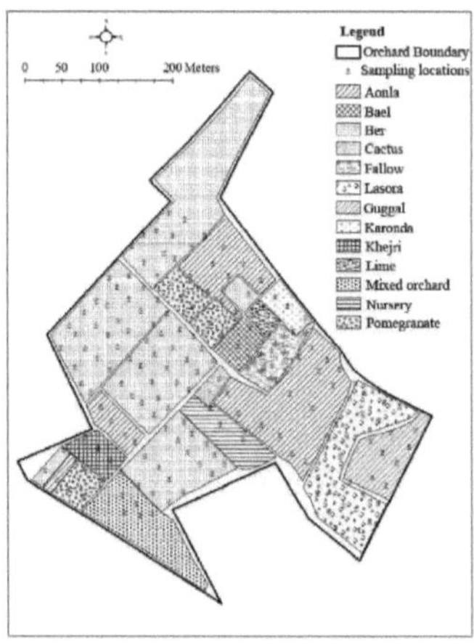

Fig. 25.1: Location of the horticultural orchard along with sampling points
Source: Santra *et al.*, 2016

Spatial data of soil properties in the horticultural orchard

Soil samples were collected from 175 locations within the orchard in several transects at regular interval of about 20 m following a grid network (Fig. 25.1). Disturbed soil samples were collected from surface (0-15 cm) and subsurface (15-30 cm) layers using khurpi and soil tube auger. Geographical coordinates of each sampling locations were recorded using hand held geographical positioning systems (GPS) (Garmin, Etrex Vista). Collected soil samples were air dried and then passed through 2 mm sieve. Processed soil samples were analysed in laboratory to determine soil organic carbon (SOC) content, pH, electrical conductivity (EC) and available K content. SOC contents of soil samples were determined through Walkley Black method whereas pH and EC were determined through standard laboratory procedure.

Table 25.1: Different fruit tree plantations in the horticultural orchard at ICAR-CAZRI Jodhpur

Species	Common name	No. of accessions	Area (ha)	Spacing (row × plant) (m.)	Year of plantation	Agronomic practices adopted
Emblica officinalis	'*Aonla*'	07	2.83	6 × 6	1986	Rainfed system
Aegel marmelos	'*Bael*'	07	0.27	6 × 6	1997	Supplementary basin irrigation
Ziziphus spp.	'*Ber*'	22	5.85	6 × 6	1978	Sole plantation of 22 cvs. under rainfed system along with few plantations in low lying areas
Opuntia ficus-indica	Cactus pear	10	0.12	3 × 3	1998	Supplementary basin irrigation
Fallow block	-	-	0.39	-	-	-
Cordia myxa	'*Lasora*'	17	1.33	5 × 5	2001-03	Supplementary basin irrigation and manual leaf removal in winter
Commiphora wightii	Guggal	-	0.67	4 × 4	2002	Rainfed system
Carissa carandas	'*Karonda*'	07	0.24	5 × 5	2000-2003	Supplementary basin irrigation
Prosopis cineraria	'*Khejri*'	13	0.29	5 × 4	2004	Supplementary basin irrigation
Citrus aurantifolia	Lime	04	0.12	6 × 6	2000-2003	Supplementary basin irrigation
Mixed orchard (*Ficus carica, Citrus aurantifolia and Ziziphus* spp)	Fig, lime and '*ber*'	-	0.94	5 × 5	2004	Rainfed system and pruning in summer
Nursery block	-	-	0.51	-	-	Raising of seedlings of ber and other horticultural plants
Punica granatum	Pomegranate	24	0.77	5 × 5	1986 &2000	Under drip and basin irrigation system
Total			14.33			

Descriptive statistics of SOC content, pH and EC of the orchard is given in Table 25.2. SOC content was found as low as 0.09% with an average of 0.30-0.33% for surface and subsurface soil layers. At few locations, SOC content was found as high as 1.41% and 1.28%, at surface and subsurface layers, respectively. In general it has been observed that SOC content was higher in surface layer than subsurface layer. However, after checking the data distribution, SOC content was found skewed towards lower values and thus log-transformation was carried out. Descriptive statistics of log-transformed SOC (%) are given in Table 25.2. Soil pH in the orchard was observed to vary from 7.07 to 8.99 with an average value of about 8.23. Average soil EC was found 0.18-0.20 mS cm^{-1}, however at few locations it was observed as high as 1.24 mS cm^{-1}. Soil EC was also observed skewed towards lower values and hence log-transformation was carried out, which are given in Table 25.2.

Histogram plot of log transformed SOC (%) are given in Fig. 25.2. About 65-70% of the samples have log[SOC(%)] content between -2.0 to -1.0 in both surface and subsurface layers.

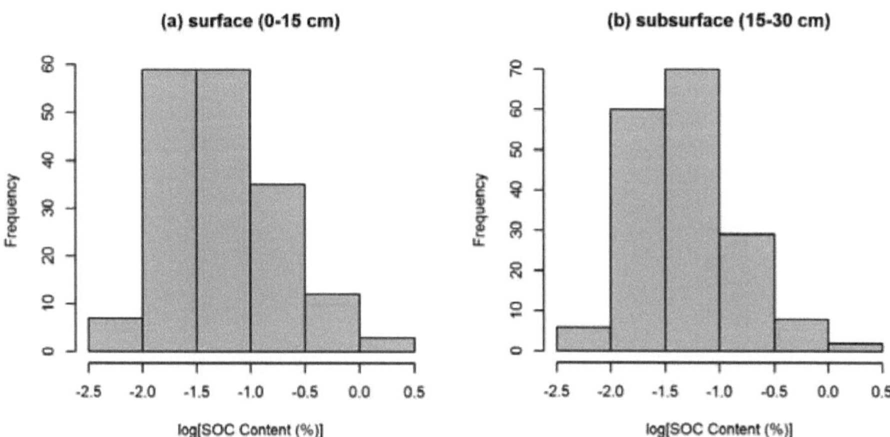

Fig. 25.2: Histogram plot of log-transformed SOC contents

SOC contents under different fruit trees

It was noted that irrespective of species SOC concentration decreased with soil depth. Significant influence of fruit species on SOC stock has also been observed, which is expected because of the difference in leaf biomass litter content as well as their characteristics across fruit species. Highest SOC stock has been observed in block plantation of ber under low lying conditions, where runoff water is generally accumulated for a short period during monsoon season. Under rainfed situation, SOC stock followed the order as *bael* > mixed orchard > pomegranate > *khejri* > *aonla* > *ber* > lemon > *gonda* > cactus pear > *guggal* > *karonda.* It has also been observed that when ber was intercropped with moong bean, SOC stock was higher than the system intercropped with pearl millet or senna or even without intercropping.

Table 25.2: Descriptive statistics of soil properties in the orchard

Soil properties	Soil layer	Original				Log-transformed			
		Min	Max	μ	σ	Min	Max	μ	σ
Soil organic carbon (%)	0-15 cm	0.09	1.41	0.33	0.21	-2.41	0.34	-1.25	0.52
	15-30 cm	0.09	1.28	0.30	0.18	-2.41	0.25	-1.32	0.48
pH	0-15 cm	7.07	8.99	8.20	0.38				
	15-30 cm	7.12	8.92	8.26	0.37				
Electrical conductivity (mS cm^{-1})	0-15 cm	0.05	1.24	0.20	0.17	-2.96	0.22	-1.84	0.60
	15-30 cm	0.06	1.16	0.18	0.15	-2.85	0.15	-1.94	0.57

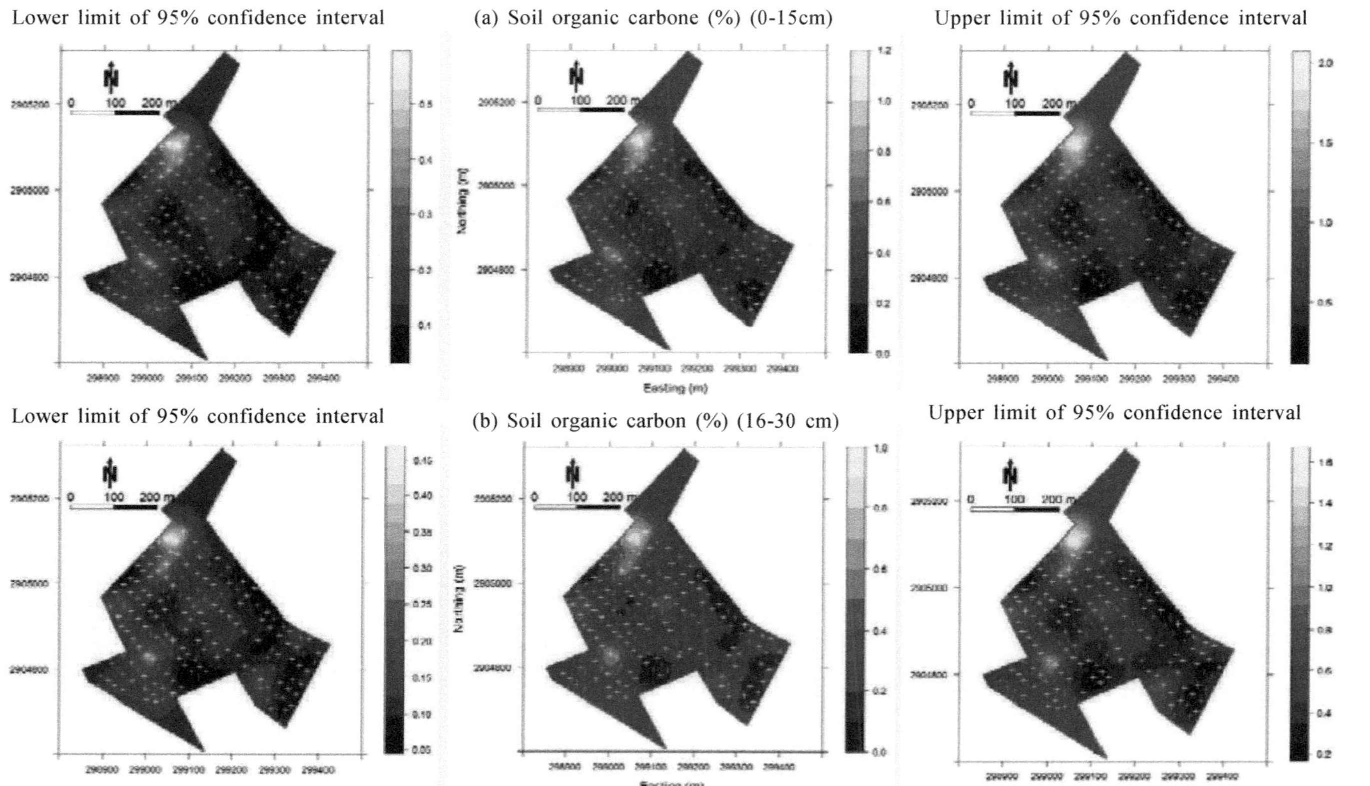

Fig. 25.3: Predicted surface map of SOC contents (%) in the orchard through ordinary kriging; (a) surface soil layer (0-15 cm) along with lower and upper limit of 95% confidence interval and (b) subsurface soil layer (15-30 cm) along with lower and upper limit of 95% confidence interval

Surface maps of SOC contents for both surface and subsurface layers are shown in Fig. 25.3. Except few patches of higher SOC contents at North-eastern and Eastern part of the orchard, most areas have SOC content <0.4%. It has also been observed that SOC content at surface layer (0-15 cm) is slightly higher than subsurface layer (15-30 cm), which may be noted from the legend bar of the respective maps. Along with the prediction map, 95% confidence interval maps for each soil layer are presented in Fig. 25.3, which gives an idea about the uncertainty of estimation at each predicted location of the orchard.

SOC stock density across fruit blocks plantations of the orchard

SOC stock density of different fruit block plantations in the orchard is calculated along with their uncertainty of estimation. Although, the ordinary kriging estimate and co-kriging estimate were found slightly different, the later was used here as the representative SOC stock density because of its better performance in cross validation. It has been observed that SOC stock density was highest 19.52 Mg ha^{-1}, in *Prosopis cineraria* ('khejri') block having 0.29 ha area. It was followed by mixed orchard of *Ficus carica* (fig) + *Citrus aurantifolia* (lime) + *Ziziphus spp* ('*ber*'), block of *Aegel marmelos* ('*bael*'), block of *Ziziphus spp.*('*ber*') and block of *Punica granatum* (pomegranate) all having SOC stock density >16 Mg ha^{-1}. Lowest SOC stock density was observed in *Carissa carandas* (*'karonda'*) block (9.01 Mg ha^{-1}) followed by *Commiphora wightii* (guggal) block (9.77 Mg ha^{-1}) and *Cordia myxa* ('*lasora*') block (11.01 Mg ha^{-1}). The better SOC stock density in blocks of *Prosopis cineraria*, mixed orchard of *Ficus carica* + *Citrus aurantifolia* + *Ziziphus spp*, *Aegel marmelos*, *Ziziphus spp.*and *Punica granatum* was due to higher litter addition potential from its canopy biomass and easy decomposition of their leaf residues. In case of *Carissa carandas* and *Commiphora wightii*, leaf biomass portion is comparatively less to significantly contribute to litter addition on soil and thus low SOC stock density was expected. For *Cordia myxa*, although the leaf biomass component is high, leaf residues of it are not easily decomposable because of high lignin content but if it decomposes properly, very good quality organic compost may be produced which will be sufficient for that tree (Meghwal et al. 2014). Confidence interval of SOC stock estimation at 90% and 95% levels are also presented in Fig 25.3. For the whole orchard, total SOC stock estimate was 14.48 Mg ha^{-1}whereas through arithmetic averaging technique, it was 12.74 Mg ha^{-1}. Therefore, difference between two estimates was observed, which is also noted for SOC estimates of different fruit block plantations.

Uncertainty of SOC stock estimate of the whole orchard as 90% confidence interval was found 6.83-26.96 Mg ha^{-1}. Bulk density of surface soils from few measured locations within the orchard was found to range from 1.47 to 1.58 Mg m^{-3} with an average value of 1.54 Mg m^{-3}, which was very close to the average bulk density of top most soil horizon in arid western India as reported in soil survey data (Shyampura et al. 2002). Therefore, a bulk density of 1.54 Mg m^{-3} was considered while calculating the SOC stock from the SOC concentration. Uncertainty of bulk density in terms of standard deviation for these two surface horizons was found 0.11 Mg m^{-3}, thus 90% confidence interval of bulk density assumption was 1.36-1.72 Mg m^{-3}. Therefore, uncertainty in the SOC stock estimate was contributed by uncertainties involved in spatial estimation of SOC content and bulk density. A detailed discussion on uncertainty of SOC stock estimation has been reported in Shrumpf et al. (2011). Confidence interval of SOC stock estimate of the whole orchard as well as for different fruit block plantations was found narrow as compared to arithmetic averaging estimates. For example, 90% confidence interval of SOC stock estimate of the orchard through arithmetic averaging technique was 0.26-28.16 Mg ha^{-1}, which is quite wider than the spatial averaging technique. However, it was noted that when number of sampling points in a block plantation was very low (<10) the uncertainty component of arithmetic averaging estimate was narrow as compared to spatial averaging estimate because sample size was not enough to represent the population.

Conclusions

Digital maps of soil organic carbon content in the horticultural orchard of ICAR-Central Arid Zone Research Institute were prepared through kriging approaches. SOC stock density of whole orchard as well as of different fruit block plantations of the orchard were also calculated from SOC concentration map through spatial averaging technique along with uncertainty of the estimates. Comparison of these estimates with non-spatial estimates showed better accuracy and lower uncertainty of spatial averaging technique. Among different fruit block plantations of the orchard, the block with *Prosopis cineraria* ('khejri') plantation was found having higher SOC stock density than others. Block plantations of *Ziziphus* sp., *Punica granatum* and mixed plantation of *Ficus carica* + *Citrus aurantifolia* were also found with comparatively higher SOC stock density than the blocks of *Commiphora wightii*, *Carisa carandas* and *Cordia myxa*. Uncertainty in spatial averaged estimates was found narrow as compared to arithmetic averaging estimates. Apart from low uncertainty, the major advantage of spatial characterization is the availability of SOC estimates along with uncertainty at each possible location of the orchard, which may help in location-specific nutrient management.

References

Jenny, H. and Rayachaudhari, S.P. 1960. Effect of climate and cultivation on nitrogen and organic matter reserve in Indian soils. Indian Council of Agricultural Research, New Delhi.

Narain, P., Kar, A., Ram, B., Joshi D.C. and Singh, R.S. 2000. Wind erosion in Western Rajasthan. ICAR-Central Arid Zone Research Institute, Jodhpur: pp. 1-36.

Post, W.M., Emanuel, W.R., Zinke, P.J., and Stangenberger, A.G. 1982. Soil carbon pools and world life zones. Nature 298: 156–159.

Ram, B. and Lal, G. 1997. Land use and agriculture in India arid ecosystem. In: Desertification Control in the Arid Ecosystem of India for Sustainable Development. Agrobotanical Publisher (India) Bikaner: 159-173

Singh, A., Santra, P., Kumar, M., Panwar, N. R. and Meghwal, P.R., 2016. Spatial assessment of soil organic carbon and physicochemical properties in a horticultural orchard at arid zone of India using geostatistical approaches. Environmental Monitoring and Assessment, DOI: 10.1007/s10661-016-5522-x.

Thompson, J.A. and Kolka, R.K. 2005. Soil carbon storage estimation in a forested watershed using quantitative soil landscape modeling. Soil Science Society of America Journal 69: 1086–1093.

United Nations 2011. Global drylands: a UN system-wide response. Full Report. United Nations Environment Management Group. pp. 131.

Watson, R.T., Noble, I.R., Bolin, B., Ravindramath, N.H., Verardo, D.J., and Dokken, D.J. 2000. Land use, land use change, and forestry (p. 375). Cambridge: Cambridge University Press.

Wiesmeier, M., Steffens, M., Kolbl, A. and Kogel-Knabner, I. 2009. Degradation and small scale spatial homogenization of top soils in intensively grazed steppes of Northern China. Soil and Tillage Research 104: 299–310.

Appendix

Soil series data of surface soil horizon in arid western India covering Rajasthan and Gujarata

Easting and northing data are according to the LCC coordinate system (projection information: +proj=lcc +lat_1=12.4729444 +lat_2=35.17280555 +lat_0=24 +lon_0=80 +x_0=4000000 +y_0=4000000 +datum=WGS84 +units=m +no_defs)

Sr. No	ID	Soil series	long	lat	Easting (m)	Northing (m)	Soil depth (cm)	Sand (%)	Clay (%)	OC (g/kg)	pH	EC (dS/m)
1	R0011	Shergarh	72.2823	26.34378	3244324	4275353	160	84.5	5.7	1	8	0.14
2	R0021	Tanot	70.24269	27.91453	3056729	4458247	250	97	2	0.8	8.4	0.2
3	R0031	Lathi	71.57806	27.04759	3180020	4355714	123	90.7	5.9	1.8	8.2	0.13
4	R0041	Lohawat	72.50969	26.99002	3270302	4344337	120	77.5	6	1.8	8.2	0.2
5	R0051	Sri Dungargarh	74.01729	28.02181	3421840	4449535	65	92	3.5	1	8.6	1.4
6	R0061	Bhiyar	71.45142	26.34171	3163050	4279824	30	90.8	4.8	0.4	8.1	0.25
7	R0071	Ballewa	72.83484	26.09937	3297026	4245958	170	82.8	9.2	1.1	8	0.4
8	R0081	Gajner	73.03885	27.95441	3326994	4446545	75	90.6	5.2	0.9	8.2	0.28
9	R0091	Balesar	71.10098	26.11014	3127196	4256823	80	93.8	4	0.2	7.7	0.24
10	R0101	Bhadasar	70.96469	27.03878	3120319	4358457	45	75.6	10.7	4	8.6	0.66
11	R0111	Devikot	70.07031	26.79876	3031540	4338270	90	82.4	8.3	1.4	8.1	0.26
12	R0121	Lanela	70.93832	27.1265	3118369	4368147	95	77.8	10.2	1	8.1	0.3
13	R0131	Bap	72.35448	27.38666	3257538	4388248	80	58.3	21.9	1.1	7.7	0.23
14	R0141	Sankara	71.51625	26.7742	3172220	4326388	65	78.3	12.8	0.7	8.2	0.24
15	R0151	Baru	71.97678	27.10833	3219173	4360046	55	92.6	4.1	0.7	8.1	0.18
16	R0161	Jamsar	73.2989	28.13969	3353076	4465488	167	88.8	9.2	0.6	7.8	0.51
17	R0171	Lunkaransar	73.72526	28.56169	3396256	4509538	120	63	17.1	3.4	8.1	0.14
18	R0181	Lalgarh	73.26182	28.091	3349245	4460360	50	92.4	6.3	1	8.5	0.12
19	R0191	Ajjasar	71.80031	27.29909	3203226	4381759	100	93.8	3.6	0.9	8.2	0.22
20	R0201	Jaitaran	73.91745	26.28659	3404048	4261298	150	87.2	8.8	2	8.5	0.03
21	R0211	Rawatsar	74.37558	29.2721	3461886	4584327	150	86	12.4	1	7.7	0.03
22	R0221	Sanderao	73.15805	25.30848	3324539	4158532	102	40.2	29.6	7	8.1	0.13

23	R0231	Balotra	72.2479	25.85185	3238019	4222143	150	42	25	5	8.3	0.31
24	R0241	Chhajai	74.24327	27.17273	3439890	4356260	50	76	15.7	1	8.5	1.2
25	R0251	Mokala	73.90612	26.62436	3404526	4298041	85	71.9	16.3	4	8.2	0.24
26	R0261	Sumerpur	73.05395	25.23312	3313866	4150857	75	17.1	25.4	2.2	8.2	0.16
27	R0271	Chawan	73.04006	26.04389	3316846	4238925	73	78.6	17.2	4.6	8.1	0.17
28	R0281	Pachpadra	72.25719	25.93896	3239450	4231546	70	75.3	15.1	3.3	8.2	0.28
29	R0291	Osian	72.89767	26.73668	3306677	4314856	120	90.7	5.9	1.8	8.9	0.13
30	R0301	Gachipura	74.42764	26.95401	3456877	4331761	120	90.5	7.3	2	7.4	0.13
31	R0311	Degana	74.32498	26.83964	3446376	4319728	120	91.3	7.5	1.1	7.4	0.2
32	R0321	Samdari	72.56058	25.23962	3265212	4154050	120	83.4	8.2	0.4	8.2	0.2
33	R0331	Anupgarh	73.20104	29.2146	3349289	4583013	113	68.2	12.9	3	8.2	0.24
34	R0341	Ladnu	74.3995	27.66486	3457211	4409173	120	88.6	9.3	0.7	7.4	0.09
35	R0351	Navalgarh	75.27142	27.87686	3542455	4429148	130	95.3	3.7	0.6	8.4	0.09
36	R0361	Karanpur	73.42065	29.85547	3373559	4651924	130	89.1	8	0.8	8.7	0.11
37	R0371	Anandgarh	72.91515	28.79322	3319639	4538447	160	88.9	9.1	0.1	7.8	0.5
38	R0381	Prithvirajpur	73.90772	29.73236	3419323	4636400	135	72.5	10.9	1.6	8	0.5
39	R0401	Suratgarh	73.90948	29.32746	3417580	4592218	125	50.6	32.5	2.7	7.9	0.2
40	R0411	Fategarh	74.23306	29.53929	3449451	4614025	126	30.8	35.8	4	8.5	0.23
41	R0441	Goyali	72.84146	24.90268	3291071	4116056	87	73.5	14.9	5	8.1	1.1
42	R0451	Palsana	75.23583	27.49901	3537622	4388146	120	79.5	13.2	3	7.7	0.46
43	R0481	Pachaori	75.8912	28.15796	3603301	4457875	120	82.4	12.5	3.7	7.6	0.3
44	R0501	Kanwat	75.65987	27.59121	3579072	4396848	100	90.8	6.8	1	7.5	0.02
45	R0531	Pisangon	74.39324	26.66894	3452293	4300909	92	52.6	27.8	3	9.2	0.4
46	R0831	Khatu	75.39561	27.36734	3552656	4373305	120	86.8	6.5	3	8.2	0.2
47	R0851	Sambhar	75.18975	27.91788	3534709	4433881	120	80	11.4	1.7	9.2	6.1
48	R0861	Nawan	75.00748	27.01755	3513613	4336550	120	83.8	5.1	2.3	9.8	4.1
49	R0871	Mithri	74.89224	27.07222	3502606	4342896	125	81.8	9.2	1.7	7.9	0.25
50	G0011	Bhardia	70.23719	23.34164	3021784	3962407	40	84	15.4	1.4	8	1.54
51	G0021	Desalpar	70.68446	23.7717	3069618	4005975	110	43.8	42.4	8	8.1	0.2
52	G0031	Vejapur	70.9961	23.52826	3099011	3977567	105	58.7	32.9	3.6	8.7	0.3

53	G0041	Adesar	71.03191	23.56468	3102841	3981286	105	74.6	21.9	4	8.7	0.3
54	G0051	Balasar	70.66525	23.83645	3068168	4013118	137	85.5	9.9	2	8.9	0.4
55	G0061	BhedraMota	68.91907	23.42896	2890711	3981629	42	84	15.4	1.4	7.5	1.54
56	G0071	Kharirohar	70.22636	23.08656	3018780	3934846	120	81.6	9.2	1.5	8.7	0.2
57	G0081	Kotada	68.88499	23.48783	2887810	3988271	110	65.7	17.6	3	8.3	0.06
58	G0091	Kunri	68.55258	23.60109	2855619	4003193	35	69.5	14.7	2.1	7.7	0.15
59	G0101	Luni	69.78369	22.85381	2972583	3912787	160	61.6	23.9	5.1	8.3	0.2
60	G0111	Ramania	68.78384	23.12053	2874538	3949310	70	63.4	17	1.9	7.4	0.11
61	G0121	Wadawa	69.79411	23.18747	2976252	3948854	45	61	19.5	5	7.5	0.23
62	G0131	Amiliara	70.46549	23.22583	3043773	3948292	105	73.3	13.8	6.1	8	0.2
63	G0141	Bhujpar	70.04568	23.31556	3002426	3960924	150	77.3	19.1	2	8.7	0.6
64	G0151	Jarpara	69.63534	22.83876	2957571	3912247	140	76.8	14.5	1.9	8.8	0.3
65	G0161	Jogarimata	70.00881	22.93498	2995809	3919958	70	72	19.3	4.8	8.2	0.2
66	G0171	Kanthkot	70.5022	23.46886	3049227	3974376	91	86.16	12.2	1.5	7.8	0.19
67	G0181	Kidana	70.1071	23.03248	3006414	3929823	120	84.6	9.6	3.3	8	0.2
68	G0191	Lakhpat	68.72251	23.82919	2874551	4026515	150	41.1	41.6	2.9	8.7	0.2
69	G0201	Padana	70.2662	23.2629	3024095	3953674	100	77.9	13.4	3.4	8.8	0.2
70	G0211	Padhdhar	69.81857	23.01853	2977376	3930376	150	68	25	2.8	8.3	0.5
71	G0221	Rampar	70.03484	23.01333	2999022	3928260	36	65.1	22.4	1	8	0.1
72	G0231	Bhadreswar	69.86016	22.88875	2980533	3916017	95	78.7	14.6	0.7	8	0.2
73	G0241	Bhimdevka	70.7726	23.32472	3075231	3956958	159	39.7	39.3	5.3	8.5	0.2
74	G0251	Bhojarda	69.80091	23.61145	2980263	3994727	150	56.8	16.2	6	7.7	0.5
75	G0261	Fatehgarh	70.84461	23.76065	3085512	4003729	113	88.6	7.9	6.3	8	5.5
76	G0271	Hajipir	69.18322	23.68407	2919227	4007196	170	27.3	28.5	6	8	2.2
77	G0281	Jangi	70.56294	23.20527	3053382	3945406	85	65	17.2	3.8	8.2	0.3
78	G0291	Motichirai	70.28279	23.21958	3025431	3948867	175	53.6	30.6	3	8.9	0.3
79	G0301	Nara	69.15761	23.65662	2916444	4004421	150	16	42.1	4.2	7.2	6.2
80	G0351	Aniyali	70.85363	23.03727	3081316	3925276	135	55.8	25	7	8.6	2.25
81	G0411	Ghundada	70.83224	22.70781	3076843	3889710	24	51.5	34.2	7	8	0.05
82	G0531	Lajai	70.78154	22.71231	3071777	3890531	88	28.8	58.4	5	8.2	0.12
83	G0601	Mitana	70.72511	22.5991	3065295	3878634	29	11.2	53.2	3.2	8.5	0.28

84	G0611	Motimarad	70.39131	22.19404	3028708	3836993	56	16.8	60.7	7	8.1	0.02
85	G0641	Pipli	70.86806	22.85262	3081464	3905171	43	59.9	18.4	6.9	8.1	0.2
86	G0801	Bhogat	69.24415	21.99954	2911349	3824304	26	41.5	36.1	3.6	8.2	3.8
87	G0811	Charkala	69.1222	22.19994	2900710	3846957	135	14.9	47	8.7	8	33
88	G0831	Dhinki	69.0716	22.19997	2895606	3847356	59	36.4	38.6	4	7.5	0.34
89	G0851	Gaga	69.17696	22.14492	2905774	3840571	127	72.4	18.1	3	8	2.2
90	G0881	Mulvel	69.12655	22.34523	2902365	3862658	21	36.4	36	5.2	7.7	0.2
91	G0901	Nageshwar	69.07167	22.34133	2896804	3862665	140	59.4	27.1	2	8.5	0.27
92	G0931	Shivrajpur	68.95349	22.32171	2884731	3861470	31	41.4	36.6	4	8.4	0.3

Colour Plates

Chapter 1: Fundamentals of Geostatistics

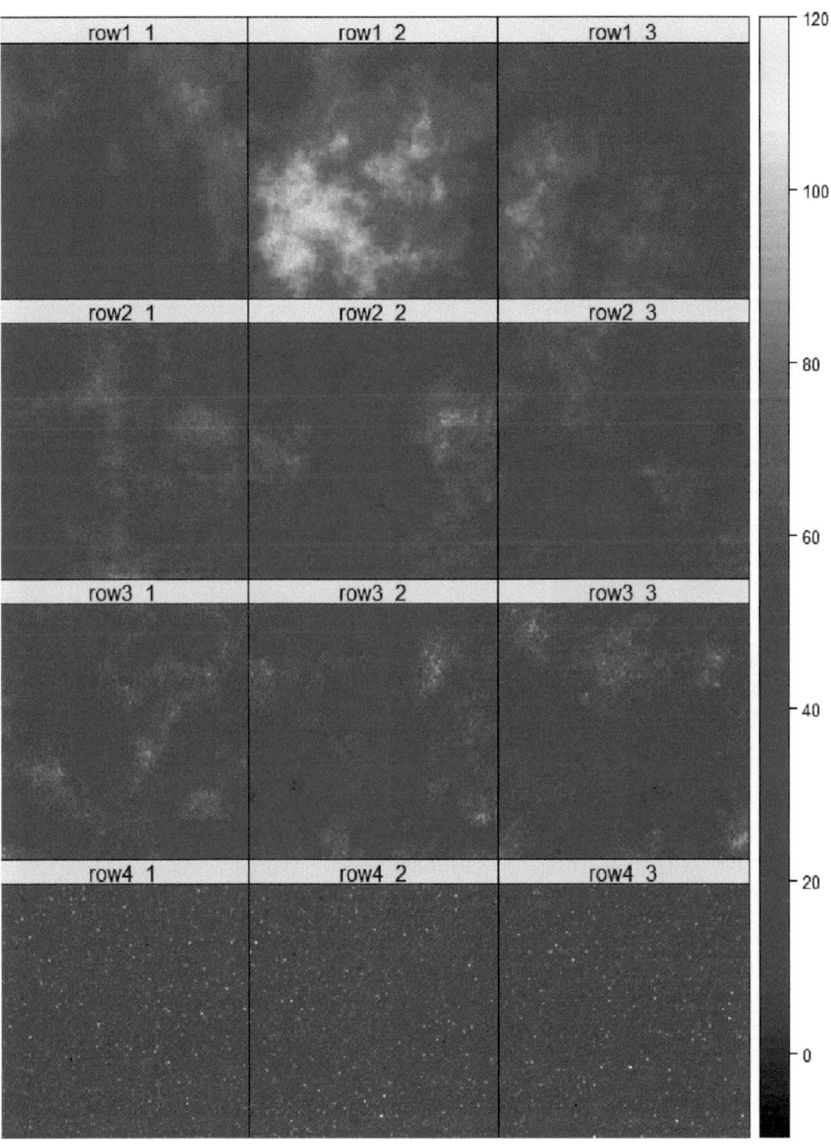

Fig. 1.5: Twelve realisations of a random field with a constant mean of 50 and standard deviation 15. The degree of spatial correlation is strongest in the top row and decreases from the top to the bottom row. Realisations within a row are all different but have the same spatial structure.

Chapter 6: Spatial Data Handling and Plotting in R

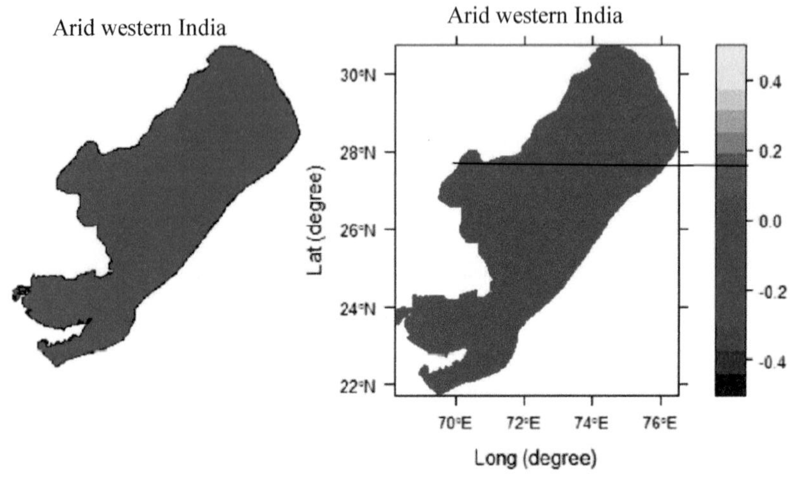

Fig. 6.5: Plot of spatial point data using plot() function at left hand side and using spplot () function at right hand side

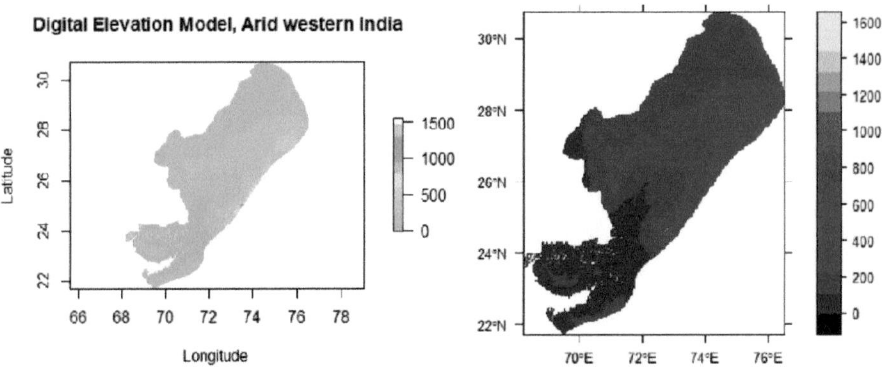

Fig. 6.6: Plot of spatial grid data on digital elevation model using *plot*() function at left hand side and using *spplot*() function at right hand side

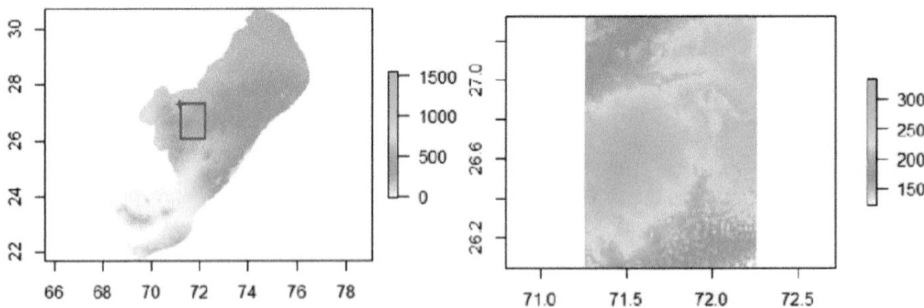

Fig. 6.7: Cropping of DEM using *drawExtent*() function in R